T0191644

Advances in Intelligent Systems and Computing

Volume 534

Series editor

Janusz Kacprzyk, Polish Academy of Sciences, Warsaw, Poland
e-mail: kacprzyk@ibspan.waw.pl

About this Series

The series "Advances in Intelligent Systems and Computing" contains publications on theory, applications, and design methods of Intelligent Systems and Intelligent Computing. Virtually all disciplines such as engineering, natural sciences, computer and information science, ICT, economics, business, e-commerce, environment, healthcare, life science are covered. The list of topics spans all the areas of modern intelligent systems and computing.

The publications within "Advances in Intelligent Systems and Computing" are primarily textbooks and proceedings of important conferences, symposia and congresses. They cover significant recent developments in the field, both of a foundational and applicable character. An important characteristic feature of the series is the short publication time and world-wide distribution. This permits a rapid and broad dissemination of research results.

Advisory Board

More information about this series at http://www.springer.com/series/11156

Shin-ya Kobayashi · Andrzej Piegat
Jerzy Pejaś · Imed El Fray
Janusz Kacprzyk
Editors

Hard and Soft Computing for Artificial Intelligence, Multimedia and Security

 Springer

Editors
Shin-ya Kobayashi
Graduate School of Science and Engineering
Ehime University
Ehime
Japan

Andrzej Piegat
West Pomeranian University of Technology
in Szczecin
Szczecin
Poland

Jerzy Pejaś
West Pomeranian University of Technology
in Szczecin
Szczecin
Poland

Imed El Fray
West Pomeranian University of Technology
in Szczecin
Szczecin
Poland

Janusz Kacprzyk
Polish Academy of Sciences
Systems Research Institute
Warsaw
Poland

ISSN 2194-5357   ISSN 2194-5365   (electronic)
Advances in Intelligent Systems and Computing
ISBN 978-3-319-48428-0   ISBN 978-3-319-48429-7   (eBook)
DOI 10.1007/978-3-319-48429-7

Library of Congress Control Number: 2016954703

Printed on acid-free paper

This Springer imprint is published by Springer Nature
The registered company is Springer International Publishing AG
The registered company address is: Gewerbestrasse 11, 6330 Cham, Switzerland

Preface

This volume contains a collection of carefully selected, peer-reviewed papers presented at the Advanced Computer System 2016 (ACS 2016) Conference organized by the Faculty of Computer Science and Information Technology, West Pomeranian University of Technology in Szczecin, Poland, in cooperation with the Faculty of Mathematics and Information Science, Warsaw University of Technology, Faculty of Physics and Applied Computer Science, AGH University of Science and Technology in Cracow, Poland, the Institute of Computer Science, Polish Academy of Sciences in Warsaw, Poland, and—last but not least—Ehime University in Matsuyama, Japan.

The ACS 2016 is a very special event because it is the twentieth one of this conference series. Since the first ACS 1994, which has just been a relatively small event organized by a small group of enthusiasts who have noticed that various issues related to broadly perceived advanced computer systems have become one of the major challenges faced by both science and technology. Over the years we have been able to clearly see that the very concept of an *advanced computer system* has been changing, in some cases more rapidly, in some cases quite slowly. However, the organizers have always showed an ingenious ability to find topics which have become attractive and challenging, and then gather so many people from all over the world. This has implied a considerable growth of the ACS Conference, in the sense of a growing number of participants, coming from increasingly many countries, but also in the sense of a growing number of co-organizers because more and more top academic and research institutions have highly appreciated the great job done by the Faculty of Computer Science and Information Technology West Pomeranian University of Technology in Szczecin, Poland, which has initiated the ACS conference series, and has been the main organizer since the very beginning. The involvement of foreign academic and research institutions, notably the Ehime University of Matsuyama, Japan, has also been a major factor in the internationalization of this conference series.

This volume is an account of the contributions submitted to the ACS 2016. The papers have been peer reviewed and the authors have introduced valuable suggestions and remarks of the reviewers. The volume is divided into parts which correspond to the main lines of the conference. For the convenience of the readers, we will now briefly summarize the contents of the papers accepted.

Part I, *Artificial Intelligence*, is concerned with some basic conceptual, theoretic and applied issues of artificial intelligence, as well as computational intelligence, that constitute a foundation of virtually all tools and techniques covered by the topics of the Conference.

In their paper, *On fuzzy RDM-arithmetic*, A. Piegat and M. Landowski present a novel concept of a horizontal membership function (HMF, for short), and its related fuzzy relative distance measure (fuzzy RDM) based arithmetic. Then, they compare it with the standard fuzzy arithmetic (SF arithmetic). The fuzzy RDM-arithmetic possesses many important and strong mathematical properties which, as opposed to the SF-arithmetic that delivers, in general, approximate, partial fuzzy solutions, makes it possible to obtain complete fuzzy solution sets. The authors explain how to implement the RDM-arithmetic and show some applications.

M. Pietras (*Hidden Markov Models with affix based observation in the field of syntactic analysis*) introduces some new Hidden Markov Models (HMMs) with N-gram observation based on words bound morphemes (affixes) used in natural language text, with a focus on syntactic classification. The curtailment of the consecutive gram's affixes presented, decreases the accuracy in observation, but reveals statistically significant dependencies so that a considerably smaller size of the training data set is required. Then, an impact of affix observation on the knowledge generalization and an improved word mapping are also discussed. The main issue is the evaluation of the HMMs in the field of syntactic analysis for English and Polish languages based on the Penn and Składnica treebanks. In total, 10 HMMs differing in the structure of observation are compared, and advantages of the approach and the particular structures are presented.

In *Opinion acquisition: an experiment on numeric, linguistic and color coded rating scale comparison* by O. Pilipczuk and G. Cariowa, the authors present the problem of acquiring an opinion of a person using different rating scales. A particular attention is given to the scale devised using a visible color spectrum and to the selection of the optimum number of colors in the scale. A comparison of the effectiveness of the color-coded scale, word scale and some selected numerical scales is performed in the process of student's opinion acquisition with the opinions collected by questionnaire and interviews. The authors compare the average time of giving answers and the cognitive load (mental effort), and describe some important problems occurring in the question answering process. They show that the opinion is the most difficult to acquire using the word scale, while the results are the easiest using the color scale and the -10 to $+10$ numerical scale.

W. Rogoza and M. Zabłocki (*A weather forecasting system using intelligent BDI multiagent-based group method of data handling*) propose the concept of analysis of complex processes based on a multiagent platform is presented. The analysis, rooted in the Group Method of Data Handling (GMDH) is employed to construct a model of the process analyzed. The model is used to predict the future development of the process. The author employs a multiagent platform composed of BDI agents which provides an intelligent distributed computational environment. Moreover, different evaluation criteria are considered, and examples of applications are shown.

In *Comparison of RDM Complex Interval Arithmetic and Rectangular Complex Arithmetic*, M. Landowski presents a comparison of the RDM complex interval arithmetic with the rectangular complex arithmetic. The RDM complex interval arithmetic is multidimensional and this property gives a possibility to find full solutions of problems with complex interval variables. To show the application of the RDM complex interval arithmetic, some examples with complex variables are solved using both the RDM and rectangular complex interval arithmetics.

In *Homogeneous Ensemble Selection - Experimental Studies*, R. Burduk and P. Heda present a new dynamic ensemble selection method. The method proposed uses information from the so-called decision profiles which are formed from the outputs of the base classifiers. In order to verify the algorithms, a number of experiments have been carried out on several publicly available benchmark data sets. The proposed dynamic ensemble selection is experimentally compared against all base classifiers and the ensemble classifiers based on the sum and decision profiles based methods. As base classifiers, the authors use a pool of homogeneous classifiers.

W. Rogoza (*Deterministic method for the prediction of time series*) considers the problem that is important to virtually all business analysts who are frequently forced to make decisions using data on a certain business process obtained within a short time interval. Under these conditions the analyst is not in a position to use traditional statistical methods and should be satisfied with just a few experimental samples. The paper deals with the new method for the prediction of time series based on system identification. The new method proposed shows flexibility and accuracy when the analyzed process is regular in a sense, and it is useful for the prediction of time series in a short-term perspective.

L. Chmielewski, A. Orłowski and M. Janowicz (*A Study on Directionality in the Ulam Square with the Use of the Hough Transform*) use a version of the Hough transform in which the direction of the line is represented by a pair of co-prime numbers for the investigation of the directional properties of the Ulam spiral. The method reveals the detailed information on the intensities of the lines which can be found in the square and on the numbers of primes contained in these lines. This makes it possible to make quantitative assessments related to the lines. The analysis, among others, confirms the fact that is well known from observations that one of the diagonal directions is more populated with lines than the other one. The results are compared to those obtained for a square containing randomly located points with a density close to that for the Ulam square of a corresponding size. Besides its randomness, such a square has also a directional structure resulting from the square shape of the pixel lattice. This structure does not depend significantly on the size of the square. The analysis reveals that the directional structure of the Ulam square is both quantitatively and qualitatively different from that of a random square. A larger density of lines in the Ulam square along one of the diagonal directions in comparison to the other one is confirmed.

Part II *Design of Information and Security Systems* deals with various aspects and approaches to the formulation, analysis and solution of a crucial and challenging issue of information systems in general and security systems in particular.

N. Borgest and M. Korovin (*Ontological approach towards semantic data filtering in the interface design applied to the interface design and dialogue creation for the 'Robot-aircraft designer' informational system*) discuss an approach to data compression in the *robot-airplane designer* system which is a system for an automated airplane design. The proposed approach is tested on a prototype of a demo of the *robot-airplane designer* system which is capable of solving the task within a given amount of time, with some presentation limitations. The importance of providing the user with just enough information to help him or her make a correct decision, without a data and information overload is emphasized. The solution proposed is based on an ontological approach in which the required bits of information are extracted from a knowledge base with regard to the user's level of competence and personal preferences. The project data, including statistical information, design decision-making strategies and values for the target values are stored in a number of databases, interconnected using a thesaurus.

G. Śmigielski, R. Dygdała, H. Zarzycki and D. Lewandowski (*Real-time system of delivering water-capsule for firefighting*) consider the method of explosive-produced water aerosol, delivered by a helicopter to a location near a fire area and then released and detonated, as an alternative and efficient technique of large-scale fire extinguishment. The effectiveness and efficiency of this technique depends on quality of the control system—its determinism, speed of computation of the moment of capsule's release and reliability of components. The article presents a proposed solution to the design of such system, with design assumptions, selection of the integration step size in numerical method, structure of the real-time system and a practical verification of the system.

K. Yamaguchi, T. Inamoto, K. Endo, Y. Higami, and S. Kobayashi (*Evaluation of Influence Exerted by a Malicious Group's Various Aims in the External Grid*) are concerned with some aspects related to the grid computing systems. Basically, though the external grid realizes high performance computing, it is necessary to guarantee the robustness of its functioning against malicious behaviors of the computers. Though in the literature a technique to protect program codes against such behaviors has been proposed, only one type of malicious behavior has been considered to evaluate the effectiveness and efficiency of the approach proposed though in reality malicious behaviors may vary according to the purpose of malicious groups. The goal of this paper is to present a new approach to guarantee the safety of the external grid in a quantitative way, and the authors evaluate the effectiveness of concealing processes against several types of malicious behaviors.

In *Subject-specific methodology in the frequency scanning phase of SSVEP-based BCI* by I. Rejer and Ł. Cieszyński, the authors are concerned with some issues related to the Steady State Visual Evoked Potentials (SSVEPs) often used in Brain Computer Interfaces (BCIs). Since they may differ across subjects, the authors propose a new SSVEP-based BCI in which the user should always start the session with the frequency-scanning phase. During this phase, the stimulation frequencies evoking the most prominent SSVEPs are determined. It is proposed that not only the stimulation frequencies specific for the given user should be chosen in the scanning phase but also the methodology used for the SSVEP detection. The paper reports the results of a survey aimed at finding whether using subject specific methodology for identifying stimulation frequencies would increase the number of frequencies found. Three factors are analyzed: the length of time window used for the power spectrum calculation, a combination of channels, and the number of harmonics used for the SSVEP detection. The experiment (performed with 6 subjects) shows the mean drop in the number of SSVEPs detected with any other but the best combination of factors to be very large for all subjects (from 31.52 % for subject S3 to 51.76 % for subject S4).

A. Grocholewska-Czuryło (*S-boxes cryptographic properties from a statistical angle*) presents the design of a new, strong S-box to be incorporated as the non-linear element of the PP-2 block cipher designed recently at the Poznan University of Technology. A statistical analysis of the cryptographic criteria characterizing a group of S-boxes generated by inverse mapping with random search elements is presented. The statistical tests used in this research are not pure randomness checks but are also related to the most important real cryptographic criteria like the non-linearity, SAC and collision avoidance.

In *Wavelet transform in detection of the subject specific frequencies for SSVEP-based BCI* by I. Rejer, the author considers one of the paradigms often used to build a brain computer interface (BCI), namely the paradigm based on steady state visually evoked potentials (SSVEPs). In a SSVEP-based BCI the user is stimulated with a set of light sources flickering with different frequencies. In order to ensure the best performance of the BCI built according to this paradigm, the stimulation frequencies should be chosen individually for each user. Usually, during the frequency-scanning phase the user-specific stimulation frequencies are chosen according to the power of the corresponding SSVEPs. However, not only the power should be taken into account while choosing the stimulation frequencies, and the second very important factor is the time needed to develop the SSVEP. The wavelet transform (WT) seems to be an excellent tool for dealing with this task since it provides not only information about the frequency components represented in the signal but also about the time of the occurrence. A procedure, based on WT, is proposed that can be used to determine the user-specific frequencies with respect to the synchronization time and its strength.

T. Klasa and I. El Fray (*Data scheme conversion proposal for information security monitoring systems*) are concerned with the information security monitoring in a highly distributed environment which requires the gathering and processing of data describing the state of its components. For a proper interpretation, these data should be acquired in a proper form, and numerous meta languages and description schemes are available, but usually only one or few of them is supported by a given data source. A set of those schemes supported by a given device or program is defined by its manufacturer, and due to the utilization of proprietary formats, usually it is impossible to apply a single scheme to all data sources. As a consequence, it is necessary to apply a data conversion scheme, transforming various incompatible messages to a chosen data scheme, supported by the main repository and the analytic subsystem. Only then it is possible to process data to determine the current state of security of the whole information system. The issues mentioned above and considered and some new solutions are proposed.

In *Non-Standard Certification Models for Pairing Based Cryptography* by T. Hyla and J. Pejaś, the authors are concerned with some important issues related to certification. Namely, in the traditional Public Key Infrastructure (PKI), a Certificate Authority (CA) issues a digitally signed explicit certificate binding a user's identity and a public key to achieve this goal. The main purpose of introducing an identity-based cryptosystem and certificateless cryptosystem is to avoid high costs of the management of certificates. In turn, the goal of introducing an implicit certificate-based cryptosystem is to solve the certificate revocation problem. The certificate and pairing based cryptography is a new technology and so far it mainly exists in theory and is just being tested in practice. This is in contrast to the PKI-based cryptography which has been an established and is widespread technology. New types of cryptographic schemes require new non-standard certification models supporting different methods of public keys' management, including theirs generation, certification, distribution and revocation. In this paper the authors take a closer look at the most prominent and widely known non-standard certification models, and discuss their most relevant properties and related issues. Also, they survey and classify the existing non-standard certification models proposed for digital signature schemes that use bilinear pairings. The authors discuss and compare them with respect to some relevant criteria.

Part III, *Multimedia Systems*, contains original contributions dealing with many aspects of broadly perceived multimedia.

A. Cariow and G. Cariowa (*An algorithm for the Vandermonde matrix-vector multiplication with reduced multiplicative complexity*) present a new algorithm for computing the Vandermonde matrix-vector product. Its main ideas boil down to the use of Winograd's formula for the inner product computation. The multiplicative complexity of the proposed algorithm is less than that of the textbook (naïve) method of calculation. If the textbook method requires MN multiplications and M(N-1) additions, the proposed algorithm needs only M+N(M+1)/2 multiplications at the cost of extra additions compared to the naïve method. From the point of view of its hardware implementation on the VLSI chip, when the implementation cost of the multiplier is significantly greater than the implementation cost of the adder, the new algorithm is generally more efficient than the naïve algorithm. When the order of the Vandermonde matrix is relatively small, this algorithm will have a smaller multiplicative complexity than some well-known fast algorithm for the same task.

J. Peksiński, G. Mikolajczak and J. Kowalski (*The use of the objective digital image quality assessment criterion indication to create panoramic photographs*) present a new method for creating panoramic photographs. The method uses their own matching which is based on the analysis of the indication of a popular digital image assessment quality measure, the Universal Quality Index. The result of applying the suggested algorithm is an effective and efficient match of the sequence of partial digital photographs which make the panoramic photograph.

In *Human Face Detection in Thermal Images Using an Ensemble of Cascading Classifiers* by P. Forczmański, the problem of thermal imagery in the context of face detection is dealt. The purpose is to propose and investigate a set of cascading classifiers learned on thermal facial portraits. In order to attain this, an own database is employed which consists of images from an IR thermal camera. The classifiers employed are based on the AdaBoost learning method with three types of low-level descriptors, namely the Haar like features, the histogram of oriented gradients, and the local binary patterns. Results of experiments on images taken in controlled and uncontrolled conditions are promising.

R. Mantiuk (*Accuracy of high-end and self-built eye-tracking systems*) is concerned with eye tracking which is a promising technology for human–computer interactions though rather rarely used in practical applications. The author argues that the main drawback of the contemporary eye trackers is their limited accuracy. However, there is no standard way of specifying this accuracy which leads to underestimating the accuracy error by eye tracker manufacturers. In this work a subjective perceptual experiment is performed of measuring the accuracy of two typical eye trackers: a commercial corneal reaction-based device mounted under a display and a head-mounted do-it-yourself device of the author's own construction. During the experiment, various conditions are taken into consideration including the viewing angle, human traits, visual fatigue, etc. The results indicate that the eye tracker accuracy is observer-dependent and measured gaze directions exhibit a large variance. Interestingly enough, the perceptually measured accuracy of the low-cost do-it-yourself device is close to the accuracy of the professional device.

R. Staniucha and A. Wojciechowski (*Mouth features extraction for emotion analysis*) deal with the analysis of face emotions which is one of the fundamental techniques that might be exploited in a natural human–computer interaction process and thus is one of the most studied topics in the current computer vision literature. In consequence, the extraction of face features is an indispensable element of the face emotion analysis as it influences a decision-making performance. The paper concentrates on extraction of the mouth features which, next to the eye region features, become one of the most representative face regions in the context of emotion retrieval. In the paper an original gradient-based mouth feature extraction method is presented. Its high performance (exceeding 90 % for selected features) is also verified for a subset of the Yale images database.

D. Oszutowska-Mazurek and P. Mazurek (*Sensitivity of Area-Perimeter Relation for Image Analysis and Image Segmentation Purposes*) are concerned with the image analysis with the use of fractal estimators which is important for the description of grayscale images. The sensitivity of the Area Perimeter Relation (APR) using the Brodatz texture database and the Monte Carlo approach is analyzed in this paper. The APR curve obtained APR is approximated using a polynomial and two parameters of the polynomial are applied as the discrimination parameters. A few techniques for the evaluation of APR are applied. The results show the possibility of the discrimination using a single or two polynomial parameters even for a few textures. The quality of the discrimination (separation between texture classes) can be improved if a larger window analysis sizes is applied.

In *Vocal tract resonance analysis using LTAS in the context of the singer's level of advancement* by E. Półrolniczak and M. Kramarczyk, the authors present results of signal analysis of the recorded singing voice samples. The performed analysis is focused on the presence of resonances in the singing voices. The LTAS (Long-Term Average Spectrum) is estimated over the vocal samples, and then analyzed to extract the valuable information to conclude about the quality of the singer's voices. The study is part of a broader research on singing voice signal analysis. The results may contribute to the development of diagnostics tools for the computer analysis of singer's and speaker's voices.

J. Bobulski (*Parallel Facial recognition System Based on 2D HMM*) dealt with some important issues related to facial recognition. With the constantly growing amount of digital data, in virtually all applications, applying increasingly efficient systems for processing them is required quite often and an increase of the performance of individual processors has already reached its upper limit, multiprocessor systems are a necessity. To fully take advantage of the potentials of such systems, it is necessary to use parallel computing. The system for face recognition requires high-computational power, which is one of potential applications of the computations parallelization, especially for large databases. The purpose of this paper is to develop a parallel system of face recognition based on two-dimensional hidden Markov models (2D HMMs). The obtained results show that compared to sequential calculations, the best effects are obtained for the parallelization of tasks, and the acceleration for the training mode is by the factor of 3.3 and for test mode is 2.8.

M. Kubanek, Filip Depta and D. Smorawa (*System of Acoustic Assistance in Spatial Orientation for the Blind*) develop a prototype of an electronic device which navigates a blind person by means of sound signals. Sounds are meant to provide the blind with a simplified map of the object depth in their path. What makes the work innovative is the use of the Kinect sensor applied to scan the space in front of the user as well as the set of algorithms designed to learn and generate the acoustic space which also take into account the tilt of the head. The results of experiments indicate that a correct interpretation of the sound signals is obtained. The tests conducted on the people prove that the concept developed is highly effective and efficient.

Part IV, *Software Technologies*, contains very relevant contributions which show some effective and efficient software solutions that can be used for the implementation of virtually all kinds of novel algorithms and systems proposed in the papers included in this volume.

A. Luntovskyy (*Performance and Energy Efficiency in Distributed Computing*) discusses some performance-to-energy models and tradeoffs in distributed computing exemplified by clusters, grids and clouds. The performance models are examined. A very relevant problem of energy optimization is considered for the data centers. It is advocated that a better tradeoff of *performance-to-energy* can be reached using advanced *green* technologies as well as in the so-called Internet of Things (IoT) environment.

O. Koval, L. Globa and R. Novogrudska (*The Approach to Web Services Composition*) present an approach to the composition of Web services based on their meta descriptions. The process of Web services sequence formation is depicted. Such sequences of Web services describe the execution of certain user's task. The method of user's tasks scenario formation is proposed that allows to dynamically define an ordered sequence of Web services required to run a specific user's tasks. The scenario formation for the real user's task 'Calculation of the strength for the power components of magnetic systems' is represented, showing the applicability and efficiency of new approach proposed.

W. Bielecki and M. Palkowski (*Loop Nest Tiling for Image Processing and Communication Applications*) dealt with the loop nest tiling which is one of the most important loop nest optimization techniques. They present a practical framework for an automatic tiling of affine loop nests to reduce time of application execution which is crucial for the quality of image processing and communication systems. The new framework is derived via a combination of the Polyhedral and Iteration Space Slicing models and uses the transitive closure of loop nest dependence graphs. To describe and implement the approach in the source-to-source TRACO compiler, loop dependences are presented in the form of tuple relations. The applicability of the framework to generate tiled code for image analysis, encoding and communication program loop nests from the UTDSP Benchmark Suite is shown. Experimental results demonstrate the speed-up of optimized tiled programs generated by means of the approach implemented in TRACO.

W. Bielecki and P. Skotnicki (*Tile Merging Technique to Generate Valid Tiled Code by Means of the Transitive Closure*) present a novel approach for the generation of a parallel tiled code of arbitrarily nested loops whose original inter-tile dependence graphs contain cycles. The authors demonstrate that the problem of cyclic dependences can be reduced to the problem of finding strongly connected components of an inter-tile dependence graph, and then solved by merging tiles within each component. The technique proposed is derived via a combination of the Polyhedral Model and Iteration Space Slicing frameworks. The effectiveness and efficiency of the generated code is verified by means of well-known linear algebra kernels from the PolyBench benchmark suite.

In *Performance Evaluation of Impact of State Machine Transformation and Run-Time Library on a C# Application* by A. Derezińska and M. Szczykulski, the authors discuss some issues related to the Model-Driven Development (MDD) of software applications, notably the. UML models which are transformed into code and combined with a run-time library, the mapping of state machine concepts, including concurrent behavior issues, which can be realized in various ways. The authors discuss several problems of call and time event processing and their impact of an application performance. In experiments, different solutions were evaluated and quantitatively compared. They are applied in the refactoring of FXU—a framework for C# code generation and application development based on UML classes and state machine models.

L. Fabisiak (*The method of evaluating the usability of the website based on logs and user preferences*) discusses the development of usability assessment services based on data from the internal structure of the websites. The author considers the problem of evaluating the usability of the websites on the basis of: selecting appropriate criteria, determine their significance, selection decision support methods and users preferences. The website user preferences are assumed to be variable in time and are usually different from those included in the requirements when the site is designed and launched. Once created the website may lose its usability due to variable needs of its users. The aging of services, developments in software and hardware in computer science, technological and life style changes, change of trends and fashion, varying conditions related to the users' behavior, etc., can imply a need for a new analysis and development of new methods to evaluate the usability of websites. The author shows the analysis of the usability of websites based on the data contained in the logs and accounting for changing user preferences based on the history of the use of services.

A. Konys (*Ontology-Based Approaches to Big Data Analytics*) discuses some important issues related to a new reality which basically boils down to the fact that the access to relevant information is obviously one of the determining factors which directly influences the quality of decision-making processes. However, since huge amounts of data have been accumulated in a large variety of sources in many different formats, this makes the use of data difficult. The Web of Data is considered to provide great opportunities for ontology-based services. The combination of ontology-based approaches and Big Data may help solve some problems related to the extraction of meaningful information from various sources. This paper presents a critical analysis of some selected ontology-based approaches to Big Data analytics as well as the proposal of a new procedure for ontology-based knowledge discovery.

Many people deserve our thanks. First of all, we wish to cordially thank the authors and participants because without their hard work to prepare such good papers, the conference would have not been successful and would have not attracted such a big audience from all over the world. In this context, special thanks are also due to the reviewers who have done an excellent job and whose extremely valuable remarks and suggestions have greatly improved the scientific excellence of papers.

Since, as we have already mention the ACS 2016 is already the twentieth conference in this series, with the first Advanced Computer System conference (at that time known as conference on Applications of Computer Systems) held during December 16–17, 1994, in Szczecin, Poland, we wish to thank many people who have made an invaluable contribution. First of all, special appreciation and thanks are due to Professor Jerzy Sołdek who has initiated the conference series and has been since the very beginning in 1994 encouraging and supporting the organizers of the subsequent events which has been a key factor in the conference success.

It has been a tradition since the first conference that the organizers have always invited top specialists in the fields, and many top scientists and scholars have presented plenary talks over the years which have always provide much inspiration for future research and for both young and experienced participants. To mention a few, thanks are due to Profs. Anna Bartkowiak (Poland), Liming Chen (France),

Alexandr Dorogov (Russia), Alexandre Dolgui (France), Rolf Drechsler (Germany), Gisella Facchinetti (Italy), Larysa Globa (Ukraine), Olli-Pekka Hilmola (Finland), Akira Imada (Belarus), Shinya Kobayashi (Japan), Ivan Lopez Arevalo (Mexico), Andriy Luntovskyy (Germany), Kurosh Madani (France), Karol Myszkowski (Germany), Witold Pedrycz (Canada), Andrzej Piegat (Poland), Khalid Saeed (Poland), Alexander Schill (Germany), Sławomir Wierzchoń (Poland) and Toru Yamaguchi (Japan).

In keeping with ACS mission over last 20 years, this twentieth anniversary conference, ACS 2016, is a very special event providing comprehensive state-of-the-art summaries from keynote speakers as well as a look forward towards future research priorities. Thanks are due to the this year invited speakers, Professors Anna Bartkowiak from the University of Wroclaw (Poland), Gisella Facchinetti from the University of Salento (Italy), Jerzy August Gawinecki from the Military University of Technology (Poland), Larisa Globa from the National Technical University of Ukraine (Ukraine), Akira Imada from Brest State Technical University (Belarus), Khalid Saeed from Bialystok University of Technology and Warsaw University of Technology (Poland), Arkadiusz Orłowski from Warsaw University of Life Sciences SGGW (Poland), and Jacek Pomykała from the University of Warsaw (Poland).

We would like to thank all members of the International Programme Committee who have helped us shape the scope and topics of the conference and have provided us with much advice and help.

Moreover, we want to express a gratitude to all of the organizers from the Faculty of Computer Science and Information Technology, West Pomeranian University of Technology in Szczecin for their enthusiasm and hard work, notably Ms. Hardej, Secretary of the Conference, and all other members of Programme Committee Secretaries including Mykhailo Fedorov, Tomasz Hyla, Agnieszka Konys, Witold Maćków, Marcin Pietrzykowski and Karina Tomaszewska.

And last but not least, we wish to thank Dr. Tom Ditzinger, Dr. Leontina di Cecco and Mr. Holger Schaepe from SpringerNature for their dedication and help to implement and finish this large publication project on time maintaining the highest publication standards.

August 2016   *The Editors*

Organization

Advanced Computer System 2016 (ACS 2016) is organized by the West Pomeranian University of Technology in Szczecin, Faculty of Computer Science and Information Technology (Poland), in cooperation with Warsaw University of Technology, Faculty of Mathematics and Information Science (Poland), AGH University of Science and Technology, Faculty of Physics and Applied Computer Science (Poland), Ehime University (Japan) and Polish Academy of Sciences IPI PAN (Poland).

Programme Committee Chairs

Włodzimierz Bielecki	West Pomeranian University of Technology in Szczecin, Poland
Andrzej Piegat	West Pomeranian University of Technology in Szczecin, Poland
Jacek Pomykała	Warsaw University, Poland
Khalid Saeed	Bialystok University of Technology, Warsaw University of Technology, Poland
Jerzy Pejaś	West Pomeranian University of Technology in Szczecin, Poland
Imed El Fray	West Pomeranian University of Technology in Szczecin, Poland

Honorary Chairs

Witold Pedrycz	University of Alberta, Canada
Jerzy Sołdek	West Pomeranian University of Technology in Szczecin, Poland

Programme Committee Secretaries

Mikhailo, Fedorov	West Pomeranian University of Technology in Szczecin, Poland
Sylwia Hardej	West Pomeranian University of Technology in Szczecin, Poland
Tomasz Hyla	West Pomeranian University of Technology in Szczecin, Poland
Agnieszka Konys	West Pomeranian University of Technology in Szczecin, Poland
Witold Maćków	West Pomeranian University of Technology in Szczecin, Poland
Marcin Pietrzykowski	West Pomeranian University of Technology in Szczecin, Poland
Karina Tomaszewska	West Pomeranian University of Technology in Szczecin, Poland

International Programming Committee

Artificial Intelligence

Anna Bartkowiak	Wroclaw University, Poland
Krzysztof Ciesielski	Polish Academy of Sciences, Poland
Gisella Facchinetti	University of Salento, Italy
Akira Imada	Brest State Technical University, Belarus
Janusz Kacprzyk	Systems Research Institute, Polish Academy of Sciences, Poland
Piotr Andrzej Kowalski	AGH University of Science and Technology and SRI Polish Academy of Sciences, Poland
Jonathan Lawry	University of Bristol, UK
Kurosh Madani	Paris-XII University, France
Andrzej Piegat	West Pomeranian University of Technology, Poland
Elisabeth Rakus-Andersson	Blekinge Institute of Technology, School of Engineering, Sweden
Izebela Rejer	West Pomeranian University of Technology, Poland
Leszek Rutkowski	Czestochowa University of Technology, Poland
Zenon Sosnowski	University of Finance and Management in Białystok, Poland
Jan Węglarz	Poznan University of Technology, Poland
Sławomir Wierzchoń	Institute of Computer Science, Polish Academy of Sciences, Poland

Antoni Wiliński   West Pomeranian University of Technology, Poland
Toru Yamaguchi   Tokyo Metropolitan University, Japan

Software Engineering

Włodzimierz Bielecki   West Pomeranian University of Technology, Poland
Leon Bobrowski   Bialystok Technical University, Poland
Larisa Globa   National Technical University of Ukraine, Ukraine
Janusz Górski   Technical University of Gdansk, Poland
Andriy Luntovskyy   BA Dresden University of Coop. Education, Germany
Andrzej Niesler   Wroclaw University of Economics, Poland
Marcin Paprzycki   Systems Research Institute, Polish Academy of
                         Sciences, Poland
Valery Rogoza   West Pomeranian University of Technology, Poland
Vaclav Snašel   Technical University of Ostrava, Czech Republic

Multimedia Systems

Andrzej Cader   Academy of Humanities and Economics in Lodz,
                         Poland
Ryszard S. Choraś   University of Technology and Life Sciences, Poland
Bernard Dumont   European Commission, Information Society and Media
                         Directorate General
Dariusz Frejlichowski   West Pomeranian University of Technology, Poland
Pawel Forczmański   West Pomeranian University of Technology, Poland
Michelle Joab   LIRMM, Université Montpellier 2, France
Andrzej Kasiński   Poznan University of Technology, Poland
Mariusz Kubanek   Czestochowa University of Technology, Poland
Przemysław Mazurek   West Pomeranian University of Technology, Poland
Khalid Saeed   Bialystok University of Technology, Poland
Albert Sangrá   Universitat Oberta de Catalunya, Spain
Władysław Skarbek   Warsaw University of Technology, Poland
Ryszard Tadeusiewicz   AGH University of Science and Technology, Poland

Design of Information and Security Systems

Costin Badica   Craiova, Romania
Zbigniew Banaszak   Warsaw University of Technology, Poland
Grzegorz Bocewicz   Koszalin University of Technology, Poland
Robert Burduk   Wroclaw University of Technology, Poland

Aleksandr Cariow	West Pomeranian University of Technology, Poland
Nabendu Chaki	Calcutta University, India
Krzysztof Chmiel	Poznan University of Technology, Poland
Nicolas Tadeusz Courtois	CP8 Crypto Lab, SchlumbergerSema, France
Albert Dipanda	Le Centre National de la Recherche Scientifique, France
Jos Dumortier	K.U. Leuven University, Belgium
Imed El Fray	West Pomeranian University of Technology, Poland
Oleg Fińko	Kuban State University of Technology, Russia
Jerzy August Gawinecki	Military University of Technology, Poland
Władysław Homenda	Warsaw Univ of Technology, Poland
Jason T.J. Jung	Yeungnam University, Korea
Shinya Kobayashi	Ehime University, Japan
Zbigniew Adam Kotulski	Polish Academy of Sciences, Poland
Mieczysław Kula	University of Silesia, Poland
Mirosław Kurkowski	Cardinal Stefan Wyszyński University in Warsaw, Poland
Javier Lopez	University of Malaga, Spain
Arkadiusz Orłowski	Warsaw University of Life Sciences SGGW, Poland
Özgür Ertuğ	Gazi University, Turkey
Paweł Pawlewski	Poznań University of Technology, Poland
Jerzy Pejaś	West Pomeranian University of Technology in Szczecin, Poland
Jacek Pomykała	Warsaw University, Poland
Josef Pieprzyk	Macquarie University, Australia
Vincent Rijmen	Graz University of Technology, Austria
Kurt Sandkuhl	University of Rostock, Germany
Marian Srebrny	Institute of Computer Science, Polish Academy of Sciences, Poland
Peter Stavroulakis	Technical University of Crete, Greece
Janusz Stokłosa	Poznan University of Technology, Poland

Referees

Zbigniew Banaszak
Anna Bartkowiak
Włodzimierz Bielecki
Janusz Bobulski
Grzegorz Bocewicz
Robert Burduk
Aleksandr Cariow

Nabendu Chaki
Krzysztof Chmiel
Imed El Fray
Oleg Fińko
Pawel Forczmański
Jerzy August Gawinecki
Larysa Globa

Contents

Design of Information and Security Systems

Multimedia Systems

Software Technologies

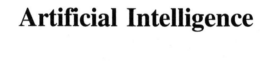

Artificial Intelligence

On Fuzzy RDM-Arithmetic

Andrzej Piegat[1]($\boxtimes$) and Marek Landowski[2]

[1] West Pomeranian University of Technology,
Zolnierska Street 49, 71-210 Szczecin, Poland
apiegat@wi.zut.edu.pl
[2] Maritime University of Szczecin, Waly Chrobrego 1-2, 70-500 Szczecin, Poland
m.landowski@am.szczecin.pl

Abstract. The paper presents notion of horizontal membership function (HMF) and based on it fuzzy, relative distance measure (fuzzy RDM) arithmetic that is compared with standard fuzzy arithmetic (SF arithmetic). Fuzzy RDM-arithmetic possess such mathematical properties which allow for achieving complete fuzzy solution sets of problems, whereas SF-arithmetic, in general, delivers only approximate, partial solutions and sometimes no solutions of problems. The paper explains how to realize arithmetic operations with fuzzy RDM-arithmetic and shows examples of its application.

Keywords: Fuzzy arithmetic · Granular computing · Fuzzy RDM arithmetic · Horizontal membership function · Fuzzy HMF arithmetic · Multidimensional fuzzy arithmetic

1 Introduction

Operations of SF-arithmetic mostly are realized on fuzzy numbers (F-numbers) and on fuzzy intervals (F-intervals), [2,3,5–8]. Any trapezoidal F-interval A is fully characterized by the quadruple (a, b, c, d) of real numbers occuring in the special canonical form (1),

$$A(x) = \begin{cases} (x-a)/(b-a) & \text{for } x \in [a, b) \\ 1 & \text{for } x \in [b, c] \\ (d-x)/(d-c) & \text{for } x \in (c, d] \\ 0 & \text{otherwise} \end{cases} \quad (1)$$

Let $A = (a, b, c, d)$ be used as a shorthand notation of trapezoidal fuzzy intervals. When $b = c$ in (1), A is usually called triangular fuzzy number (TF-number). F-number is a special case of F-interval. Conventional interval can be defined as follows [10]: closed interval denoted by $[a, b]$ is the set of real numbers given by (2),

$$[a, b] = \{x \in \mathbb{R} : a \le x \le b\} \quad (2)$$

Figure 1 shows conventional interval, fuzzy interval, and fuzzy number.

© Springer International Publishing AG 2017
S. Kobayashi et al. (eds.), *Hard and Soft Computing for Artificial Intelligence, Multimedia and Security*, Advances in Intelligent Systems and Computing 534, DOI 10.1007/978-3-319-48429-7_1

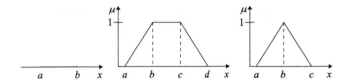

Fig. 1. Conventional interval (a, b), fuzzy interval (a, b, c, d), fuzzy number (a, b, c).

Fuzzy set A can be defined as sum of its μ-cuts, also called α-cuts. Definition of μ-cut is as follows [7]: Given a fuzzy set A defined on $\mathbb{R}$, and a real number $\mu \in [0, 1]$, the crisp set $A_\mu = \{x \in \mathbb{R} : A(x) \geq \mu\}$ is called μ-cut of A. The crisp set $S(A) = \{x \in \mathbb{R} : A(x) > 0\}$ is called the support of A. When $\max_{x \in \mathbb{R}} A(X) = 1$, A is called a normal fuzzy set.

Figure 2 shows μ-cut of F-interval on the level $\mu = 0.5$.

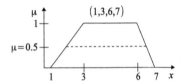

Fig. 2. Fuzzy interval about $(3, 6)$ characterized by quadruple $(1, 3, 6, 7)$ and its μ-cut $A_{0.5} = \{x \in \mathbb{R} : A(x) \geq 0.5\}$.

Because fuzzy set A can be defined as set of its μ-cuts A_μ then arithmetic operations $(+, -, \cdot, /)$ realized on fuzzy sets A and B can be understood as arithmetic operations on intervals. Hence, interval arithmetic (I-arithmetic) can be basis for F-arithmetic. In the practice mostly used I-arithmetic is Moore's arithmetic called also standard interval arithmetic (SI-arithmetic), [10]. Further on realization of basic operations of this arithmetic will be presented. If we have two intervals $[a_1, a_2]$ and $[b_1, b_2]$ then basic operations are realized according to (3) and (4).

$$[a_1, a_2] \oplus [b_1, b_2] = [a_1 \oplus b_1, a_2 \oplus b_2] \tag{3}$$

$$[a_1, a_2] \otimes [b_1, b_2] = [\min(a_1 \otimes b_1, a_1 \otimes b_2, a_2 \otimes b_1, a_2 \otimes b_2), \\ \max(a_1 \otimes b_1, a_1 \otimes b_2, a_2 \otimes b_1, a_2 \otimes b_2)] \tag{4}$$

where: $\oplus \in \{+, -\}$, $\otimes \in \{\cdot, /\}$ and $0 \notin [b_1, b_2]$ if $\otimes = /$.

If an arithmetic operation on F-intervals A and B is to be realized then operations defined by (3) and (4) have to be performed for each μ-cut, $\mu \in [0, 1]$. Figure 3 shows subtraction of two identical F-intervals $X - X$ where $X = (1, 2, 4, 5)$.

In the case of SF-arithmetic subtraction of two identical F-intervals does not result in crisp zero but in fuzzy zero $(\tilde{0})$.

$$X - X \neq 0, \ X - X = \tilde{0} = (-4, -2, 2, 4) \tag{5}$$

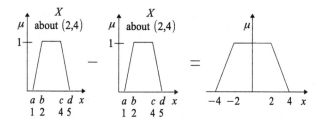

Fig. 3. Visualization of subtraction of two identical intervals $X - X$ determined by quadruple $(1, 2, 4, 5)$.

This result seems rather illogical, because F-interval X represents only one true value x that really occurred in a system. However, it is not known precisely but only approximately as $(1, 2, 4, 5)$. Hence, the difference should be equal to crisp zero. Let us consider now fuzzy equation $A - X = C$, where $A = (1, 2, 4, 5)$ and $C = (7, 8, 9, 10)$. Solving this equation with SF-arithmetic gives strange result $X = (-5, -, 5, -6, -6)$ shown in Fig. 4 in which $x_{min} > x_{max}$.

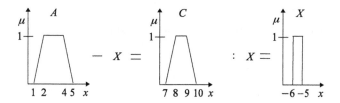

Fig. 4. Paradoxical and incomplete solution $X = [-6, -6, -5, -5]$ of fuzzy equation $A - X = C$ with $A = (1, 2, 4, 5]$ and $C = (7, 8, 9, 10)$ in which $x_{min} > x_{max}$.

One can also try to solve the equation $A - X = C$ using other forms of it, e.g.: $A - X - C = 0$, $A = C + X$, $A - C = X$. The form $A - X - C = 0$ gives paradoxical result $X = (-2, -4, -7, -9)$ in which $x_{min} > x_{max}$. The form $A = C + X$ gives result $X = (-6, -6, -5, -5)$ which is not fuzzy but conventional interval. The form $A - C = X$ gives result $X = (-9, -7, -4, -2)$ being inverse form of the result $X = (-2, -4, -7, -9)$ delivered by the form $A - X - C = 0$. This paradoxical phenomenon of different results achieved from different forms of one and the same equation has been described in many publications, e.g. in [4]. To "enable" solving of equations of type $A - X = C$ apart of the usual calculation way of the difference $A - B = X$ a second way called Hukuhara difference (H-difference) has been in SF-arithmetic introduced. It is calculated from equation $A = B + X_H$ [11]. Thus, officially two ways of difference calculation exist in SF-arithmetic. In SF-arithmetic many next paradoxes exist which are reason of its limited application. SF-arithmetic can solve only part of real problems but the rest lies outside its possibilities. What are reasons of this situation?

Reason 1 is assumption and conviction that result X of arithmetic operation on two 2D fuzzy intervals A and B also is a 2D fuzzy-interval. However, this is not true. Result X of such operation exist not in 2D-space but in 3D-space what will be shown further on. Each next F-interval added to the operation increases dimension of the result. This state of matter is diametrically different from the state in crisp, conventional arithmetic.

Reason 2. In SF-arithmetic and SI-arithmetic calculations are performed only with interval borders. Interiors of intervals do not take part in calculations. And after all calculations should be made on full and complete sets and only on their borders.

Reason 3. In SF-arithmetic as calculation result is accepted not complete solution set but partial, incomplete solution set. In interval arithmetic being basis of SF-arithmetic S.P. Shary [16] introduced 3 different notions of solutions of linear equation systems: the united solution set, the tolerable solution set, the controlled solution set. There also exists notion of the interval algebraic solution [9]. Basing on [9] definitions of the above solution sets for the case of equation $A - X = C$ are as below.

The united solution set $\sum_{\exists\exists}(A, C)$ is the set of solutions of real systems $a - x = c$ with $a \in A$ and $c \in C$, i.e.,

$$\sum\nolimits_{\exists\exists}(A, C) = \{x \in \mathbb{R} | \exists_{a \in A} \exists_{c \in C}(a - x = c)\} \tag{6}$$

The tolerable solution set $\sum_{\forall\exists}(A, C)$ is the set of all real values x such that for every real value $a \in A$ the real difference $a - x$ is contained in the interval vector C, that is;

$$\sum\nolimits_{\forall\exists}(A, C) = \{x \in \mathbb{R} | \forall_{a \in A} \exists_{c \in C}(a - x = c)\} \tag{7}$$

The controlled solution set $\sum_{\exists\forall}(A, C)$ of all real values $x \in \mathbb{R}$, such that for any $c \in C$ we can find the corresponding value $a \in A$ satisfying $a - x$;

$$\sum\nolimits_{\exists\forall}(A, C) = \{x \in \mathbb{R} | \exists_{a \in A} \forall_{c \in C}(a - x = c)\} \tag{8}$$

The united, tolerable and controlled solution sets are not complete algebraic solutions of the equation $A - X = C$ because there exists also point solutions being outside these sets. The full solution set was called "interval algebraic solution". Notion of it [9] adapted for the equation $A - X = C$ is as below.

The interval algebraic solution of interval equation $A - X = C$ is an interval X, which substituted in the equation $A - X$, using interval arithmetic, results in C, that is (9).

$$A - X = C \tag{9}$$

According to [9] "the interval algebraic solutions do not exist in general in the ordinary intervals space (the space without improper intervals)". However, let us remark that in definition of "the interval algebraic solution" as solution "an interval X" is understood, i.e. the same mathematical object as intervals A and C occurring in the equation $A - X = C$. As will be shown further on algebraic solution of expression $A - X = C$ exists and it can be called complete solution. It can be achieved with use of RDM fuzzy arithmetic what will be shown further on. However, this solution is not an interval but a multidimensional granule existing in 3D-space.

Reason 4. SF-arithmetic does not possess certain important mathematical properties which are necessary in solving more complicated problems as fuzzy equation systems. In particular, SF-arithmetic does not possess the inverse element $(-X)$ of addition and $(1/X)$ of multiplication. Hence properties (10) are true (except for degenerate intervals), 0 means crisp zero and 1 means crisp 1.

$$X + (-X) \neq 0, \ X \cdot (1/X) \neq 1 \tag{10}$$

In SI- and SF-arithmetic also subdistributivity law and cancellation law for multiplication does not hold in general, (11) and (12).

$$X(Y + Z) \neq XY + XZ \tag{11}$$

$$XZ = YZ \nRightarrow X = Y \tag{12}$$

2   RDM Variables and Horizontal Membership Functions

Figure 1 shows an interval. It is a model of x-value that is not precisely but only approximately known and the knowledge about it is expressed in the form $x \in [a, b]$. If such a model is used in calculations of SI-arithmetic then only the interval borders a and b take part in the calculations. The whole interior does not take part. The Relative-Distance-Measure (RDM) allows for participation also the interval interior in calculations. The RDM model of interval is given by [10] and shown in Fig. 5.

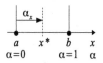

Fig. 5. Visualization of the RDM interval model.

In Fig. 5 x^* means the true precise value of variable x that has occurred in a real system but that is not precisely known. This true value can be expressed by (13).

$$x^* = a + (b - a)\alpha_x, \ \alpha_x \in [0, 1] \tag{13}$$

However, for simplicity notation (14) will be used.

$$x = a + (b - a)\alpha_x \tag{14}$$

The RDM-variable α_x has here meaning of the relative distance of the true value x^* from the interval beginning a. Thus, it can be interpreted as a local

coordinate. A fuzzy interval (a, b, c, d) is shown in Fig. 1. For such intervals vertical MFs can be used. Vertical models express the vertical dependence $\mu = f(x)$.
Vertical MF of fuzzy interval (a, b, c, d) is given by (15).

$$\mu(x) = \begin{cases} (x - a)/(b - a) & \text{for } x \in [a, b) \\ 1 & \text{for } x \in [b, c] \\ (d - x)/(d - c) & \text{for } x \in (c, d] \\ 0 & \text{otherwise} \end{cases} \tag{15}$$

The model (15) is a model of the fuzzy interval borders only. It does not
model the interval interior. This fact limits usefulness of the vertical model in
calculations. A question can be asked: would it be possible to create a horizontal model of fuzzy interval in the form $x = f^{-1}(\mu)$? At first glance it seems
impossible because such dependence would be ambiguous and hence would not
be function. However, let us consider a horizontal cut of the MF shown in Fig. 6
on level μ. The cut is a usual 1D interval and has two boundaries $x_L(\mu)$ and
$x_R(\mu)$ that are expressed by (16).

$$x_L(\mu) = a + (b - a)\mu, \quad x_R(\mu) = d - (d - c)\mu \tag{16}$$

The RDM variable α_x with its increase transforms the left boundary $x_L(\mu)$
in the right one $x_R(\mu)$, Fig. 6.

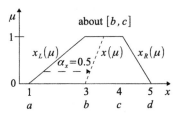

Fig. 6. Visualization of the horizontal approach to fuzzy membership functions.

The contour line $x(\mu, \alpha_x)$ of constant α_x-values in the interior of the MF
(Fig. 6) is expressed by (17).

$$x(\mu, \alpha_x) = x_L(\mu) + [x_R(\mu) - x_L(\mu)]\alpha_x, \quad \mu, \alpha_x \in [0, 1] \tag{17}$$

The contour line $x(\mu, \alpha_x)$ is set of points lying at equal relative distance α_x
from the left boundary $x_L(\mu)$ of the MF in Fig. 6. A more precise form (18)
of (17) can be called horizontal MF (HMF).

$$x = [a + (b - a)\mu] + [(d - a) - (d - c + b - a)\mu]\alpha_x, \quad \mu, \alpha_x \in [0, 1] \tag{18}$$

The horizontal MF $x = f(\mu, \alpha_x)$ is function of two variables and exists in
3D-space, Fig. 7. It is unique and expresses the 2D MF with its interior shown

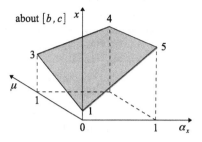

Fig. 7. The horizontal membership function $x = (1 + 2\mu) + (4 - 3\mu)\alpha_x$, $\mu, \alpha_x \in [0, 1]$, corresponding to vertical function shown in Fig. 6.

in Fig. 1. The HMF defines a 3D information granule, Fig. 7, and hence can be denoted as x^{gr}.

Formula (18) describes the trapezoidal MF. However, it can be adapted to triangular MF by setting $a = b$ and to rectangular MF by $a = b$ and $c = d$. Boundaries of these functions are here linear. To derive formulas for nonlinear boundaries, e.g. of Gauss type, formulas for the left $x_L(\mu, \alpha_x)$ and for the right boundary $x_R(\mu, \alpha_x)$ should be determined and set in (17). Concept of the horizontal MF was elaborated by A. Piegat [12–15,17].

3   RDM Fuzzy Arithmetic with Horizontal Membership Functions

Let $x^{gr} = f(\mu, \alpha_x)$ be a horizontal MF representing a fuzzy interval X (19) and $y^{gr} = f(\mu, \alpha_y)$ be a horizontal MF representing a fuzzy interval Y (20).

$$X : x^{gr} = [a_x + (b_x - a_x)\mu] + [(d_x - a_x) - (d_x - c_x + b_x - a_x)\mu]\alpha_x, \ \mu, \alpha_x \in [0, 1] \quad (19)$$

$$Y : y^{gr} = [a_y + (b_y - a_y)\mu] + [(d_y - a_y) - (d_y - c_y + b_y - a_y)\mu]\alpha_y, \ \mu, \alpha_y \in [0, 1] \quad (20)$$

Addition of two independent fuzzy intervals, (21).

$$X + Y = Z : x^{gr}(\mu, \alpha_x) + y^{gr}(\mu, \alpha_y) = z^{gr}(\mu, \alpha_x, \alpha_y), \ \mu, \alpha_x, \alpha_y \in [0, 1] \quad (21)$$

For example, if X is trapezoidal MF $(1, 3, 4, 5)$, (22),

$$x^{gr}(\mu, \alpha_x) = (1 + 2\mu) + (4 - 3\mu)\alpha_x \quad (22)$$

and Y is trapezoidal MF $(1, 2, 3, 4)$, (23),

$$y^{gr}(\mu, \alpha_y) = (1 + \mu) + (3 - 2\mu)\alpha_y \quad (23)$$

then $z^{gr}(\mu, \alpha_x, \alpha_y)$ is given by (24),

$$z^{gr}(\mu, \alpha_x, \alpha_y) = (2 + 3\mu) + (4 - 3\mu)\alpha_x + (3 - 2\mu)\alpha_y, \ \mu, \alpha_x, \alpha_y \in [0, 1] \quad (24)$$

The 4D-solution (24) exists in the space which cannot be seen. Therefore we can be interested in its low dimensional representations. Frequently, the 2D-representation in the form of span $s(z^{gr})$ is determined. It can be found with known methods of function examination (25).

$$s(z^{gr}) = [\min_{\alpha_x,\alpha_y} z^{gr}(\mu,\alpha_x,\alpha_y), \max_{\alpha_x,\alpha_y} z^{gr}(\mu,\alpha_x,\alpha_y)] \qquad (25)$$

In the case of discussed example, extrema of (25) lie not inside but on boundaries of the result domain. The minimum corresponds to $\alpha_x = \alpha_y = 0$ and the maximum to $\alpha_x = \alpha_y = 1$. Span of the 4D-result granule (24) is given by (26).

$$s(z^{gr}) = [2 + 3\mu, 9 - 2\mu], \ \mu \in [0,1] \qquad (26)$$

The span (26) is not the addition result. The addition result has the form of 4D-function (24). The span is only a 2Dinformation about the maximal uncertainty of the result.

Subtraction of two independent fuzzy intervals, (27).

$$X - Y = Z : x^{gr}(\mu,\alpha_x) - y^{gr}(\mu,\alpha_y) = z^{gr}(\mu,\alpha_x,\alpha_y), \ \mu,\alpha_x,\alpha_y \in [0,1] \quad (27)$$

For example, if X and Y are trapezoidal MF (22) and (23) then the result is given by (28).

$$z^{gr}(\mu,\alpha_x,\alpha_y) = \mu + (4 - 3\mu)\alpha_x - (3 - 2\mu)\alpha_y, \ \mu,\alpha_x,\alpha_y \in [0,1] \qquad (28)$$

If we are interested in the span representation $s(z^{gr})$ of the 4D-subtraction result, then it can be determined from (29).

$$s(z^{gr}) = [\min_{\alpha_x,\alpha_y} z^{gr}, \max_{\alpha_x,\alpha_y} z^{gr}] = [-3 + 3\mu, 4 - 2\mu], \ \mu \in [0,1] \qquad (29)$$

The span (29) of z^{gr} (28) corresponds to $\alpha_x = 0$, $\alpha_y = 1$ for $\min z^{gr}$ and $\alpha_x = 1$, $\alpha_y = 0$ for $\max z^{gr}$.

Multiplication of two independent fuzzy intervals, (30).

$$X \cdot Y = Z : x^{gr}(\mu,\alpha_x) \cdot y^{gr}(\mu,\alpha_y) = z^{gr}(\mu,\alpha_x,\alpha_y), \ \mu,\alpha_x,\alpha_y \in [0,1] \qquad (30)$$

For example, if X and Y are trapezoidal MF (22) and (23) then the multiplication result z^{gr} is expressed by (31).

$$\begin{aligned} z^{gr}(\mu,\alpha_x,\alpha_y) &= x^{gr} \cdot y^{gr} \\ &= (1 + 2\mu) + (4 - 3\mu)\alpha_x] \cdot [(1 + \mu) + (3 - 2\mu)\alpha_y], \ \mu,\alpha_x,\alpha_y \in [0,1] \end{aligned} \qquad (31)$$

Formula (31) describes the full 4D-result of the multiplication. If we are interested in the 2D simplified representation of this result in the form of a span $s(z^{gr})$ then formula (32) should be used.

$$s(z^{gr}) = [\min_{\alpha_x,\alpha_y} z^{gr}, \max_{\alpha_x,\alpha_y} z^{gr}] = [(1 + 2\mu)(1 + \mu), (5 - \mu)(4 - \mu)], \ \mu \in [0,1] \quad (32)$$

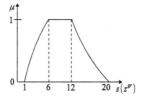

Fig. 8. MF of the span representation of the 4D multiplication result (31).

Figure 8 shows the MF of the span representation of the multiplication result. **Division X/Y of two independent fuzzy intervals, $0 \notin Y$, (33).**

$$X/Y = Z : x^{gr}(\mu, \alpha_x)/y^{gr}(\mu, \alpha_y) = z^{gr}(\mu, \alpha_x, \alpha_y), \ \mu, \alpha_x, \alpha_y \in [0, 1] \qquad (33)$$

For example, if X and Y are trapezoidal MF (22) and (23) then the division result z^{gr} is given by (34).

$$z^{gr}(\mu, \alpha_x, \alpha_y) = x^{gr}/y^{gr} = \frac{(1 + 2\mu) + (4 - 3\mu)\alpha_x}{(1 + \mu) + (3 - 2\mu)\alpha_y}, \ \mu, \alpha_x, \alpha_y \in [0, 1] \qquad (34)$$

The span representation $s(z^{gr})$ of the result (34) is expressed by (35) and is shown in Fig. 9.

$$s(z^{gr}) = \left[\min_{\alpha_x, \alpha_y} z^{gr}, \ \max_{\alpha_x, \alpha_y} z^{gr} \right] = \left[\frac{1 + 2\mu}{4 - \mu}, \frac{5 - \mu}{1 + \mu} \right], \ \mu \in [0, 1] \qquad (35)$$

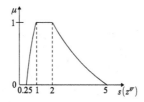

Fig. 9. Span representation of the 4D division result (34).

The solution granule of the division (34) is 4-dimensional, so it cannot be presented in its full space. However, it can be shown in a simplified way, in the $X \times Y \times Z$ 3D-space, without μ-coordinate. Figure 10 presents surfaces for constant $\mu = 0$ and $\mu = 1$ values.

As Fig. 10 shows, the solution granule (34) is uniform. This results from the fact that the divisor does not contain zero. Division results can be discontinuous and multigranular in more complicated cases.

What happens when uncertain denominator Y of division X/Y contains zero? Then the solution is multi-granular. Such situation does not occur in conventional crisp arithmetic.

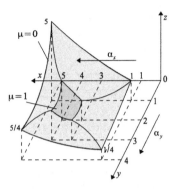

Fig. 10. Simplified view of the 4D-solution granule (34) $z^{gr}(x, y, z)$ in 3D-space $X \times Y \times Z$, without μ-coordinate

Let us now apply RDM fuzzy arithmetic to solve equation $A - X = C = (1, 2, 4, 5) - X = (7, 8, 9, 10)$ which previously has been "solved" with use of SF-arithmetic. Using Eq. (18) HMF of A is achieved in form of (36) and HMF of C in form of (37).

$$a^{gr} = (1 + \mu) + (4 - 2\mu)\alpha_a, \ \mu, \alpha_a \in [0, 1] \tag{36}$$

$$c^{gr} = (7 + \mu) + (3 - 2\mu)\alpha_c, \ \mu, \alpha_c \in [0, 1] \tag{37}$$

The solution x^{gr} can be found similarly as in crisp number arithmetic according to (38).

$$x^{gr} = a^{gr} - c^{gr} = [(1+\mu) + (4-2\mu)\alpha_a] - [(7+\mu) + (3-2\mu)\alpha_c], \ \mu, \alpha_a, \alpha_c \in [0, 1] \tag{38}$$

$$a^{gr} - x^{gr} = c^{gr} = a^{gr} - (a^{gr} - c^{gr}) = c^{gr} \tag{39}$$

One can easily check that after substitution of the solution x^{gr} in the solved equation $A - X = C$ (39) is achieved which gives the result $c^{gr} = c^{gr}$, which means that the solution (38) is the algebraic solution of the equation according to [9]. One can easily check that solution (38) is complete by substituting various possible triple combinations of variables $(\mu, \alpha_a, \alpha_c)$. E.g. for $\mu = 0$, $\alpha_a = 1/5$ and $\alpha_c = 2/5$ we have $a = 1.8$, $c = 8.2$ and $x = -6.4$. It is one of point solutions of equation $A - X = C$ because $a - x = 1.8 - (-6.4) = c = 8.2$. With various combinations $(\mu, \alpha_a, \alpha_c)$ one can generate each triple (a, x, c) satisfying equation $a - x = c$.

4   Mathematical Properties of Multidimensional RDM Fuzzy Arithmetic

Commutativity. For any fuzzy intervals X and Y, Eqs. (40) and (41) are true.

$$X + Y = Y + X \tag{40}$$

$$XY = YX \tag{41}$$

Associativity. For any fuzzy intervals X, Y and Z, Eqs. (42) and (43) are true.

$$X + (Y + Z) = (X + Y) + Z \tag{42}$$

$$X(YZ) = (XY)Z \tag{43}$$

Neutral element of addition and multiplication. In multidimensional RDM FA, there exist additive and multiplicative neutral elements such as the degenerate interval 0 and 1 for any interval X, Eqs. (44) and (45).

$$X + 0 = 0 + X = X \tag{44}$$

$$X \cdot 1 = 1 \cdot X = X \tag{45}$$

Inverse elements. In MD RDM FA, fuzzy interval $-X : -x^{gr} = -[a + (b - a)\mu] - [(d - a) - \mu(d - a + b - c)]\alpha_x$, $\mu, \alpha_x \in [0, 1]$, is an additive inverse element of fuzzy interval $X : x^{gr} = [a + (b - a)\mu] + [(d - a) - \mu(d - a + b - c)]\alpha_x$, $\mu, \alpha_x \in [0, 1]$.

If parameters of two fuzzy intervals X and Y are equal: $a_x = a_y$, $b_x = b_y$, $c_x = c_y$, $d_x = d_y$, then the interval $-Y$ is the additive inverse interval of X, when also inner RDM-variables are equal: $\alpha_x = \alpha_y$. It means full coupling (correlation) of both uncertain values x and y modelled by intervals.

Assuming that $0 \notin X$, a multiplicative inverse element of the fuzzy interval X is equal in MD RDM FA $\frac{1}{X} / \frac{1}{x^{gr}} = \frac{1}{[a+(b-a)\mu]+[(d-a)-\mu(d-a+b-c)]\alpha_x}$, $\mu, \alpha_x \in [0, 1]$.

If parameters of two fuzzy intervals X and Y are equal: $a_x = a_y$, $b_x = b_y$, $c_x = c_y$, $d_x = d_y$, then the interval $1/Y$ is the multiplicative inverse interval of X only when also inner RDM-variables are equal: $\alpha_x = \alpha_y$. It means full coupling (correlation) of both uncertain values x and y modelled by intervals. Such full or partial correlation of uncertain variables occurs in many real problems.

Subdistributivity law. The subdistributivity law holds in MD RDM FA (46).

$$X(Y + Z) = XY + XZ \tag{46}$$

The consequence of this law is a possibility of formulas transformations. They do not change the calculation result.

Cancellation law for addition and multiplication. Cancellation laws (47) and (48) hold in MD RDM FA:

$$X + Z = Y + Z \Rightarrow X = Y \tag{47}$$

$$XZ = YZ \Rightarrow X = Y \tag{48}$$

5   Application Example of RDM Fuzzy Arithmetic in Solving Differential Equation

Solving fuzzy differential equation (FD-equation) is difficult task that has been considered since many years and has not been finished until now. In the example both SF-arithmetic and RDM fuzzy arithmetic will be applied to solve a FD-equation (49). This equation is a type of benchmark because it has been discussed in few important papers on FD-equation solving methods, e.g. in [1].

Example. Find the solution of fuzzy differential equation (49) taken from [1].

$$\begin{cases} \dot{X}(t) = -X(t) + W\cos t \\ X(0) = (-1, 0, 1) \end{cases} \tag{49}$$

where $W = (-1, 0, 1)$.

The solution of Eq. (49) for $t \geq 0$ expressed in the form of μ-solution sets [1], for $\mu \in [0, 1]$, is given by (50).

$$x_\mu(t) = 0.5(\sin t + \cos t)[W]_\mu + ([X(0)]_\mu - 0.5[W]_\mu)\exp(-t) \tag{50}$$

The solution (50) obtained by standard fuzzy (SF-) arithmetic for $[W]_\mu = [\mu - 1, 1 - \mu]$ and $[X(0)]_\mu = [\mu - 1, 1 - \mu]$ is given by (51). This standard solution exists in 3D-space.

$$x_\mu^{SFA}(t) = 0.5(\sin t + \cos t)[\mu - 1, 1 - \mu] + ([\mu - 1, 1 - \mu] - 0.5[\mu - 1, 1 - \mu])\exp(-t) \tag{51}$$

In this case fuzzy numbers $[W]_\mu = [X(0)]_\mu = [\mu - 1, 1 - \mu]$ are equal. Figure 11(a) presents in 2D-space border values of the SFA solution (51).

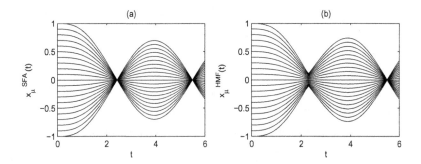

Fig. 11. 2D presentation of border values of the FD-equation (49) obtained with use of standard fuzzy arithmetic (a), and horizontal membership function FA (b), for $\mu \in [0 : 0.1 : 1]$, $t \in [0 : 0.1 : 6]$.

The fuzzy numbers in the form of horizontal membership functions are presented by (52) and (53), where α_W, α_X are RDM variables defined in 3D-space.

$$[W]_\mu = -1 + \mu + 2(1 - \mu)\alpha_W, \ \alpha_W \in [0, 1] \tag{52}$$

$$[X(0)]_\mu = -1 + \mu + 2(1 - \mu)\alpha_X, \; \alpha_X \in [0, 1] \tag{53}$$

In terms of SF-arithmetic both fuzzy numbers W and $X(0)$ are equal ($W = (-1, 0, 1)$ and $X(0) = (-1, 0, 1)$). But it does not mean that real values $w \in W$ and $x(0) \in X(0)$ that have occurred in the real system also have been equal. E.g., it is possible that $w = -0.5$ ($-0.5 \in (-1, 0, 1)$) and $x(0) = 0.3$ ($0.3 \in (-1.0.1)$). RDM fuzzy arithmetic thanks to RDM variables α_W and α_X enables modeling of such situations. If $\alpha_W = \alpha_X$ then $w = x(0)$. When $\alpha_W \neq \alpha_X$ then $w \neq x(0)$.

The μ-solution sets of FD-equation (49) using horizontal membership function is presented by (54). Let us notice that HMF-solution exists in 5D-space.

$$
\begin{aligned}
x_\mu^{HMF} = {} & 0.5(\sin t + \cos t)(-1 + \mu + 2(1 - \mu)\alpha_W) \\
& + [(-1 + \mu + 2(1 - \mu)\alpha_X) - 0.5(-1 + \mu + 2(1 - \mu)\alpha_W)] \exp(-t)
\end{aligned}
\tag{54}
$$

where $\mu \in [0, 1]$, $\alpha_W \in [0, 1]$, $\alpha_X \in [0, 1]$.

Figure 11(b) presents the 5D-solution (54) in 2D-space showing only border values of the HMF-solution.

To check that the solution achieved with use of HMF-functions is correct and complete let us take the testing point $x_\mu^0(t)$ from the HMF-solution (54). It can be proved that the testing point does not belong to the SF-arithmetic solution (51). Let us take the point $x_\mu^0(2.5) = 0.2245$, where $t = 2.5$, $\mu = 0$, $[W]_{\mu=0} = -1$, $[X(0)]_{\mu=0} = 1$, $\alpha_W = 0$ and $\alpha_X = 1$. The value $x_\mu^0(2.5) = 0.2245 \in x_\mu^{HMF}$ belongs to the HMF-solution but it does not belong to the SFA-solution $x_\mu^0(2.5) = 0.2245 \notin x_\mu^{SFA}$.

Solution of the Eq. (49) equals (50). Derivative of the solution (50) is (55).

$$[\dot{X}(t)]_\mu = 0.5(\cos t - \sin t)[W]_\mu - ([X(0)]_\mu - 0.5[W]_\mu)\exp(-t) \tag{55}$$

The left-part of Eq. (49) is the derivative (55). The derivative (55) for $t = 2.5$, $\mu = 0$, $[W]_{\mu=0} = -1$ and $[X(0)]_{\mu=0} = 1$ equals $[\dot{X}(2.5)]_{\mu=0} = 0.5767$. The right-part for $x_\mu^0(2.5) = 0.2245$, $t = 2.5$, $\mu = 0$, $[W]_{\mu=0} = -1$ also equals 0.5767, so the testing point $x_\mu^0(2.5) = 0.2245$ objectively belongs to the HMF-solution of Eq. (49).

In conclusion of the example one can say that standard fuzzy arithmetic can give only a part of full solution. The SF-arithmetic did not find the correct solution of Eq. (49) in the case $w \neq x(0)$. However, when using horizontal membership function with RDM variables α_W and α_X the fuzzy numbers W and $X(0)$ can be independent and w can be different from $x(0)$.

6   Conclusions

The paper has shown a new model of membership function called vertical MF and its application in RDM fuzzy arithmetic. RDM fuzzy arithmetic possess important mathematical properties which SF-arithmetic has not. These properties enable transformation of equations in the process of solving them.

Further on, they increase possibilities of fuzzy arithmetic in solving real problems. Examples of such problems were shown in the paper. Because uncertainty is prevalent in reality, RDM fuzzy arithmetic becomes important tool of solving real problems.

References

1. Abbasbandy, S., Nieto, J.J., Alavi, M.: Tubing of reacheable set In one dimensional fuzzy differentia inclusion. Chaos, Solitons & Fractals **26**(5), 1337–1341 (2005)
2. Dubois, D., Prade, H.: Operations on fuzzy numbers. Int. J. Syst. Sci. **9**(6), 613–626 (1978)
3. Dutta, P., Boruah, H., Ali, T.: Fuzzy arithmetic with and without using α-cut method: a comparative study. Int. J. Latest Trends Comput. **2**(1), 99–107 (2011)
4. Dymova, L.: Soft Computing in Economics and Finance. Springer, Heidelberg (2011)
5. Hanss, M.: Applied Fuzzy Arithmetic. Springer, Heidelberg (2005)
6. Kaufmann, A., Gupta, M.M.: Introduction to Fuzzy Arithmetic. Van Nostrand Reinhold, New York (1991)
7. Klir, G.J., Pan, Y.: Constrained fuzzy arithmetic: Basic questions and answers. Soft Comput. **2**(2), 100–108 (1998)
8. Kosinski, W., Prokopowicz, P., Slezak, D.: Ordered fuzzy numbers. Bull. Polish Acad. Sci. Ser. Sci. Math. **51**(3), 327–338 (2003)
9. Markov, S.M., Popova, E.D., Ullrich, C.: On the solution of linear algebraic equations involving interval coefficients. In: Margenov, S., Vassilevski, P. (eds.) Iterative Methods in Linear Algebra, II. IMACS Series in Computational and Applied Mathematics, vol. 3, pp. 216–225 (1996)
10. Moore, R.E., Kearfott, R.B., Cloud, M.J.: Introduction to Interval Analysis. Society for Industrial and Applied Mathematics, Philadelphia (2009)
11. Pedrycz, W., Skowron, A., Kreinovich, V.: Handbook of Granular Computing. John Wiley & Sons, Chichester (2008)
12. Piegat, A., Landowski, M.: Two interpretations of multidimensional RDM interval arithmetic - multiplication and division. Int. J. Fuzzy Syst. **15**(4), 486–496 (2013)
13. Piegat, A., Plucinski, M.: Fuzzy number addition with the application of horizontal membership functions. Scient. World J., Article ID: 367214, 1–16 (2015). Hindawi Publishing Corporation
14. Piegat, A., Landowski, M.: Horizontal membership function and examples of its applications. Int. J. Fuzzy Syst. **17**(1), 22–30 (2015)
15. Piegat, A., Landowski, M.: Aggregation of inconsistent expert opinions with use of horizontal intuitionistic membership functions. In: Atanassov, K.T., et al. (eds.) Novel Developments in Uncertainty Representation and Processing. AISC, vol. 401, pp. 215–223. Springer, Heidelberg (2016). doi:10.1007/978-3-319-26211-6_18
16. Shary, S.P.: On controlled solution set of interval arithmetic of interval algebraic systems. Interval Comput. **6**, 66–75 (1992)
17. Tomaszewska, K., Piegat, A.: Application of the horizontal membership function to the uncertain displacement calculation of a composite massless rod under a tensile load. In: Wiliński, A., Fray, I., Pejaś, J. (eds.) Soft Computing in Computer and Information Science. AISC, vol. 342, pp. 63–72. Springer, Heidelberg (2015). doi:10.1007/978-3-319-15147-2_6

Hidden Markov Models with Affix Based Observation in the Field of Syntactic Analysis

Marcin Pietras[(✉)]

Computer Science and Information Technology,
West Pomeranian University of Technology, Żołnierska 49, Szczecin, Poland
mpietras@wi.zut.edu.pl

Abstract. This paper introduces Hidden Markov Models with N-gram observation based on words bound morphemes (affixes) used in natural language text processing focusing on the field of syntactic classification. In general, presented curtailment of the consecutive gram's affixes, decreases the accuracy in observation, but reveals statistically significant dependencies. Hence, considerably smaller size of the training data set is required. Therefore, the impact of affix observation on the knowledge generalization and associated with this improved word mapping is also described. The focal point of this paper is the evaluation of the HMM in the field of syntactic analysis for English and Polish language based on Penn and Składnica treebank. In total, a 10 HMM differing in the structure of observation has been compared. The experimental results show the advantages of particular configuration.

Keywords: Hidden Markov Models · N-gram · Syntactic analysis · Natural Language Processing · Treebank · Part-Of-Speech · Data clustering

1 Introduction

The mathematically rich Hidden Markov Models (HMM) are widely used in natural language processing, especially in Part-Of-Speech (POS) tagging. In recent years, various HMMs designated for this purpose has been presented [1–3]. In order to solve NLP challenges, the researchers introduced N-grams analysis [4] with the assumption that text or language structure (such as grammar or syntax) can be recognized by using occurrence probabilities for particular words (i.e. unigram - single word) or sequences of words (i.e. N-gram - sequence of N consecutive words) in the text. However, with increasing complexity of the N-gram observations (associated with length of words sequence), much more training data has to be provided to the model in order to obtain statistically significant learning results [5]. Nonetheless, in the most cases training data will still be insufficient to cover all possible sequences of words, which implies that the model may misinterpret the meaning of the observed text. Furthermore, certain expressions (in database) may be used only occasionally and be heavily dependent on the current geo-language trends and situation. Therefore, the HMM based system should also support an observation's uncertainty handling. The study presented in [6, 7] addressed the problem of new instances classifications which were not included in the training corpus. The database deficiency to some extent can be overcome by the

S. Kobayashi et al. (eds.), *Hard and Soft Computing for Artificial Intelligence, Multimedia and Security*, Advances in Intelligent Systems and Computing 534, DOI 10.1007/978-3-319-48429-7_2

smoothing methods presented well in [8]. In addition, in the [9] are shown the advantages of morphological features extraction in order to handle unknown information. In contrary to the whole words N-gram based observation the study [10] introduces more advanced method that is based on morphological analysis and is dividing the word into morphemes (morpheme-word representation). However, in this paper presented concept put the main focus on the bound morphemes features detection, which is an intermediary approach between morphological analysis and dictionary words mapping. Therefore, the observation model includes mainly the word affixes, with the naive assumption of its constant length. This assumption leads to creation of the observation window (both for Prefix and Suffix of the word), which can be extended (to give a more accurate observation) or reduced (which leads to a statistical generalization). Applying this modification allows to adjust the observation window individually for every gram of observation. This approach was evaluated in the field of syntactic tagging which involves text classification and disambiguation depending on the characteristics of the recognized expression and its surroundings [11]. Performance of the HMM for text syntactic analysis is strongly dependent on the observation complexity and training database. As previously mentioned, database that would cover all words combination is an enormous challenge. The HMM creation and training based on limited treebank [12, 13] database enforce an alternative (to the dictionary word mapping methods) approaches in order to increase proper text syntactic analysis. Another issue related with observations and model structure complexity is significantly larger resources utilization. Consequently, usage of model at some rate of complexity would simply be impractical for many applications because of too time-consuming computation effort. As a one of possible solution a special "affix oriented" observation is introduce in order to reduce model complexity (by strongly decreasing the emission matrix) and to improve unknown expression recognition.

2 Methodology

This paper does not assume any restrictions concerning the Markov Models. All states are by default fully connected and the transitions probabilities are established by the evaluation of training database. The transitions between states are carried out at regular discrete intervals. Process of transition between states is defined by probabilities of states occurrence obtained from training database. In general, the transitions probabilities may depend on the whole process so far. For the first order HMM states probabilities are reduced to the transition from a previous state only. Formally, HMM is described with the same notation as in [14].

The information related to the model structure, the observation type as well as the textual data preparation and the HMM training process for each HMM presented in this paper are described in details.

In order to create and train HMMs served "HMM-Toolbox" application developed specifically for this project. The HMM-Toolbox is equipped with, among others, the GUI to allow manual states tagging and HMM parameters modifications. The core algorithms associated with HMM computation (such as Viterbi path, Forward-Backward, Baum-Welch and HMM factory) are provided by Jahmm [15] library.

2.1   Treebank Database

The treebank as a parsed and tagged text corpus that annotates syntactic sentence structure provides ready to use database for the purpose of HMM training and further verification. For this reason all states in HMM are convergent with tags occurred in treebank database and moreover states path and observations sequence are also extracted from it. Two treebanks were utilized: Penn Treebank for English language and Składnica treebank for Polish language. In both treebanks lower level syntactic tags (Parts-Of-Speech) and higher level syntactic classes (phrase/dependency classes) can be distinguishing. Hence, two types of HMMs are designed for the each treebank.

3   Complexity of M-order Model with N-gram Observation

The independence assumption states that the output observation at time t is dependent only on the current state and is conditionally independent of previous observations. A number of studies show that this assumption becomes a significant deficiency of the HMM [16]. Nonetheless, the model could be improved in terms of Expectation Maximization for used data by applying observation extension. This technique is well known in computational linguistic and refers to N-gram modification [17]. The N-gram observation is created by combining N past observations together and shifting them in queue with length equal to N.

Let us consider a HMM (St = 3, Obs = 11) with a state sequence equal: 122213223212, where (1-vowel, 2-consonant, 3-pause) and corresponding observation sequence equal: ABCDE_FG_HIJ. Then the bigram (2-gram) observation will look: _A, AB, BC, CD, DE, E_, _F, FG, G_, _H, HI, IJ. Analogically the trigram (3-gram) observation will look: __A, _AB, ABC, BCD, CDE, DE_, E_F, _FG, FG_, G_H, _HI, HIJ. Figure 1 visualizes states path and corresponding observations for first order HMM. Certainly, by N-gram operation the observation is enhanced by occurrence of past N expressions. Still, state sequence did not change. Nevertheless, N-gram conversion affects number of observation and observation probability distribution, thereby to estimate HMM parameters at least O·N more training data is required to obtain similar statistical validity in comparison to basic HMM. Similar operation can be performed in terms of states sequence. This technique is well known and refers to M-order Markov model [18]. The M-order chain can be created by combining M past states together. Let us consider the second order HMM (St = 9, Obs = 11), then to create a first order Markov chain each state will include combination with other state to allow precedent state occurrence memory. So basics first order states (1-vowel, 2-consonant, 3-pause) will be transformed in second order model by full states combination: 11, 12, 13, 21, 22, 23, 31, 32, 33, (where e.g. state 12 correspond to vowel-consonant; state 22 correspond to consonant-consonant). Regardless to the order of Markov model the observation sequence stay the same:

11-A, 12-B, 22- C, 22- D, 21- E, 13-_, 32- G, 23- _, 32- H, 21-I, 12-J.

Analogically, for N-gram derivate, e.g. trigram observations:

11-__A, 12-_AB, 22-ABC, 22-BCD, 21-CDE, 13-DE_, 32-E_G, 22-_FG, 23- FG_, 32-G_H, 21-_HI, 12-HIJ.

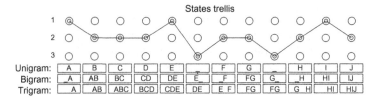

Fig. 1. First order HMM states path and corresponding observation sequence.

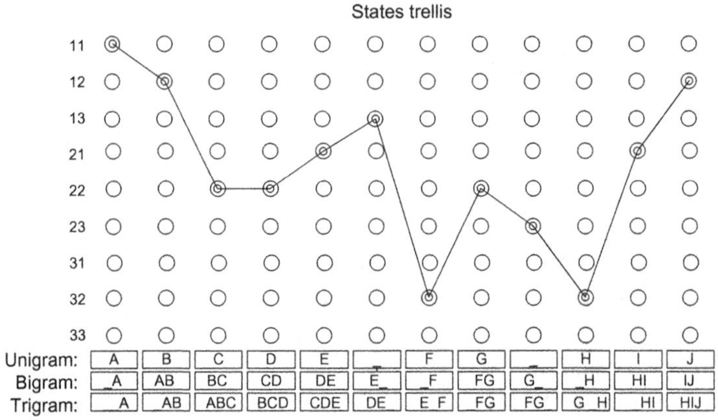

Fig. 2. Second order HMM states path and corresponding observation sequence.

Figure 2 visualizes states path and corresponding observations sequence for second order HMM.

However, higher-order Markov models are able to model only fixed length dependencies and in practice are strong limited by exponential growth in number of states with every next order. This example shows that to cover all states occurrence in M order HMM at least S·M more training data is required.

Combination of both M-order HMM with N-gram observation results in (O·N)· (S·M) more training data requirements. Both approaches modify the HMM model in terms of maximizing the probability of the data and from this point of view greater model complexity and higher demanding on training data possibly pays off with better HMM fitting to the data.

4   HMM in the Field of Syntactic Classification

The process of determining a syntactic class for each word in the analyzed text can be achieved by using HMM, where the discreet observation represent the word while the state of the Markov Model will represent the corresponding part of speech/sentence. In order to estimate all parameters correctly, it is crucial to determine what kind of features are represent by the observation. In presented models it is assumed that the word's affixes are more important for syntactic analysis then a stem. Hence, every word is represented only by its Prefix and Suffix, while the stem can be omitted. For the simplicity, both Prefix and Suffix has a fixed number of characters. Consequently, for short words, the values are redundantly overlapped (e.g. with - > "wit-ith"), while for longer words middle part of the word is cut out (e.g. approachable - > "app-ble"). The main advantage of this approach is a simplified mechanism for inclusion of new words that were not involved in the training process. A fixed number of characters in both Suffix and Prefix results a limited number of possible combinations. Hence, the vast majority of syntactic classes can be covered even at smaller training database. In other words, affix oriented expression groups more words to the same observation (e.g. "accorda ble" = "accrua ble" or "administra ble" = "admira ble"). Words that have different, unknown stem will be categorized by their Prefix and Suffix learned from training database.

4.1 N-gram Affix Oriented Observation

A basic observation is built from word's Prefix and Suffix. The N-gram observation is created by combining N past observations together and shifting them in queue with length equal to N. However, affix oriented observations can be degraded by removing letters from the inner side of the word. Correspondingly, the total number of possible combinations significantly decreases, revealing only important statistic relations. Simple sentence case study example in Table 1 shows the main idea about affix N-gram observation and its derivations.

Because of progressive clarity reduction for past elements in the gram sequence the presented example, is limited to use up to 3 grams in the observation. The progressive character of the N-gram observation modification is here utilized to limit the information carried by the past grams. However, observation window can as well have (for all or for some selected grams) a fixed length. At this point three different affix observation

Table 1. Applied affix modification. Uni-, Bi- and Tri-gram observations extracted from example sentence: "those who cannot remember the past are condemned to repeat it".

Raw	Unigram	Bigram	Trigram
those	tho-ose	___-___: tho-ose	___-___ : ___-___: tho-ose
who	who-who	th_-_se: who-who	___-___: th_-_se: who-who
cannot	can-not	wh_-_ho: can-not	t__-__e: wh_-_ho: can-not
remember	rem-ber	ca_-_ot: rem-ber	w__-__o: ca_-_ot: rem-ber
the	the-the	re_-_er: the-the	c__-__t: re_-_er: the-the
past	pas-ast	th_-_he: pas-ast	r__-__r: th_-_he: pas-ast
are	are-are	pa_-_st: are-are	t__-__e: pa_-_st: are-are
condemned	con-ned	ar_-_re: con-ned	p__-__t: ar_-_re: con-ned
to	to_-_to	co_-_ed: to_-_to	a__-__e co_-_ed: to_-_to
repeat	rep-eat	to_-_to: rep-eat	c__-__d: to_-_to: rep-eat
it	it_-_it	re_-_at: it_-_it	t__-__o: re_-_at: it_-_it

structures for N-gram observation can be distinguished. Absolute progressive: every subsequent gram is degraded more than its predecessor until the loss of whole information. Offset progressive: every subsequent gram is degraded more than its predecessor until reach some level of information which is mandatory. Constant affix: here the level of observation clarity is adjusted for all grams.

4.2   Model Observation Evaluation

When using an affixes based observation a question about the accuracy of word mapping arises. If length of affixes is large enough all words are directly mapped, which means that even very similar words are distinguished from each other. However, if the length is too small, too many different words will be grouped together and the

Table 2. Dictionary coverage regarding to different Prefix-Suffix characters length.

Nr	Prefix	Suffix	Observations for English	English dictionary coverage [%]	Observations for Polish	Polish dictionary coverage [%]
1	1	1	685	1.07	967	1,03
2	1	2	5191	8.12	5879	6,29
3	1	3	18152	28.40	18952	20,29
4	1	4	35291	55.22	36264	38,82
5	1	5	48180	75.39	54232	58,06
6	2	1	4518	7.06	6681	7,15
7	2	2	17710	27.71	20469	21,91
8	2	3	36086	56.46	39447	42,23
9	2	4	49322	77.17	56025	59,98
10	2	5	56689	88.70	70834	75,84
11	3	1	20479	32.04	25185	26,96
12	3	2	38939	60.93	46657	49,95
13	3	3	51251	80.19	62991	67,44
14	3	4	57689	90.27	74076	79,31
15	3	5	61065	95.55	83169	89,04
16	4	1	40348	63.13	49335	52,82
17	4	2	52309	81.85	68175	72,99
18	4	3	58367	91.33	78218	83,74
19	4	4	61268	95.87	84347	90,31
20	4	5	62773	98.22	89550	95,88
21	5	1	52345	81.90	67430	72,19
22	5	2	58654	91.78	80607	86,30
23	5	3	61535	96.28	86370	92,47
24	5	4	62784	98.24	89962	96,32
25	5	5	63423	99.24	92882	99,44
All words:			63906	100	93396	100

accuracy of syntactic recognition will suffer or even go to zero. Table 2 presents dictionary coverage regarding to different Prefix-Suffix character length [19, 20].

As it is shown in Table 2 the length of the affixes affect word representation and word grouping. For Prefix and Suffix with the length of 2 characters each, the clarity of word representation is about 28 % (17710 different affixes represents 63906 words), while for affixes with single character each gives 1 % of clarity of word representation (685 different affixes represents 63906 words). Similar situation is for Polish language.

For completely new expression (misspelled or not included in the training database) some additional precautions method may be applied. Primarily by detecting unknown expression a Levenshtein distance [21] can be measured in order to find closest observation that approximately match this expression. For N-gram observation the Levenshtein distance should be calculated for each word observation in N-word sequence and each obtained value should be weighted with multiplicative inverse of N. Depends on the analyzed language the Levenshtein distance can be modified in order to mitigate Suffix or Prefix impact as it is shows in [22]. If the expression does not fit to any representation then such expression is classified as unknown (___-___).

4.3   HMM Preparation

Presented process of preparation assumes that the HMM is based on affix observations which are statistically significant so that HMM will be able to determine the membership of the word to the one of the syntactic classes. At HMM initialization a number of states is established based on syntactic tags found in treebank database.

The database for Polish language (Składnica treebank) consist more than twenty thousand sentences with more than 200 thousand syntactically tagged words. The database for English language (based on Penn treebank) consist more than ten thousand sentences with more than 200 thousand syntactically tagged words. The set of all possible affix observations for given language is obtained from the proper treebank database. However, the initialization for emissions probabilities is based on Wordnet database. All counters related to observation occurrence are initialized with the value corresponding to the membership level of a given affix to the given Part-Of-Speech (Noun, Verb, Adjective or Adverb) in Wordnet. For example, counter for affix "re-al" will be initialized with values (4-strong, 1-weak): 4 for Noun class, 3 for Adjective class, 2 for Verb class and 1 for Adverb class. To overcome data sparseness problem and to improve probabilities estimation for unseen observations the additional Laplace smoothing is also applied.

5   Experiments and Results

Classification of words syntactic category has been made by calculating the Viterbi path. Correctness of syntactic class recognition (expressed in percentage) is represent by average accuracy of all HMM states. For affix observation structure two digits are assigned, the first determines the number of letters for the Prefix and the second digit determines the number of letters for the Suffix. The verification was conducted on test database. A part of unknown expression occurred in test database is also pointed out. Table 3 lists all 10 HMMs included in the experiment.

The HMM for Polish Dependency Types [23] classification and HMM for English Phrase Chunk classification are hierarchical HMMs [24] based on word's affix observation and on POS class recognized previous by HMM for Part-Of-Speech classification for a given language.

6 Discussion

Results presented in Table 3 for unigram HMMs conforms that by using affix oriented observation HMM performs well. Even for single letters affixes more than 75 % of class ware recognize correctly. The differences in the observation number may reach decimal of percents. This is due to combinatorial limitations of the model where a smaller number of letters is subject of observation. The biggest advantage of HMMs with simple and small set of observations is the computation time. For many applications (where computation latency/time is essential) complex HMM compute to slow and despite the higher accuracy more useful is a smaller model even with some classification deficiency.

Table 3. Classification results for affix based HMM in the field of syntactic analysis.

HMM description	Number of states/Observations	Size of Prefix-Suffix	Accuracy [%]	Unknown [%]
Polish Part-Of-Speech	40/836	1-1	75,89	0.07
Polish Part-Of-Speech	40/4791	1-2	83.69	0.67
Polish Part-Of-Speech	40/14703	2-2	88.40	2.87
Polish Part-Of-Speech	40/36196	3-3	93.63	10.31
Polish Dependency Types	28/11579	2-2 +POS state	82.54	7.34
English Part-Of-Speech	47/767	1-1	80.02	0.13
English Part-Of-Speech	47/3994	1-2	90.20	0.93
English Part-Of-Speech	47/9055	2-2	91.20	2.69
English Part-Of-Speech	47/15546	3-3	95.26	5.48
English Phrase Chunk	24/12915	2-2 +POS state	82.70	3.89

By affix orientation modification the knowledge generalization takes place at the observation level. Nonetheless, the clarity of the observation should be chosen carefully. Insufficient level of accuracy may lead to the loss of its statistical significance and the observation becomes useless.

Obtained results for unigram HMMs are very promising especially for Polish language. The accuracy of Part-Of-Speech classification is comparable with the results presented in [25]. Although, the Dependencies Types classification is much more difficult task still unigram HMM ware able to achieve more than 80 % of classification correctness.

Nevertheless, classification results presented in this paper refers only to unigram HMMs. Hence, the further research will concentrate on examination of HMMs based on N-gram affix oriented observations. Presumably, N-gram variant should significantly improve the classification accuracy.

7 Conclusions

In the paper affix oriented observations was introduce and comprehensively described. The analysis was performed to examine the correlation between clarity of word representation (affix size) and HMM classification accuracy. In addition, a general example for M-order HMM model N-grams was presented for concept better understanding and to point out further application of affix oriented observation in order to decrease model complexity.

Evaluated Hidden Markov Models based and trained on Penn Treebank (for English) and on Składnica Treebank (for Polish) database prove that HMMs are suited to the language processing in the field of syntactic tagging even by limited clarity of observations. The number of unknown words has significantly decreased for HMMs that less accurate observation. Furthermore, accuracy of recognizing the syntactic class remained at a similar level in comparison to models with exact observation. Depending on the requirements of the programs which utilize a syntactic analysis, different HMMs derivatives can be useful. If very high accuracy is required, a complex N-gram HMM with additional support for unknown words recognition should be applied. However, if processing speed is a priority and accuracy plays a secondary role then HMM based on simplified observation (single character affixes) will be more appropriate.

References

1. Kupiec, J.: Robust part-of-speech tagging using a hidden Markov model. In: Computer Speech and Language, pp. 225–242 (1992)
2. Goldwater, S., Griffiths, T.: A fully Bayesian approach to unsupervised part-of-speech tagging. In: Proceedings of the 45th Annual Meeting of the Association of Computational Linguistics, Prague, Czech Republic, pp. 744–751. Association for Computational Linguistics, June 2007
3. Gao, J., Johnson, M.: A comparison of Bayesian estimators for unsupervised hidden Markov model pos taggers. In: Proceedings of the 2008 Conference on Empirical Methods in Natural Language Processing, pp. 344–352 (2008)

4. Lioma, C.: Part of speech n-grams for information retrieval. Ph.D. thesis, University of Glasgow (2008)
5. Brants, T.: TnT — A statistical part of speech tagger. In: Proceedings of the 6th Applied NLP Conference(ANLP-2000), pp. 224–231 (2000)
6. Thede, S.M.: Predicting part-of-speech information about unknown words using statistical methods. In: Proceedings of the 36th Annual Meeting of the Association for Computational Linguistics - v.2, pp. 1505–1507 (1998)
7. Nakagawa, T., Kudoh, T., Matsumoto, Y.: Unknown word guessing and part-of-speech tagging using support vector machines. In: Proceedings of the Sixth Natural Language Processing Pacific Rim Symposium, pp. 325–331 (2001)
8. Jurafsky, D., Martin, J.H.: Speech and Language Processing. Prentice Hall, Upper Saddle River (2000)
9. Tseng, H., Jurafsky, D., Manning, C.: Morphological features help POS tagging of unknown words across language varieties. In: Proceedings of the Fourth SIGHAN Bakeoff (2005)
10. Luong, M.T., Nakov, P., Ken, M.Y.: A hybrid morpheme-word representation for machine translation of morphologically rich languages. In: Proceedings of the Conference on Empirical Methods in Natural Language Processing (EMNLP 2010), Cambridge, MA, pp. 148–157 (2010)
11. Adler, M.: Hebrew morphological disambiguation: an unsupervised stochastic word-based approach. Ph.D. thesis, Ben-Gurion University of the Negev, Israel (2007)
12. Taylor, A., Marcus, M., Santorini, B.: The Penn Treebank: An Overview (2003)
13. Hajnicz, E.: Lexico-semantic annotation of składnica treebank by means of PLWN lexical units. In: Proceedings of the Seventh Global WordNet Conference, Tartu, Estonia, pp. 23–31 (2014)
14. Rabiner, L.R.: A tutorial on hidden Markov models and selected applications in speech recognition. Proc. IEEE **77**(2), 257–286 (1989)
15. Jahmm, Java implementation of HMM related algorithms (2009)
16. Layton, M.: Augmented Statistical Models for Classifying Sequence Data (2006)
17. Langkilde, I., Knight, K.: The practical value of n-grams in generation. In: Proceedings of the Ninth International Workshop on Natural Language Generation, Niagara-on-the-Lake, Ontario, pp. 248–255 (1998)
18. Lee, L.-M., Lee, J.-C.: A study on high-order hidden Markov models and applications to speech recognition. In: Ali, M., Dapoigny, R. (eds.) IEA/AIE 2006. LNCS (LNAI), vol. 4031, pp. 682–690. Springer, Heidelberg (2006)
19. Fellbaum, C.: WordNet: An Electronic Lexical Database. MIT Press, Cambridge (1998)
20. Maziarz, M., Piasecki, M., Szpakowicz, S.: Approaching plWordNet 2.0. In: Proceedings of the 6th Global Wordnet Conference, Matsue, Japan (2012)
21. Levenshtein, A.: Binary codes capable of correcting deletions, insertions and reversals. Sov. Phys. Dokl. **10**(8), 707–710 (1966)
22. Pietras, M.: Sentence sentiment classification using fuzzy word matching combined with fuzzy sentiment classifier. Electrical Review - Special issue, Poland (2014). doi:10.15199/48.2015.02.26
23. Wróblewska, A.: Polish dependency parser trained on an automatically induced dependency bank. Ph.D. dissertation, Institute of Computer Science, Polish Academy of Sciences, Warsaw (2014)
24. Fine, S., Singer, Y., Tishby, N.: The hierarchical hidden Markov model: analysis and applications. Mach. Learn. Boston **32**, 41–62 (1998)
25. Kobyliński, Ł.: PoliTa: a multitagger for Polish. In: Proceedings of the Ninth International Conference on Language Resources and Evaluation, Iceland, pp. 2949–2954 (2014)

Opinion Acquisition: An Experiment on Numeric, Linguistic and Color Coded Rating Scale Comparison

Olga Pilipczuk[1(✉)] and Galina Cariowa[2]

[1] University of Szczecin, Mickiewicza 64, 71-101 Szczecin, Poland
olga.pilipczuk@wneiz.pl
[2] West Pomeranian University of Technology,
Żołnierska 49, 70-210 Szczecin, Poland
g.tariova@wi.zut.pl

Abstract. The paper presents the problem of acquiring person opinion using different rating scales. Particular attention is given to the scale created using a visible color spectrum and to the selection of the optimum number of colors in the scale. This paper aims to compare the effectiveness of the color coded scale, word scale and selected numerical scales during the process of students opinion acquisition. Opinions were collected by questionnaire and interview. The authors compare the average time of giving answers and the cognitive load (mental effort), and describe the problems occurring in the question-answer process. It was found that the opinion is most difficult to acquire using the word scale, while the results are most effective with a color and −10 + 10 numerical scale.

Keywords: Rating scale · Color coding · Opinion acquisition · Cognitive load

1 Introduction

Knowledge acquisition is the process by which the knowledge engineer attempts to extract information from a person: opinions on an object or a phenomenon, the evaluation of an object or a phenomenon, comments, notes, etc. [1]. Information from a person is usually extracted by means of a numerical or a linguistic scale. A constraint of numerical data is that they are unable to reflect fuzzy information coming from the environment, whereas verbal information is not sufficiently precise and has a limited scale. For a long time now scientists have argued that phenomena or object descriptions using linguistic data are very labor intensive, controversial, and imprecise. Typically, five to seven linguistic quantifiers are used; this significantly limits the precision of evaluating the object or criterion and thus makes precise forecasting using verbal data impossible. In addition, for every person the same word can have different meanings.

The situation where we have to manipulate numbers and words at the same time very often appears. This indicates the need for data integration. The traditional method of numerical data "fuzzing" for their integration with linguistic data significantly reduces the precision of the result. Therefore, researchers are constantly looking for new tools that enable the integration of numerical and linguistic data and that will also increase the

© Springer International Publishing AG 2017
S. Kobayashi et al. (eds.), *Hard and Soft Computing for Artificial Intelligence, Multimedia and Security*, Advances in Intelligent Systems and Computing 534, DOI 10.1007/978-3-319-48429-7_3

accuracy of the data. To this end, among the most interesting modern knowledge acquisition tools are the multiple linguistic scales presented by [2]. These authors introduce the extended linguistic hierarchy to manage multiple linguistic scales in a precise way. The model is based on the linguistic hierarchies, but the extension defines a new way to build a multi-granular evaluation framework with more flexibility [3].

Many knowledge acquisition methods and techniques were adapted from cognitive methods or from other disciplines such as anthropology, ethnography, counseling, education, and business management [4–7]. The results of such studies indicate that during expert opinion acquisition the intuitive approach should be used to manipulate the information on the cognitive level. Another problem in the acquisition of expert knowledge is the appropriate reflection of feelings and emotions, which sometimes are more important than quantifiable data. Even when using words we cannot always determine our perception. We can use color in powerful ways to enhance the meaning and clarity of data displays, but only when we understand how it works, what it does well, and how to avoid problems that often arise when it's used improperly [8].

2 Related Works

Color coding is a fundamental visualization method for representing scalar values [9]. The coding problem is precisely described in works [10, 11]. Well-realized encoding can make features and patterns in data easy to find, select and analyze [9]. There are many possible ways of coloring the values of a single continuous variable, but they all involve the mapping of a set of values to a color scale, an ordered arrangement of colors [9].

At the moment, a lot of different measurement color coding scales are used in practice. One of the cognitive tools of knowledge acquisition is a visible color spectrum scale. Pain scales can be used as a tool for opinion acquisition on a cognitive level [12]. This scale measures a patient's pain intensity or other features and is based on self-reported philological data. Although different pain scales often have been used interchangeably, it is not known whether interchanging them is appropriate [13].

The color spectrum scale is also used in many other situations and fields of science. For example, such a scale is used in geographical research to represent the level of temperature, salinity, density, and frequency of the ocean biosphere [14]. In making medical diagnosis such a scale is used to indicate the surface plasmon resonance angle during the study of the live cell membrane or in echocardiography images [15]. In physics, a spectrum scale is used, for example, to show the correlation between different spectral components during the investigation of a single electron transport through semiconductor nanostructure [16].

Many researches have been conducted to compare the effectiveness of different rating scales [17, 18]. However, there is a lack of studies on the effectiveness of a color coding scale application and on the determination of which scale – word, number, or color – more accurately reflects the people opinions. Therefore, we sought to determine the optimal number of colors contained in the visible color spectrum scale and organized a number of experiments to compare the linguistic, numerical, and color rating scales. Expert reviews were collected by interview or questionnaire.

There are several ways to calculate the effectiveness of visualization: subjective (or self-report) measures such as rating scales, performance-based measures such as secondary tasks and physiological measures such as pupillary dilation (eye-tracking) [19].

Cognitive load (CL) is defined as an individual's cognitive capacity which is used work on a task, to learn, or to solve a problem [20]. The measurement of CL is highly complicated and problematic activity [21]. Many studies adapt a scale developed by Paas [22] or improve it [23]. Leppink et al. [24] have published an method to separately measure the different types of CL: intrinsic, extraneous, and germane load.

Many authors treated the cognitive load as amount of mental effort. Mental effort refers to the amount of cognitive capacity that is actually allocated to accommodate the demand imposed by the task; thus mental effort reflects the actual cognitive load [25].

Methods and techniques used to measure mental effort are the same as in case of visualization effectiveness measurement. The subjective rating scale method has is the most widespread technique [26]. However, some studies reveal the disadvantages of subjective method [27].

Finely, Paas reviewed these studies and concluded that rating scale techniques "are easy to use; do not interfere with primary task performance; are inexpensive; can detect small variations in workload (that is, sensitivity); are reliable; and provide decent convergent, construct, and discriminate validity" [23].

The subjective mental effort rating scales according to Sweller [28] can be used to measure extraneous load and intrinsic load. Ayres [29] also argues that subjective rating scales can be used to measure intrinsic load, if the germane and extraneous load are kept constant. One of most widespread subjective mental effort rating scale was proposed by Kalyuga, Chandler and Sweller [30]. The scale consists of a seven points from 'extremely easy' to 'extremely difficult'. We applied this scale to our research, due to its easiness and clarity. The following section presents the experiments to compare the different rating scale types. We tried to confirm the assumption that the color coding scale more effectively reflects the views of the people than other scales do, analyzing the cognitive load and response time.

3   Experiments

To calculate the effectiveness of selected rating scales we consider the students opinion on their age and on the weather conditions separately. Our aim was to compare the scales effectiveness adapted to different data types (age-amount in numbers, weather – sum of the definitions and phenomenon).

Our choice was caused by the fact that the climate conditions involve large and heterogeneous datasets with spatial and temporal references [31]. This makes climate research an important application area for color coding visualization [32, 33]. However, visualization of weather and climate data is remain challenging for climate research scientists, who are usually not visualization experts [31].

Nevertheless, a color coded weather warning system has been introduced by the UK Met Office to alert the public to the predicted severity of storms and winds, highlighting areas in red, amber, yellow or green to advise people on what weather conditions to expect. The color coded scales also are used to describe the weather in pilot guides systems.

The experiments runs as follows. The first experiment aimed to find the optimal number of colors in the color coded scale and to determine whether it is effective to obtain students opinions on description of numerical and abstract objects. We selected and compared color coded scales contained from 15 to 25 colors from violet to red. All the scales were center concentrated with green center point (See Figs. 1, 2, 3, 4, 5 and 6). We calculated the effectiveness of these scales on the basis of the response time and the cognitive load (mental effort).

The second experiment aimed to compare the effectiveness of numeric −10 + 10 scale, numeric 1–21 scale and 21-word scale with the effectiveness of color scale contains optimal number of colors, obtained in previous experiment. It was noted during preliminary tests that many students the green color intuitively serves as a center of the scale, which is why each scale includes an odd number of colors.

The experiment examined the opinions of 70 IT students. First, they evaluated their age, using color coded scales (Figs. 1, 2, 3, 4, 5 and 6).

Next, they were asked to calculate the time and answer the following questions:

- "Do you have a problem with giving answers?" The possible answers were as follows:

  1. Application response does not cause me any problems.
  2. Yes, the scale is too short.
  3. Yes, the scale is too long.
  4. Another problem: ....
  5. I have a problem with understanding the scale, but I cannot determine what it is.

- "Which scale do you prefer?"

The obtained results suggest (Table 1) that the more suitable for students are 19- and 21-color scales. The Fig. 1 shows that for further experiments 21-color scale was the most suitable one.

min   max

Fig. 1. Color coded scale with 15 colors (Color figure online)

min   max

Fig. 2. Color coded scale with 17 colors (Color figure online)

min   max

Fig. 3. Color coded scale with 19 colors (Color figure online)

min   max

Fig. 4. Color coded scale with 21 colors (Color figure online)

min   max

Fig. 5. Color coded scale with 23 colors (Color figure online)

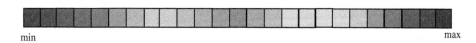

min   max

Fig. 6. Color coded scale with 25 colors (Color figure online)

Table 1. The results of the color coded scales comparison

	Average time	Preference	Number of problems
Figure 1	8,3	14	16
Figure 2	7,7	1	12
Figure 3	5,5	7	10
Figure 4	5,3	8	12
Figure 5	6	6	17
Figure 6	6,2	32	18

During the second experiment students were asked to evaluate the "attractiveness" of the day's weather conditions using 21 color, numeric and word scales presented in Fig. 7 and, after that answer the question about how hard was to give the answer. The scale which contained a seven points from 'extremely easy' to 'extremely difficult' was used. The response time in seconds was checked for each of the presented scales separately (Fig. 8).

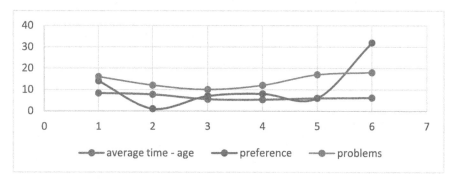

Fig. 7. The visualization of results of color coded scales comparison (Color figure online)

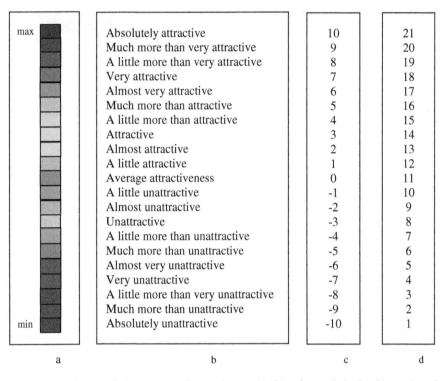

max	Absolutely attractive	10	21
	Much more than very attractive	9	20
	A little more than very attractive	8	19
	Very attractive	7	18
	Almost very attractive	6	17
	Much more than attractive	5	16
	A little more than attractive	4	15
	Attractive	3	14
	Almost attractive	2	13
	A little attractive	1	12
	Average attractiveness	0	11
	A little unattractive	-1	10
	Almost unattractive	-2	9
	Unattractive	-3	8
	A little more than unattractive	-4	7
	Much more than unattractive	-5	6
	Almost very unattractive	-6	5
	Very unattractive	-7	4
	A little more than very unattractive	-8	3
	Much more than unattractive	-9	2
min	Absolutely unattractive	-10	1
a	b	c	d

Fig. 8. The scales used during second experiment (a) 21-color coded, (b) 21-word scale, (c) −10 + 10 numerical scale, (d) 1–21 numerical scale (Color figure online)

4   Result and Discussion

Table 1 shows the results of the first experiment. Sixteen students declared a problem in giving answers based on a scale of 15 colors. For them, the scale was too short; it should contain more of the green color. Some of the students can not notice the

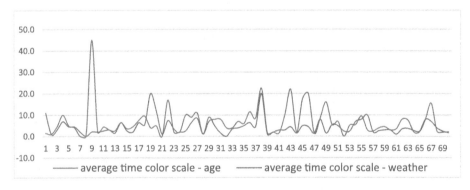

Fig. 9. The results of time comparison. (Color figure online)

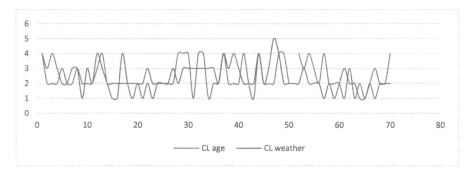

Fig. 10. The results of cognitive load comparison (Color figure online)

difference between the color tones. The results obtained using scales of 19 and 21 colors looked almost identical and was the best. However the 21-color scale obtain a little better points. The average time of response had the increased trend. Furthermore, the scale containing 25 colors had a little success. The group of 32 students preferred this scale, but the rest of indexes get the non-impressive values.

The Figs. 9 and 10 show the results of color scales comparison based on average response time and CL estimation. As long as CL results were similar, the response time differ slightly more in favor to weather conditions.

The second experiment allowed to prove the high effectiveness of the color coded scales. The response time varied from 1 to even 50 s depending on scale. The Fig. 11 presents the results of average response time estimation for all of the scales independently: the best are for color scales and $-10 + 10$ scale, the worst - are for word scale.

The mental effort appeared at the same way. The score are assigned as follows: (1) the $-10 + 10$ scale, (2) the color scales, (3) the word and 1–21 scale (Fig. 12).

The main disadvantage of the color scales was noticed the lack of expressly underlined center.

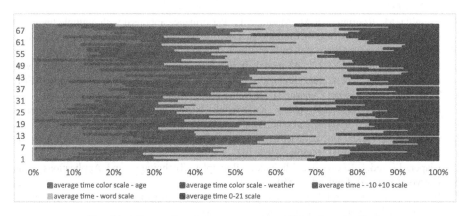

Fig. 11. The results of time comparison (Color figure online)

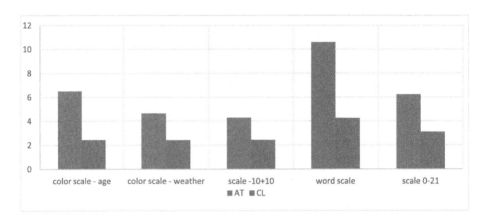

Fig. 12. The rating scales effectiveness according to average response time and cognitive load (mental effort). (Color figure online)

5   Conclusion

It was found that the opinion is most difficult to acquire using the word scale, while the results are most effective with a color and −10 + 10 numerical scale. Despite of the our expectations some problems appear. The main problem in color scale using the definition of center point. It can be obtained by integration of color scale and −10 + 10 scale with 0 center point.

References

1. Regoczei, S.B., Hirst, G.: Knowledge and knowledge acquisition in the computational context. In: Hoffman, R.R. (ed.) The Psychology of Expertise, pp. 12–25. Springer, New York (1992)
2. Espinilla, M., Rodriguez, R.M., Martinez, L.: Performance appraisal with multiple linguistic scales. In: de Andres, R. (ed.) Intelligent Decision Making Systems, vol. 2, pp. 433–443. PRESAD Research Group (2010)
3. Herrera, F., Martinez, L.: A model based on linguistic 2-tuples for dealing with multigranularity hierarchical linguistic context in multi-expert decision-making. IEEE Trans. Fuzzy Syst. Man Cybern. **31**, 227–234 (2001). doi:10.1109/3477.915345
4. Boose, J.H., Gaines, B.R. (eds.): Knowledge Acquisition Tools for Expert Systems, Knowledge Based System, 2nd edn. Academic Press, San Diego (1988)
5. Cooke, N.J.: Varieties of knowledge elicitation techniques. Int. J. Hum.-Comput. Stud. **41**, 801–849 (1994). doi:10.1006/ijhc.1994.1083
6. Diaper, D. (ed.): Knowledge Elicitation: Principles, Techniques, and Applications. Ellis Horwood Limited, England (1989)
7. Hoffman, R.R.: The problem of extracting the knowledge of experts from the perspective of experimental psychology. AI Mag. **8**, 53–67 (1987)

8. Stephen Few Practical Rules for Using Color in Charts, Perceptual Edge, Visual Business Intelligence Newsletter, allowed at February 2008. http://www.perceptualedge.com/articles/visual_business_intelligence/rules_for_using_color.pdf
9. Tominski, C., Fuch, G., Schumann, H.: Task-driven color coding. 2008 12th International Conference on Information Visualisation, IV 2008, pp. 373–380 (2008)
10. Stone, M.C.: A Field Guide to Digital Color. A.K. Peters, Natick (2003)
11. Stone, M.C.: Color in information display. In: Tutorial, IEEE Visualization Conference, Sacramento, USA, October 2007
12. Whitworth, M.: A Review of the Evaluation of Pain Using a Variety of Pain Scales. https://cme.dannemiller.com/articles/activity?id=318
13. Freeman, K., Smyth, C., Dallam, L., Jackson, B.: Pain measurement scales: a comparison of the visual analogue and faces rating scales in measuring pressure ulcer pain. J. Wound Ostomy Continence Nurs. **28**(6), 290–296 (2001)
14. Zappa, C.J., Ho, D.T., McGillis, W.R., Banner, M.L., Dacey, J.W.H., Bliven, L.F., Ma, B., Nystuen, J.: Rain-induced turbulence and air-sea gas transfer. J. Geophys. Res. **114** (2009). doi:10.1029/2008JC005008
15. Yoshifuku, S., Chen, S., McMahon, E., Korinek, J., Yoshikawa, A., Ochiai, I., Sengupta, P., Belohlavek, M.: Parametric detection and measurement of perfusion defects in attenuated contrast echocardiographic images. J. Ultrasound Med. Official J. Am. Inst. Ultrasound Med. **26**(6), 739–748 (2007)
16. Ubbelohde, N., Fricke, Ch., Flindt, Ch., Hohls, F., Haug, R.J.: Measurement of finite-frequency current statistics in a single-electron transistor. Nat. Commun. **3**, 612 (2012). doi:10.1038/ncomms1620
17. Couper, M.P., Tourangeau, R., Conrad, F.G., Singer, E.: Evaluating the effectiveness of visual analog scales: a web experiment. Soc. Sci. Comput. Rev. **24**, 227–245 (2006). doi:10.1177/0894439305281503
18. Hyun, Y.: Nonlinear Color Scales for Interactive Exploration (2001). http://www.caida.org/~youngh/colorscales/nonlinear.html. Accessed Apr 2008
19. De Waard, D.: The measurement of drivers' mental workload. Ph.D. thesis, University of Groningen, Haren, The Netherlands (1996)
20. Sweller, J., Ayres, P., Kalyuga, S. (eds.): Cognitive Load Theory. Springer, New York (2011)
21. Kirschner, P., Ayres, P., Chandler, P.: Contemporary cognitive load theory research. Comput. Hum. Behav. **27**, 99–105 (2011)
22. Paas, F.: Training strategies for attaining transfer of problem-solving skill in statistics. J. Educ. Psychol. **84**, 429–434 (1992)
23. Paas, F., Tuovinen, J., Tabbers, H., van Gerven, P.: Cognitive load measurement as a means to advance cognitive load theory. Educ. Psychol. **38**(1), 63–71 (2003)
24. Leppink, J., Paas, F., van der Vleuten, C., van Gog, T., van Merriënboer, J.: Development of an instrument for measuring different types of cognitive load. Behav. Res. Methods (2013). doi:10.3758/s13428-013-0334-1
25. Huanga, Weidong, Eadesb, Peter, Hongb, Seok-Hee: Measuring effectiveness of graph visualizations: a cognitive load perspective. Inf. Vis. **8**, 139–152 (2009). doi:10.1057/ivs.2009.10
26. Hendy, K.C., Hamilton, K.M., Landry, L.N.: Measuring subjective workload: when is a one scale better than many? Hum. Factors **35**(4), 579–601 (1993)
27. Gopher, D., Braune, R.: On the psychophysics of workload: why bother with subjective measures? Hum. Factors **26**, 519–532 (1984)
28. Sweller, J., van Merriënboer, J., Paas, F.: Cognitive architecture and instructional design. Educ. Psychol. Rev. **10**, 251–296 (1998)

29. Ayres, P.: Using subjective measures to detect variations of intrinsic cognitive load within problems. Learn. Instr. **16**, 389–400 (2006)
30. Kalyuga, S., Chandler, P., Sweller, J.: Managing split-attention and redundancy in multimedia learning. Appl. Cogn. Psychol. **13**, 351–371 (1999)
31. Tominski, C., Donges, J.F., Nocke, T.: Information visualization in climate research. In: 2011 15th International Conference on Information Visualisation (IV), pp. 298–305 (2011)
32. Nocke, T., Heyder, U., Petri, S., Vohland, K., Wrobel, M., Lucht, W.: Visualization of Biosphere Changes in the Context of Climate Change. In: Wohlgemuth, V. (ed.) Information Technology and Climate Change – 2nd International Conference IT for Empowerment. trafo Wissenschaftsverlag, pp. 29–36 (2009)
33. Ladstädter, F., Steiner, A.K., Lackner, B.C., Pirscher, B., Kirchengast, G., Kehrer, J., Hauser, H., Muigg, P., Doleisch, H.: Exploration of climate data using interactive visualization. J. Atmos. Oceanic Technol. **27**(4), 667–679 (2010). doi:10.1175/2009JTECHA1374.1

A Weather Forecasting System Using Intelligent BDI Multiagent-Based Group Method of Data Handling

W. Rogoza$^{(\boxtimes)}$ and M. Zabłocki

Faculty of Computer Science and Information Technology, West Pomeranian University of Technology, Szczecin, ul. Żołnierska 52, 71-210 Szczecin, Poland
{wrogoza,mzablocki}@wi.zut.edu.pl

Abstract. In the article the concept of analysis of complex processes based on the multi-agent platform is presented. The analysis based on Group Method of Data Handling is employed to construct a model of the analysed process. The model is used to predict future development of the process. Employed multi-agent platform composed of BDI agents provides intelligent distributed computational environment. The description of this approach, case study examining various prediction model criterion and evaluationally summary of the approach is provided in the article.

Keywords: Multi-agent system · Processes analysis · BDI agent · Java Agent Development Framework · Group Method of Data Handling

1 Introduction

The analysis of processes of different kind has two main objectives: (a) detecting the nature of the phenomenon represented by the sequence of observation and (b) forecasting (predicting future values of the processes). Both of these goals require to identify and describe, in a more or less formal manner, elements of the processes. Once established pattern can be applied to other data (i.e. used in the theory of the studied phenomenon, for example seasonal goods prices).

In our approach we apply mathematical and computer instrumentation to build a model, and further, on the basis of this model perform a prediction of future development of the process. To build mathematical model of the analysed system, the Group Method Of Data Handling (GMDH) algorithm can be used. The statistical method based on building polynomial models allows us to study the relationships between data values. Following the steps of GMDH algorithm, the complex task of constructing the model require selection of input variables, estimation of polynomial coefficients using the least squares and usage of obtained polynomial during the construction of next layer if it is desirable.

In this paper we concentrate on some computational aspects of analysis of complex processes by distributed systems. As a domain of application of our

© Springer International Publishing AG 2017
S. Kobayashi et al. (eds.), *Hard and Soft Computing for Artificial Intelligence, Multimedia and Security*, Advances in Intelligent Systems and Computing 534, DOI 10.1007/978-3-319-48429-7_4

approach we have chosen the weather phenomena. The complex nature of this problem require sophisticated solution - the use of artificial intelligence (AI). In this case, the obvious choice is the solution based on multi-agent system.

The multi-agent approach divides highly complex computational process and delegates it into few computational agents. Agents as an intelligent software units in our approach are based on belief-desire-intention (BDI) model [1]. BDI architecture is the most common logical representation of agent inference model. It is based on logical theory, which defines the mentality of beliefs, desires and intentions with modal logic. One of the most well known implementations of this architecture is Procedural Reasoning System [2].

A community of intelligent agents cooperates with each other based on multi-agent paradigm which provides among others, full autonomy of the agents and highly efficient communication protocol, namely brings collective intelligence to the system [3].

The system proposed in this publication is implemented on the basis of multi-agent platform JADE [4]. The JADE supports multi-agent software development in compliance with the FIPA specifications for interoperable intelligent multi-agent systems [5]. The platform is designed to simplify the development of multi-agent systems by providing a comprehensive set of system services and agents. It has the tools for testing and debugging large and complex multi-agent systems. The multi-agent technology has wide usage in a domain of distributed computing and especially in grid systems [6].

2   Related Work

Polynomial-based learning machines based on the group method of data handling (GMDH) model continually enjoy great usage [7]. Thy have proven to be a powerful and effective way to find sufficient models for many applications.

Presented in [8] a group method of data handling (GMDH)-type neural network and evolutionary algorithm (EAs) are used for modelling the effects of the seven decision variables on the total cost rate and exergy efficiency of the system. The hybrid feedback version of symilar algorithm, proposed in [9], is used for the medical image diagnosis of liver cancer.

In scope of protection of the bridge abutment in waterways against scour phenomena in hydraulic engineering fields GMDH algorithm is used to model a relationship between the abutment scour depth due to thinly armored bed and the governing variables [10]. In [11] GMDH algorithm is used to extract rules and polynomials from an electric load pattern to forecast future electric load. In [12], a hybrid group method of data handling (GMDH) optimized with Genetic algorithm (GA) has been applied for prediction of optimum consumed power.

Papers [13–16] describes an agent platform that implements the BDI (belief-desire-intention) architecture named BDI4JADE. It is probably the most general implementation of the BDI architecture for the JADE multiagent platform. This work was an inspiration for our implementation of the BDI architecture for the JADE multiagent platform. Our implementation is based on the reasoning process described in [13–16]. But our solution uses BDI Formal Model Ontology implemented on the basis of FIPA Specification [17,18]. The comparison of this two implementations is out of the scope of this paper.

The authors of [19] propose a highly goal-oriented architecture for software agents. The solution includes mainly an extended goal model and corresponding run-time satisfaction criteria, an environment model and a failure model. When compared to our solution, the latter is much simpler. Those presented in [19], offer three types of goal with the subgoals, softgoals and relations. It is a much more general purpose solution.

The multi-agent system, proposed in [20], employees a BDI model in each agent to support the communication and reasoning process. In this regard, our solution is very similar. The publication [21] is an example of a multi-agent system based on the BDI agents in a market environment, which gives the agents the ability to achieve a common goal with very little information.

Agents are commonly viewed as software programs capable to autonomous computations that perceive their environment through sensors and act upon it through their effectors. The idea behind multi-agent system is to break down a complex problem into smaller simpler problems handled by several agents. Such system consists of nine agents capable of acting together to satisfy a desired goal, which, in the case of [22], is power supply-demand matching.

In literature we can meet with different types of architecture for the multi-agent systems. The type of architecture has crucial impact on multi-agent system performance, but usually highly depends on system domain.

The distributed system, proposed in [23], has a lot of agents, among them each agent is designed to implement a charticular functional unit. The back end server is composed of a case based reasoning database, a task agent and interface agent. The back end server communicates with flood disaster forecasting servers through mobile agent.

In [24] proposed distributed multi-agent system architecture is composed of controller agent responsible for the coordination between the other agents, and also it is interested in the optimization of exchange operations. Other agents are responsible for specific system functions. They cooperates with each other, but mainly with controller agent.

The proposed in [25] system is composed of cognitive Theory of Planned Behaviour-based agents. The agents are organized in a small world network that resembles connections between people in a neighbourhood or a street. Each agent-household in the network is connected to eight other agents-neighbours. Connection strengths between agents vary from 0.3 to 1 representing both strong and weak connections while the total number of connections remains constant.

3   Group Method of Data Handling

Evolution strategies have a significant place in the development of modern methodologies in the area of system identification. Among these methodologies, the Group Method of Data Handling (GMDH) has proven itself as a heuristic self-organizing technique in which complex models are generated using the analysis of evolution of the object under consideration. The GMDH algorithm was first developed by A.G. Ivakhnenko [26–29]. The algorithm is dedicated to the development of the mathematical model of an object to predict future states of the object using information on its states at the current and maybe earlier instants of time. Thus the idea of the algorithm implies that time is presented as a discrete variable and that the future states are determined in the form of predicted state variables (named the output variables) as functions of known state variables defined at the current or maybe earlier instants of time (named the input variables). It is obvious that the numbers of input and output state variables are the same in the problem of behaviour prediction. Assume that the set of state variables consists of M independent variables. Denoting output variables by y and input variables by x, and using the Kolmogorov-Gabor polynomial as a basic discrete model [26], we can write the model in the form of the following function:

$$y = a_0 + \sum_{i=1}^{M} \alpha_i x_i + \sum_{i=1}^{M} \sum_{j=1}^{M} \alpha_{ij} x_i x_j + \sum_{i=1}^{M} \sum_{j=1}^{M} \sum_{k=1}^{M} \alpha_{ijk} x_i x_j x_k + ... \quad (1)$$

where $x_i, x_j, x_k, ...$ are the input state variables, and y is any of those state variables considered as an output variable.

The challenge is to build equations of the form for each state variable using experimental data on the states of the object in the past instants of time. In other words, the challenge is reduced to the calculation of coefficients $\alpha_0, \alpha_i, \alpha_{ij}, \alpha_{ijk}, ...$ which are different for each predicted variable y using known values $x_i, x_j, x_k, ...$ in the past instants of time. To obtain coefficient α, the algorithm follows these stages:

1. Establish polynomial structure.
2. Select m input variables $x_1, x_2, ..., x_m$ affecting the output variable y, where m depends on chosen polynomial structure.
3. Estimate coefficients $\alpha_0, \alpha_1, ..., \alpha_{ijk}$, using the least squares. Coefficients can be obtained by minimizing value of external criteria.
4. Using the polynomial selected in the previous steps, construct more complex model using new input-output data, and repeat the process beginning with step 2.

In subsequent layers are formed more complex models (see Table 1). The whole procedure is shown in Fig. 1.

4   Inteligent BDI Agent

Among several approaches that propose different types of mental attitudes and their relationships, the belief-desire-intention (BDI) model is the most widespread one. Originally it was proposed by Bratman [31] and gives an

Table 1. Models formed in subsequent layers in GMDH algorithm.

Layer 1	Layer 2	Layer 3	...
$z_1 = f(x_1, x_2)$	$u_1 = f(x_1, x_2, z_1, z_2)$	$v_1 = f(x_1, x_2, z_1, z_2, u_1, u_2)$	...
$z_2 = f(x_1, x_3)$	$u_2 = f(x_1, x_3, z_1, z_2)$	$v_1 = f(x_1, x_3, z_1, z_2, u_1, u_2)$	...
...	...	...	...
$z_s = f(x_{M-1}, x_M)$	$z_r = f(x_{M-1}, x_M, z_{s-1}, z_s)$	$z_p = f(x_{M-1}, x_M, z_{s-1}, z_s, u_{r-1}, u_r)$	...

inspiration to folk psychological studies of mind and human behaviour [32]. The philosophical theory of the practical reasoning, defined as an intentional model that explains the human reasoning process with the following mental attitudes: beliefs, desires and intentions. The agent behaviour is attributed to those mental attitudes that play different roles in determining an agent's behaviour. The BDI model assumes that actions are derived from a process named practical reasoning. In this process, on the basis of the current situation of the agent's beliefs, a set of desires is selected to be achieved. Then agent determines how these concrete goals can be achieved by means of the available options [33]. BDI agent balances the time spent on deliberating about plans and executing them.

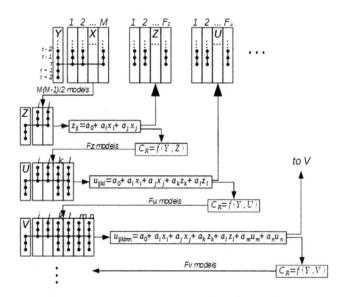

Fig. 1. GMDH algorithm visualization. Source [30], page 44.

The three basic mental attitudes are defined as follow: Beliefs are the facts that simulate the agent's environment. Desires store the information of the goals to be achieved. It is a model of agent motivations. Intentions represent the current action plan. It is a model of agent current behaviour.

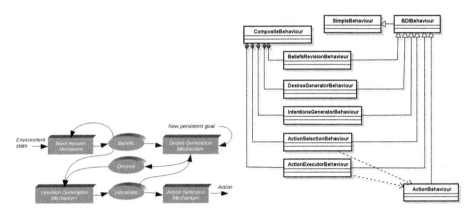

Fig. 2. The process of practical reasoning in the BDI agent.

Fig. 3. BDI life cycle behaviour class diagram.

The practical reasoning schema is presented in the Fig. 2. It is composed of four reasoning mechanisms that affect basic mental attitudes. The process of practical reasoning in the BDI agent begins by beliefs revision, namely based on environment state, the agent updates the set of beliefs. Then the set of desires is generated based on the set of intentions which the agent was not able to realize to current moment and new persistent goals. Afterwards, the agent determines a new set of intentions based on its current beliefs and desires. Finally, based on the intentions queue, the agent selects action to perform.

The reasoning life cycle is implemented in BDI agent as a CompositeBehaviour that consists of simple BDIBehaviour. As it is shown in Fig. 3, individual elements shown in Fig. 2, have its equivalent implementations.

Presented above the BDI model is a fully compliant implementation of Formal Model [17,18]. Figure 4 presents the BDI Ontology class diagram.

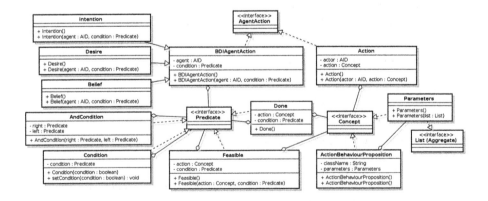

Fig. 4. BDI ontology class diagram.

5   Multiagent System

The system is based on master-slave structure organization built upon a multiagent platform. The organizational structure distinguishes two different roles for agents. The first role, called ModelAgent, is responsible for forecast computations. The second role, called DecisionAgent, makes the final decision based on forecast propositions obtained from ModelAgents. These main responsibilities consist of tasks (goals) that have to lead to execution of this responsibilities. Every task is boxed in ActionBehaviour as its implementation.

The activity diagram presented in Fig. 5 represents an overall flows of ModelAgent behaviours taking into account its choices. Behaviours placed in this activity diagram refer to following tasks: UpdateDataActionBehaviour - the acquisition of computational data from DecisionAgent (an distributor of current data); FindNewModelStructureActionBehaviour - computation of most suitable models based on the GMDH algorithm; TrainModelsActionBehaviour - computation of new models (polynomial coefficients) based on owned model structures; MakeNewForecastActionBehaviour - computation of new forecast based on owned models; SendResultActionBehaviour - sending forecast data to DecisionAgent; ReviseForecastActionBehaviour - comparison of forecast data with actual data.

DecisionAgent behaviour flows presented in form of activity diagram is depicted in Fig. 6. This behaviours corresponds to following tasks: SendDataActionBehaviour - sending current data to ModelAgent; ReceiveResultsActionBehaviour - receiving forecast data from ModelAgent; MakeDecisionActionBehaviour - makes final decision on the basis of received forecast data.

The Figs. 5 and 6 illustrate the logic relationship between the components of the system. The forecast computation process is initiated by ModelAgent, which acquires the computational data from DecisionAgent. Using the GMDH algorithm, ModelAgent, computes most suitable models. Then it uses those models to make forecast. The results are sent to the DecisionAgent. On the basis of forecast results obtained from all the ModelAgents, the DecisionAgent makes decision about the future course of analyzed process. The decision is based on

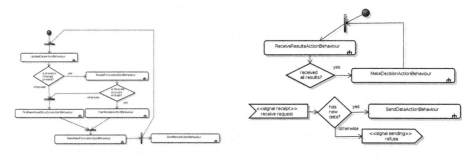

Fig. 5. ModelAgent activity diagram.   **Fig. 6.** DecisionAgent activity diagram.

the strategy that chooses the forecast of the agent, who had the best forecast[1] (BF) at the previously instants of time. Meanwhile, the ModelAgent revise forecast after obtaining data update. It compares forecast data with actual data. Based on this, it decides whether to find a new model structure or train an existing one.

An effective inter-agent communication is crucial for multiagent systems. The simplicity of this aspect increases the efficiency of the system as a whole. The Fig. 7 presents an inter-agent communication diagram. In this case the interaction between agents is reduced to minimum.

6   Multi Model Forecast Ontology

Since presented in this paper multiagent system is based on BDI concept, it uses two hierarchical ontologies. The first ontology is coupled with the implemented BDI concept of the Formal Model (see Fig. 3). It is involved in reasoning process. The second ontology extends the first one and it is responsible for inter-agent communication (see Fig. 7). This ontology is a very important element of the Multi-Model Forecast concept. Figure 8 presents the class diagram of the second ontology. Clear and meaningful ontology gives awareness of the subject for an agent, namely improves communication process.

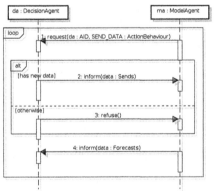

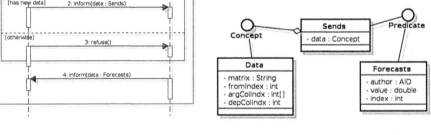

Fig. 7. Inter-agent communication diagram.

Fig. 8. Inter-agent communication ontology.

7   Case Studies

The weather forecasting system presented in this paper was tested on a dataset representing daily variations in temperature in Szczecin (Poland). As input data, the system receives three time series describing minimal, maximal, and average

---

[1] The forecasted value was closest to the actual value.

Table 2. Model evaluation criteria used by specific ModelAgents.

Agent name	testModel-Agent0	testModel-Agent1	testModel-Agent2	testModel-Agent3	testModel-Agent4	testModel-Agent5
Model evaluation criterion	Sum of absolute errors	Sum of square errors	Delta criteria	R square	Adjusted R square	Mean absolute error
Agent name	testModel-Agent6	testModel-Agent7	testModel-Agent8	testModel-Agent9	testModel-Agent10	testModel-Agent11
Model evaluation criterion	Mean square error	Mean absolute scaled error	Mean error	Root mean square error	Unbiased root mean square error	Mean percentage error
Agent name	testModel-Agent12	testModel-Agent13	testModel-Agent14	testModel-Agent15	testModel-Agent16	testModel-Agent17
Model evaluation criterion	Mean absolute percentage error	Maximum absolute error	Maximum percent error	Random walk sum of squared errors	Explained sum of squares	Amemiyas adjusted R square
Agent name	testModel-Agent18	testModel-Agent19	testModel-Agent20	testModelAgent21		
Model evaluation criterion	Random walk R square	Akaikes information criterion	Schwarz bayesian information criterion	Amemiyas prediction criterion		

temperature of each day during the observation time interval of 400 days. Following this time interval, the system made the decision on the average temperature value for the next day. Then the result was compared with the actual value. Investigations were carried out over 280 days.

In this case, an experiment was conducted where eleven ModelAgents were employed. Each ModelAgent computed the forecast on the basis of the GMHD algorithm using a special-purpose model evaluation criterion. Table 2 presents agent names and the model evaluation criteria that have been used.

Figure 9 presents the results of our experiment. The upper left chart of Fig. 9 presents the bar chart with data indicating how often each agent has the BF over 280 days of testing. The chart indicates pictorially the efficiency of each agent forecasting process efficiency. Most frequently the BF was delivered by the testModelAgent14 with the Maximum percent error model evaluation criterion. The worst result was observed in the case of the testModelAgent19 and the testModelAgent20, they delivered the best forecast only once. The upper right chart of Fig. 9 presents the ModelAgents prediction results over 280 tested days. This line chart presents strong correlation with data depicted in Fig. 9. Agents with higer scores obtained better fit with actual data. The bottom left chart of Fig. 9 presents final decisions made by the DecisionAgent based on the forecast values delivered by the ModelAgents. The bottom right chart of Fig. 9 presents a summary of the agents mean absolute percentage error (MAPE). The result obtained by the DecisionAgent illustrates the effectiveness of decision making strategy. It is situated among the best ModelAgents results. It is important to point that according to adopted decision making strategy, it is impossible for the DecisionAgent to obtain better result than the best Model Agent. Very weak

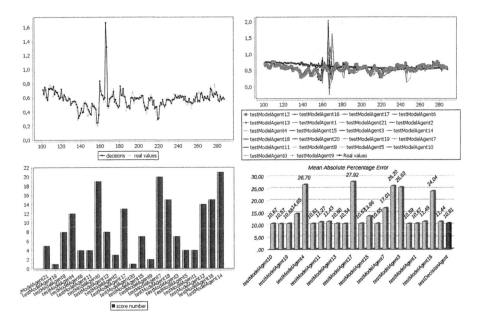

Fig. 9. Upper left: ModelAgents scores; Upper right: ModelAgents predictions results; Bottom left: Decisions made by DecisonAgent; Bottom right: Agents mean absolute percentage error value comparison.

data correlation that has been revealed during the comparison of the two bottom charts of Fig. 9 should not cause concerns. The bottom left chart shows only the frequency of BF among the ModelAgents, and it proves that the decision making strategy effectively is able to follow BF effectively.

8   Conclusions

The presented in the article case study proved the effectiveness of the approach. The approach allows to properly select best value of future development of the analysed process. The given comparison of model evaluation criteria revealed the level of their effectiveness which indicates that the majority of them is worth to be applied in GMDH algorithm.

The future study will be concentrating on further development of the concept of multi-model based prediction and improvement of proposed multi-agent framework based on BDI agents.

References

1. Subagdja, B., Sonenberg, L., Rahwan, I.: Intentional learning agent architecture. Auton. Agent. Multi-Agent Syst. **18**(3), 417–470 (2009)

2. Wobcke, W.: A logic of intention and action for regular BDI agents based on bisimulation of agent programs. Auton. Agent. Multi-Agent Syst. **29**(4), 569–620 (2015)
3. Wooldridge, M.: An Introduction to Multiagent Systems. Wiley, New York (2009)
4. Bellifemine, F., Caire, G., Grenwood, D.: Developing Multi-Agent Systems with JADE. Wiley, Chichester (2007)
5. Foundation for Intelligent Physical Agents. http://www.fipa.org
6. Rogoza, V., Zabłocki, M.: Grid computing and cloud computing in scope of JADE and OWL based semantic agents–a survey. Przegląd Elektrotechniczny **90**(2), 93–96 (2014)
7. Onwubolu, G.: GMDH-Methodology and Implementation in C. World Scientific Publishing Company (2014)
8. Khanmohammadi, S., Atashkari, K., Kouhikamali, R.: Exergoeconomic multi-objective optimization of an externally fired gas turbine integrated with a biomass gasifier. Appl. Therm. Eng. **91**, 848–859 (2015)
9. Shirmohammadi, R., Ghorbani, B., Hamedi, M., Hamedi, M., Romeo, L.M.: Optimization of mixed refrigerant systems in low temperature applications by means of group method of data handling (GMDH). J. Nat. Gas Sci. Eng. **26**, 303–312 (2015)
10. Najafzadeh, M., Barani, G., Hessami-Kermani, M.: Evaluation of GMDH networks for prediction of local scour depth at bridge abutments in coarse sediments with thinly armored beds. Ocean Eng. **104**, 387–396 (2015)
11. Kondo, T., Ueno, J., Takao, S.: Medical image diagnosis of liver cancer by hybrid feedback GMDH-type neural network using principal component-regression analysis. Artif. Life Robot. **20**(2), 145–151 (2015)
12. Koo, B., Lee, H., Park, J.: Short-term electric load forecasting based on wavelet transform and GMDH. J. Electr. Eng. Technol. **10**(3), 832–837 (2015)
13. Nunes, I., Lucena, C., Luck, M.: BDI4JADE: a BDI layer on top of JADE. In: International Workshop on Programming Multi-Agent Systems (ProMAS 2011), Taiwan (2011)
14. Nunes, I.: Improving the design and modularity of BDI agents with capability relationships. In: Dalpiaz, F., Dix, J., Riemsdijk, M.B. (eds.) EMAS 2014. LNCS (LNAI), vol. 8758, pp. 58–80. Springer, Heidelberg (2014). doi:10.1007/978-3-319-14484-9_4
15. Nunes, I., Luck, M.: Softgoal-based plan selection in model-driven BDI agents. In: The 13th International Conference on Autonomous Agents and Multiagent Systems, pp. 749–756 (2014)
16. Nunes, I.: Capability relationships in BDI agents. In: The 2nd International Workshop on Engineering Multi-Agent Systems (2014)
17. FIPA Communicative Act Library Specification. http://www.fipa.org/specs/fipa00037/
18. FIPA SL Content Language Specification. http://www.fipa.org/specs/fipa00008/
19. Morandini, M., Perini, A., Penserini, L., Marchetto, A.: Engineering requirements for adaptive systems. Requirements Eng. (2015)
20. Hilal, A.R., Basir, O.A.: A scalable sensor management architecture using BDI model for pervasive surveillance. IEEE Syst. J. **9**(2), 529–541 (2015)
21. Ren, Q., Bai, L., Biswas, S., Ferrese, F., Dong, Q.: A BDI multi-agent approach for power restoration. Paper presented at the 7th International Symposium on Resilient Control Systems, ISRCS 2014 (2014)

22. Maruf, M.N.I., Hurtado Munoz, L.A., Nguyen, P.H., Lopes Ferreira, H.M., Kling, W.L.: An enhancement of agent-based power supply-demand matching by using ANN-based forecaster. Paper presented at the 2013 4th IEEE/PES Innovative Smart Grid Technologies Europe, ISGT Europe 2013 (2013)

23. Linghu, B., Chen, F.: An intelligent multi-agent approach for flood disaster forecasting utilizing case based reasoning. Paper presented at the Proceedings - 2014 5th International Conference on Intelligent Systems Design and Engineering Applications, ISDEA 2014, pp. 182–185 (2014)

24. Elamine, D.O., Nfaoui, E.H., Jaouad, B.: Multi-agent architecture for smart microgrid optimal control using a hybrid BP-PSO algorithm for wind power prediction. Paper presented at the 2014 2nd World Conference on Complex Systems, WCCS 2014, pp. 554–560 (2014)

25. Mogles, N., Ramallo-González, A.P., Gabe-Thomas, E.: Towards a cognitive agent-based model for air conditioners purchasing prediction. Paper presented at the Procedia Computer Science, pp. 463–472 (2015)

26. Ivakhnenko, A., Ivakhnenko, G.: Problems of further development of the group method of data handling algorithms. Pattern Recogn. Image Anal. 10(2), 187–194 (2000)

27. Ivakhnenko, A., Ivakhnenko, G., Mueller, J.: Self-organization of neural network with active neurons. Pattern Recogn. Image Anal. 4(2), 185–196 (1999)

28. Ivakhnenko, A.G., Ivakhnenko, G.A., Andrienko, N.M.: Inductive computer advisor for current forecasting of Ukraine's macroeconomy. Syst. Anal. Model. Simul. (1998)

29. Ivakhnenko, G.A.: Model-free analogues as active neurons for neural networks self-organization. Control Syst. Comput. 2, 100–107 (2003)

30. Wiliński, A.: GMDH-metody grupowania argumentów w zadaniach zautomatyzowanej predykcji zachowań rynków finansowych (2009)

31. Bratman, M.E.: Intention, Plans, and Practical Reason. CSLI Publications, Stanford (1987)

32. Rao, A., Georgeff, M.: Modeling rational agents within a BDI architecture. In: Proceedings of the 2nd International Conference on Principles of Knowledge Representation and Reasoning, pp. 473–484 (1991)

33. d'Inverno, M., Kinny, D., Luck, M., Wooldridge, M.: A formal specification of dMARS. In: Agent Theories, Architectures, and Languages, pp. 155–176 (1997)

Comparison of RDM Complex Interval Arithmetic and Rectangular Complex Arithmetic

Marek Landowski[✉]

Department of Mathmatical Methods, Maritime University of Szczecin,
Waly Chrobrego 1-2, 70-500 Szczecin, Poland
m.landowski@am.szczecin.pl

Abstract. The article presents RDM complex interval arithmetic in comparison with rectangular complex arithmetic. The basic operations and the main properties of both complex interval arithmetics are described. To show the application of RDM complex interval arithmetic the examples with complex variables were solved using RDM and rectangular complex interval arithmetics. RDM means relative distance measure. RDM complex interval arithmetic is multidimensional, this property gives a possibilty to find a full solution of the problem with complex interval variables.

Keywords: Complex variable · RDM complex interval arithmetic · Rectangular complex arithmetic · Uncertainty theory

1 Introduction

The computation on the intervals is a fundamental problem of uncertainty theory and numerical computation. Well known Moore's interval arithmetic is commonly used in fuzzy arithmetic [2], Grey systems [6], granular computing [9], in decision problems [15], probabilistic arithmetic [16]. The rectangular complex arithmetic [3,10,14] is the extention of the Moore's interval arithmetic [7,8] into the complex variable, and the RDM complex interval arithmetic based on the RDM interval arithmetic [4,5,11–13].

The complex number Z, that real part and/or imaginary part are intervals, is defined as (1).

$$Z = A + iB = [\underline{a}, \overline{a}] + i\left[\underline{b}, \overline{b}\right] = \{x + iy | \underline{a} \le x \le \overline{a}, \underline{b} \le y \le \overline{b}\} \qquad (1)$$

Figure 1 presents the complex interval number $Z = [\underline{a}, \overline{a}] + i\left[\underline{b}, \overline{b}\right]$.

For complex interval numbers $Z_1 = A + iB$ and $Z_2 = C + iD$ the basic operations of rectangular complex arithmetic (also called complex interval arithmetic) are given in (2)–(5)

$$Z_1 + Z_2 = A + C + i(B + D) \qquad (2)$$

© Springer International Publishing AG 2017
S. Kobayashi et al. (eds.), *Hard and Soft Computing for Artificial Intelligence, Multimedia and Security*, Advances in Intelligent Systems and Computing 534, DOI 10.1007/978-3-319-48429-7_5

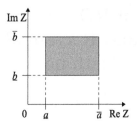

Fig. 1. Illustration of complex interval number $Z = [\underline{a}, \overline{a}] + i\left[\underline{b}, \overline{b}\right]$.

$$Z_1 - Z_2 = A - C + i(B - D) \tag{3}$$

$$Z_1 \cdot Z_2 = AC - BD + i(AD + BC) \tag{4}$$

$$Z_1/Z_2 = \frac{AC + BD}{C^2 + D^2} + i\frac{BC - AD}{C^2 + D^2}, \; 0 \notin Z_2 \tag{5}$$

The alternative for rectangular complex arithmetic is RDM complex interval arithmetic. In DRM complex interval method every interval is described by RDM variable α, where $\alpha \in [0, 1]$.

For example complex number $x + iy \in [2, 4] + i[3, 6]$ in RDM notation is described as $x + iy = 4 + 2\alpha_x + i(3 + 3\alpha_y)$, where $\alpha_x, \alpha_y \in [0, 1]$.

The complex interval number $Z_1 = A + iB = [\underline{a}, \overline{a}] + i\left[\underline{b}, \overline{b}\right]$ and $Z_2 = C + iD = [\underline{c}, \overline{c}] + i\left[\underline{d}, \overline{d}\right]$ in RDM notation are definied as (6) and (7).

$$Z_1 = \left\{a + ib : a + ib = \underline{a} + \alpha_a(\overline{a} - \underline{a}) + i\left[\underline{b} + \alpha_b(\overline{b} - \underline{b})\right], \alpha_a, \alpha_b \in [0, 1]\right\} \tag{6}$$

$$Z_2 = \left\{c + id : c + id = \underline{c} + \alpha_c(\overline{c} - \underline{c}) + i\left[\underline{d} + \alpha_d(\overline{d} - \underline{d})\right], \alpha_c, \alpha_d \in [0, 1]\right\} \tag{7}$$

Basic operations in RDM complex interval arithmetic are defined as (8)–(11)

Addition

$$\begin{aligned} Z_1 + Z_2 = \{x + iy : x + iy &= \underline{a} + \alpha_a(\overline{a} - \underline{a}) + \underline{c} + \alpha_c(\overline{c} - \underline{c}) \\ +i\left[\underline{b} + \alpha_b(\overline{b} - \underline{b}) + \underline{d} + \alpha_d(\overline{d} - \underline{d})\right] &, \alpha_a, \alpha_b, \alpha_c, \alpha_d \in [0, 1]\} \end{aligned} \tag{8}$$

Subtraction

$$\begin{aligned} Z_1 - Z_2 = \{x + iy : x + iy &= \underline{a} + \alpha_a(\overline{a} - \underline{a}) - \underline{c} - \alpha_c(\overline{c} - \underline{c}) \\ +i\left[\underline{b} + \alpha_b(\overline{b} - \underline{b}) - \underline{d} - \alpha_d(\overline{d} - \underline{d})\right] &, \alpha_a, \alpha_b, \alpha_c, \alpha_d \in [0, 1]\} \end{aligned} \tag{9}$$

Multiplication

$$\begin{aligned} Z_1 \cdot Z_2 = \{x + iy : x + iy &= [\underline{a} + \alpha_a(\overline{a} - \underline{a}) + i[\underline{b} + \alpha_b(\overline{b} - \underline{b})]] \\ \cdot [\underline{c} + \alpha_c(\overline{c} - \underline{c}) + i[\underline{d} + \alpha_d(\overline{d} - \underline{d})]] &, \alpha_a, \alpha_b, \alpha_c, \alpha_d \in [0, 1]\} \\ = \{x + iy : x + iy & \\ = [\underline{a} + \alpha_a(\overline{a} - \underline{a})] \cdot [\underline{c} + \alpha_c(\overline{c} - \underline{c})] &- [\underline{c} + \alpha_c(\overline{c} - \underline{c})] \cdot [\underline{d} + \alpha_d(\overline{d} - \underline{d})] \\ +i[[\underline{a} + \alpha_a(\overline{a} - \underline{a})] \cdot [\underline{d} + \alpha_d(\overline{d} - \underline{d})] &+ [\underline{b} + \alpha_b(\overline{b} - \underline{b})] \cdot [\underline{c} + \alpha_c(\overline{c} - \underline{c})]], \\ \alpha_a, \alpha_b, \alpha_c, \alpha_d & \in [0, 1]\} \end{aligned} \tag{10}$$

Division

$$Z_1/Z_2 = \{x + iy : x + iy = \left[\underline{a} + \alpha_a(\overline{a} - \underline{a}) + i[\underline{b} + \alpha_b(\overline{b} - \underline{b})]\right]$$
$$/ \left[\underline{c} + \alpha_c(\overline{c} - \underline{c}) + i[\underline{d} + \alpha_d(\overline{d} - \underline{d})]\right], \alpha_a, \alpha_b, \alpha_c, \alpha_d \in [0, 1]\}$$
$$= \{x + iy : x + iy = \frac{[\underline{a}+\alpha_a(\overline{a}-\underline{a})]\cdot[\underline{c}+\alpha_c(\overline{c}-\underline{c})]+[\underline{b}+\alpha_b(\overline{b}-\underline{b})]\cdot[\underline{d}+\alpha_d(\overline{d}-\underline{d})]}{[\underline{c}+\alpha_c(\overline{c}-\underline{c})]^2+[\underline{d}+\alpha_d(\overline{d}-\underline{d})]^2}$$
$$+i\frac{[\underline{b}+\alpha_b(\overline{b}-\underline{b})]\cdot[\underline{c}+\alpha_c(\overline{c}-\underline{c})]+[\underline{a}+\alpha_a(\overline{a}-\underline{a})]\cdot[\underline{d}+\alpha_d(\overline{d}-\underline{d})]}{[\underline{c}+\alpha_c(\overline{c}-\underline{c})]^2+[\underline{d}+\alpha_d(\overline{d}-\underline{d})]^2}, \alpha_a, \alpha_b, \alpha_c, \alpha_d \in [0, 1]\}, \quad (11)$$
$$0 \notin Z_2$$

For complex interval numbers $Z_1 = A + iB$ and $Z_2 = C + iD$ and the base operations $* \in \{+, -, \cdot, /\}$ span is a complex interval number defined as (12), operation $/$ is defined only if $0 \notin Z_2$.

$$s(Z_1 * Z_2) = \quad [\min\{Re(Z_1 * Z_2)\}, \max\{Re(Z_1 * Z_2)\}]$$
$$+i[\min\{Im(Z_1 * Z_2)\}, \max\{Im(Z_1 * Z_2)\}] \quad (12)$$

The solution obtained by RDM complex interval arithmetic can be presented in the form of formula, illustration, span, cardinality distribution or the center of gravity.

2 Properties of RDM Complex Interval Arithmetic and Rectangular Complex Arithmetic

For any complex interval numbers Z_1 and Z_2 in rectangular complex arithmetic and in RDM complex interval arithmetic hold the commutativity, Eqs. (13) and (14).

$$Z_1 + Z_2 = Z_2 + Z_1 \quad (13)$$

$$Z_1 \cdot Z_2 = Z_2 \cdot Z_1 \quad (14)$$

Both complex interval addition and multiplication in rectangular complex arithmetic and RDM complex interval arithmetic are associative. For any complex interval numbers Z_1, Z_2 and Z_3 are true (15) and (16).

$$Z_1 + (Z_2 + Z_3) = (Z_1 + Z_2) + Z_3 \quad (15)$$

$$Z_1 \cdot (Z_2 \cdot Z_3) = (Z_1 \cdot Z_2) \cdot Z_3 \quad (16)$$

In rectangular complex arithmetic and in RDM complex interval arithmetic there exist neutral elements of addition and multiplication, which are complex numbers, both with an imaginary part equal 0. For addition a real part of neutral element is a degravative interval 0, for multiplication a real part of neutral element is a degravative interval 1. For any complex interval number Z holds the Eqs. (17) and (18).

$$Z + 0 = 0 + Z = Z \quad (17)$$

$$Z \cdot 1 = 1 \cdot Z = Z \quad (18)$$

The complex interval number $Z = A + iB = \{a + bi : a + bi = \underline{a} + \alpha_a(\overline{a} - \underline{a}) + i[\underline{b} + \alpha_b(\overline{b} - \underline{b})], \alpha_a, \alpha_b \in [0, 1]\}$ in RDM complex interval arithmetic has an additive inverse element (19)

$$-Z = -A - iB$$
$$= \{-a - bi : -a - bi = -[\underline{a} + \alpha_a(\overline{a} - \underline{a})] - i[\underline{b} + \alpha_b(\overline{b} - \underline{b})], \alpha_a, \alpha_b \in [0, 1]\}.$$
(19)

Proof. $Z + (-Z) = \{a + bi : a + bi = \underline{a} + \alpha_a(\overline{a} - \underline{a}) + i[\underline{b} + \alpha_b(\overline{b} - \underline{b})], \alpha_a, \alpha_b \in [0, 1]\} + \{-a - bi : -a - bi = -[\underline{a} + \alpha_a(\overline{a} - \underline{a})] - i[\underline{b} + \alpha_b(\overline{b} - \underline{b})], \alpha_a, \alpha_b \in [0, 1]\} = \{a + bi - a - bi : a + bi - a - bi = \underline{a} + \alpha_a(\overline{a} - \underline{a}) + i[\underline{b} + \alpha_b(\overline{b} - \underline{b})] - [\underline{a} + \alpha_a(\overline{a} - \underline{a})] - i[\underline{b} + \alpha_b(\overline{b} - \underline{b})], \alpha_a, \alpha_b \in [0, 1]\} = 0.$ □

Also in RDM complex interval arithmetic there exists a multiplicative inverse element, so for $Z = A + iB = \{a + bi : a + bi = \underline{a} + \alpha_a(\overline{a} - \underline{a}) + i[\underline{b} + \alpha_b(\overline{b} - \underline{b})], \alpha_a, \alpha_b \in [0, 1]\}$, $0 \notin Z$, the multiplicative inverse element in RDM is (20).

$$1/Z = 1/(A + iB) = (A - iB)/(A^2 + B^2)$$
$$= \{1/(a + bi) : 1/(a + bi) = 1/[\underline{a} + \alpha_a(\overline{a} - \underline{a}) + i[\underline{b} + \alpha_b(\overline{b} - \underline{b})]], \alpha_a, \alpha_b \in [0, 1]\}$$
$$= \{x + iy : x + iy = \frac{\underline{a} + \alpha_a(\overline{a} - \underline{a}) - i[\underline{b} + \alpha_b(\overline{b} - \underline{b})]}{[\underline{a} + \alpha_a(\overline{a} - \underline{a})]^2 + [\underline{b} + \alpha_b(\overline{b} - \underline{b})]^2}, \alpha_a, \alpha_b \in [0, 1]\}.$$
(20)

Proof. $Z \cdot 1/Z$
$$= \{x + iy : x + iy = [\underline{a} + \alpha_a(\overline{a} - \underline{a}) + i[\underline{b} + \alpha_b(\overline{b} - \underline{b})]] \frac{\underline{a} + \alpha_a(\overline{a} - \underline{a}) - i[\underline{b} + \alpha_b(\overline{b} - \underline{b})]}{[\underline{a} + \alpha_a(\overline{a} - \underline{a})]^2 + [\underline{b} + \alpha_b(\overline{b} - \underline{b})]^2}, \alpha_a, \alpha_b \in [0, 1]\} = \{x + iy : x + iy = \frac{[\underline{a} + \alpha_a(\overline{a} - \underline{a})]^2 + [\underline{b} + \alpha_b(\overline{b} - \underline{b})]^2}{[\underline{a} + \alpha_a(\overline{a} - \underline{a})]^2 + [\underline{b} + \alpha_b(\overline{b} - \underline{b})]^2}, \alpha_a, \alpha_b \in [0, 1]\} = 1$$
□

In rectangular complex arithmetic for any complex interval number $Z = A + iB = [\underline{a}, \overline{a}] + i[\underline{b}, \overline{b}]$ additive inverse element in general does not exists, it exists only if real part and imaginary part are degenerated intervals, where $\underline{a} = \overline{a}$ and $\underline{b} = \overline{b}$.

Proof.
$Z - Z = [\underline{a}, \overline{a}] + i[\underline{b}, \overline{b}] - [\underline{a}, \overline{a}] - i[\underline{b}, \overline{b}] = [\underline{a} - \overline{a}, \overline{a} - \underline{a}] + i[\underline{b} - \overline{b}, \overline{b} - \underline{b}] \neq 0$ unless $\underline{a} = \overline{a}$ and $\underline{b} = \overline{b}$. □

Complex conjugate of complex number is a complex number with equal real part and the opposite of the imaginary part, for example, for complex interval number $Z = A + iB$ the complex conjugate is $\overline{Z} = A - iB$.

For complex interval number $Z = A + iB$ properties of complex conjugates (21) and (22) hold in RDM complex arithmetic.

$$Z + \overline{Z} = 2A \tag{21}$$

$$Z\overline{Z} = A^2 - B^2 \tag{22}$$

Proof. The RDM complex interval arithmetic. For $Z = A + iB = [\underline{a}, \overline{a}] + i\left[\underline{b}, \overline{b}\right]$ the complex conjugate of Z is $\bar{Z} = A - iB = [\underline{a}, \overline{a}] - i\left[\underline{b}, \overline{b}\right]$. In RDM notation Z and $\bar{Z}$ are:

$Z = \left\{a + ib : a + ib = \underline{a} + \alpha_a(\overline{a} - \underline{a}) + i\left[\underline{b} + \alpha_b(\overline{b} - \underline{b})\right], \alpha_a, \alpha_b \in [0, 1]\right\}$

$\bar{Z} = \left\{a - ib : a - ib = \underline{a} + \alpha_a(\overline{a} - \underline{a}) - i\left[\underline{b} + \alpha_b(\overline{b} - \underline{b})\right], \alpha_a, \alpha_b \in [0, 1]\right\}$

Property (21) holds in RDM interval arithmetic.

$Z + \bar{Z} = \left\{a + ib : a + ib = \underline{a} + \alpha_a(\overline{a} - \underline{a}) + i\left[\underline{b} + \alpha_b(\overline{b} - \underline{b})\right], \alpha_a, \alpha_b \in [0, 1]\right\} + \left\{a - ib : a - ib = \underline{a} + \alpha_a(\overline{a} - \underline{a}) - i\left[\underline{b} + \alpha_b(\overline{b} - \underline{b})\right], \alpha_a, \alpha_b \in [0, 1]\right\} = \{x + iy : x + iy = \underline{a} + \alpha_a(\overline{a} - \underline{a}) + i\left[\underline{b} + \alpha_b(\overline{b} - \underline{b})\right] + \underline{a} + \alpha_a(\overline{a} - \underline{a}) - i\left[\underline{b} + \alpha_b(\overline{b} - \underline{b})\right], \alpha_a, \alpha_b \in [0, 1]\} = \{x + iy : x + iy = 2[\underline{a} + \alpha_a(\overline{a} - \underline{a})], \alpha_a, \alpha_b \in [0, 1]\} = 2A.$

Also property (22) holds in RDM interval arithmetic.

$Z\bar{Z} = \left\{a + ib : a + ib = \underline{a} + \alpha_a(\overline{a} - \underline{a}) + i\left[\underline{b} + \alpha_b(\overline{b} - \underline{b})\right], \alpha_a, \alpha_b \in [0, 1]\right\}\left\{a - ib : a - ib = \underline{a} + \alpha_a(\overline{a} - \underline{a}) - i\left[\underline{b} + \alpha_b(\overline{b} - \underline{b})\right], \alpha_a, \alpha_b \in [0, 1]\right\} = \{x + iy : x + iy = [\underline{a} + \alpha_a(\overline{a} - \underline{a}) + i[\underline{b} + \alpha_b(\overline{b} - \underline{b})]] \cdot [\underline{a} + \alpha_a(\overline{a} - \underline{a}) - i[\underline{b} + \alpha_b(\overline{b} - \underline{b})]], \alpha_a, \alpha_b \in [0, 1]\} = \{x + iy : x + iy = [\underline{a} + \alpha_a(\overline{a} - \underline{a})]^2 + [\underline{b} + \alpha_b(\overline{b} - \underline{b})]^2, \alpha_a, \alpha_b \in [0, 1]\} = A^2 + B^2.$ □

Properties of complex conjugates (21) and (22) in general do not hold for rectangular complex arithmetic. Equation (21) holds only for complex interval number $A + iB = [\underline{a}, \overline{a}] + i[\underline{b}, \overline{b}]$ where imaginary part is degenerate interval $(\underline{b} = \overline{b})$, width of $[\underline{b}, \overline{b}]$ equals zero. Equation (22) is true for rectangular complex arithmetic only if product AB gives degenerate interval where width equals zero.

Proof. Let us consider complex interval number $Z = A + iB = [\underline{a}, \overline{a}] + i[\underline{b}, \overline{b}]$ and the complex conjugate $\bar{Z} = A - iB = [\underline{a}, \overline{a}] + i[-\overline{b}, -\underline{b}]$. Using rectangular complex arithmetic we have:

$Z + \bar{Z} = A + iB + A - iB = [\underline{a}, \overline{a}] + i[\underline{b}, \overline{b}] + [\underline{a}, \overline{a}] + i[-\overline{b}, -\underline{b}] = [2\underline{a}, 2\overline{a}] + i[\underline{b} - \overline{b}, \overline{b} - \underline{b}] = 2A - i(B - B) \neq [2\underline{a}, 2\overline{a}] + i0 = 2A$ unless $\underline{b} = \overline{b}$

and

$Z\bar{Z} = (A + iB)(A - iB) = [[\underline{a}, \overline{a}] + i[\underline{b}, \overline{b}]] \cdot [[\underline{a}, \overline{a}] + i[-\overline{b}, -\underline{b}]] = [\underline{a}, \overline{a}]^2 - [\underline{b}, \overline{b}][-\overline{b}, -\underline{b}] + i[[\underline{a}, \overline{a}][-\overline{b}, -\underline{b}] + [\underline{b}, \overline{b}][\underline{a}, \overline{a}]] = [\underline{a}, \overline{a}]^2 + [\underline{b}, \overline{b}]^2 + i[[\underline{a}, \overline{a}][\underline{b}, \overline{b}] - [\underline{a}, \overline{a}][\underline{b}, \overline{b}]] = A^2 + B^2 + i(AB - AB) \neq [\underline{a}, \overline{a}]^2 + [\underline{b}, \overline{b}]^2 + i[0, 0] = A^2 + B^2$ unless interval AB is degenerate interval, where width equals zero. □

3   Examples of Complex Interval Arithmetic

The faults of general interval arithmetic such as "the dependency problem" or "difficulties of solving of even simplest interval equations" [1] hold in rectangular complex arithmetic. RDM complex interval arithmetic finds the solution which is independent of the form of the function or the form of equation.

Example 1. The dependency problem.

For complex interval number $Z = [1, 3] + i[2, 5]$ find the value of function f given in the two forms (23) and (24).

$$f_1(Z) = Z^2 - Z \tag{23}$$

$$f_2(Z) = Z(Z-1). \tag{24}$$

Solution of rectangular complex arithmetic.

$f_1([1,3] + i[2,5]) = ([1,3] + i[2,5])^2 - ([1,3] + i[2,5]) = ([1,9] - [4,25]) + i([2,15] + [2,15]) - [1,3] - i[2,5] = [-24,5] + i[4,30] - [1,3] - i[2,5] = [-27,4] + i[-1,28].$

$f_2([1,3] + i[2,5]) = ([1,3] + i[2,5])([1,3] + i[2,5] - 1) = ([1,3] + i[2,5])([0,2] + i[2,5]) = [0,6] - [4,25] + i([2,15] + [0,10]) = [-25,2] + i[2,25].$

The result depends on the form of the function. In rectangular complex arithmetic $f_1(Z) \neq f_2(Z)$.

The solution of RDM complex interval arithmetic.

Let us write complex interval number Z in the RDM notation using RDM variable $\alpha_a \in [0,1]$ and $\alpha_b \in [0,1]$, (25).

$$Z = \{z : z = 1 + 2\alpha_a + i(2 + 3\alpha_b), \alpha_a, \alpha_b \in [0,1]\} \tag{25}$$

For every $z \in Z$ we have:

$f_1(z) = f_1(1 + 2\alpha_a + i(2 + 3\alpha_b)) = [1 + 2\alpha_a + i(2 + 3\alpha_b)]^2 - [1 + 2\alpha_a + i(2 + 3\alpha_b)] = (1 + 2\alpha_a)^2 + i2(1 + 2\alpha_a)(2 + 3\alpha_b) - (2 + 3\alpha_b)^2 - 1 + 2\alpha_a - i(2 + 3\alpha_b) = 1 + 4\alpha_a + 4\alpha_a^2 + 2i(2 + 3\alpha_b) + 4i\alpha_a(2 + 3\alpha_b) - (3 + 3\alpha_b)^2 - 1 - 2\alpha_a - i(2 + 3\alpha_b) = [2\alpha_a + 4\alpha_a^2 - (2 + 3\alpha_b)^2] + i[(2 + 3\alpha_b) + 4\alpha_a(2 + 3\alpha_b)].$

$f_2(z) = f_2(1 + 2\alpha_a + i(2 + 3\alpha_b)) = [1 + 2\alpha_a + i(2 + 3\alpha_b)][1 + 2\alpha_a + i(2 + 3\alpha_b) - 1] = [1 + 2\alpha_a + i(2 + 3\alpha_b)][2\alpha_a + i(2 + 3\alpha_b)] = 2\alpha_a + i(2 + 3\alpha_b) + 4\alpha_a^2 + i2\alpha_a(2 + 3\alpha_b) + i2\alpha_a(2 + 3\alpha_b) + i^2(2 + 3\alpha_b)^2 = [2\alpha_a + 4\alpha_a^2 - (2 + 3\alpha_b)^2] + i[(2 + 3\alpha_b) + 4\alpha_a(2 + 3\alpha_b)].$

RDM complex interval arithmetic gives equal solution (26) for $f_1(z)$ and $f_2(z)$. The solution is not a rectangular, see Fig. 2.

$$Re(f(Z)) = 2\alpha_a + 4\alpha_a^2 - (2 + 3\alpha_b)^2,$$
$$Im(f(Z)) = (2 + 3\alpha_b) + 4\alpha_a(2 + 3\alpha_b), \tag{26}$$
$$\alpha_a, \alpha_b \in [0,1].$$

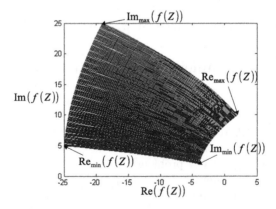

Fig. 2. Values of function f obtained by RDM complex interval arithmetic.

To find span of the solution obtained by RDM complex interval arithmetic the solution for border values of RDM variables α_a and α_b should be calculated, Table 1.

Table 1. Solution for border values of RDM-variables α_a and α_b.

α_a	0	0	1	1
α_b	0	1	0	1
$Re(f(z))$	−4	−25	2	−19
$Im(f(z))$	2	5	10	25
		$Re_{min}(f(z))$	$Re_{max}(f(z))$	
	$Im_{min}(f(z))$			$Im_{max}(f(z))$

The span of the solution obtained by RDM complex interval arithmetic is given by (27).

$$s(f(Z)) = [-25, 2] + i[2, 25]. \tag{27}$$

Example 2. Solution of simple complex interval equations.
Find the solution X of Eq. (28).

$$ZX = C \tag{28}$$

where $Z = [2, 5] + i[1, 2]$, $C = [2, 3]$.

Solution of rectangular complex arithmetic.

Equation (28) can be solved by rectangular complex arithmetic at least in two ways, as Eq. (28) or in the form of (29).

$$X = C/Z \tag{29}$$

The solution obtained from Eq. (28) is (30).

$$X = [19/24, 25/48] + i[-19/48, -5/24]. \tag{30}$$

the real part is an invert interval. The solution from Eq. (28) is (31).

$$X = [4/27, 15/5] + i[-6/5, -2/27]. \tag{31}$$

Rectangular complex arithmetic cannot solve a simple equation.

The solution of RDM complex interval arithmetic.

Let us write the complex interval number Z and interval C in RDM notation (32).

$$\begin{aligned} Z &= \{z : z = 2 + 3\alpha_a + i(1 + \alpha_b), \alpha_a, \alpha_b \in [0, 1]\} \\ C &= \{c : c = 2 + \alpha_c, \alpha_c \in [0, 1]\} \end{aligned} \tag{32}$$

For every $z \in Z$ and $c \in C$ we have:

$$[2 + 3\alpha_a + i(1 + \alpha_b)]X = 2 + \alpha_c$$
$$X = (2 + \alpha_c)/[2 + 3\alpha_a + i(1 + \alpha_b)]$$
$$X = [(2 + \alpha_c)(2 + 3\alpha_a) - i(2 + \alpha_c)(1 + \alpha_b)]/[(2 + 3\alpha_a)^2 + (1 + \alpha_b)^2]$$

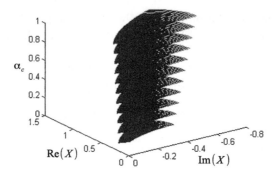

Fig. 3. Illustration of the solution of Eq. (28) obtained by RDM complex interval arithmetic, for $\alpha_c \in [0 : 0.1 : 1]$.

Table 2. Solution for border values of RDM-variables α_a, α_b and α_c.

α_a	0	0	0	1	0	1	1	1
α_b	0	0	1	0	1	0	1	1
α_c	0	1	0	0	1	1	0	1
$Re(X)$	4/5	6/5	1/2	5/13	3/4	15/26	10/29	15/29
$Im(X)$	-2/5	-3/5	-1/2	-1/13	-3/4	-3/26	-4/29	-6/29
		$Re_{\max}(X)$					$Re_{\min}(X)$	
				$Im_{\max}(X)$	$Im_{\min}(X)$			

The real part and imaginary part of solution X obtained by RDM complex interval solution are given by (33).

$$Re(X) = [(2 + \alpha_c)(2 + 3\alpha_a)]/[(2 + 3\alpha_a)^2 + (1 + \alpha_b)^2]$$
$$Im(X) = [(2 + \alpha_c)(1 + \alpha_b)]/[(2 + 3\alpha_a)^2 + (1 + \alpha_b)^2] \tag{33}$$

The solution (33) is presented in Fig. 3.

Span of the solution X is calculated by border values of RDM variables Table 2.

The span of the solution (33) obtained by RDM complex interval arithmetic is given by (34).

$$s(X) = [10/29, 6/5] + i[-3/4, -1/13]. \tag{34}$$

4   Conclusions

In the article the basic operations and properties of RDM complex interval arithmetic and rectangular complex arithmetic are described. The solved examples show that the solution obtained by rectangular complex arithmetic depends on the form of the function with complex variables where the problem is given. Also results of solved simple complex interval equation in rectangular complex arithmetic are different, the solution depends on the form of equation. The RDM complex interval arithmetic gives one solution for any form of a function or an equation. The basic difference of both arithmetics is that the RDM complex interval arithmetic gives the solution in the multidimensional space but the rectangular complex arithmetic solutions are expressed in the form of intervals. In some problems interval obtained by rectangular complex arithmetic is only a span of the solution, not a full solution.

References

1. Dymova, L.: Soft Computing in Economics and Finance. Springer, Berlin, Heidelberg (2011)
2. Hanss, M.: Applied Fuzzy Arithmetic. Springer, Berlin, Heidelberg (2005)
3. Knopp, K.: Elements of the Theory of Functions. Dover Publications, New York (1952)
4. Landowski, M.: Differences between Moore and RDM interval arithmetic. In: Proceedings of the 7th International Conference Intelligent Systems IEEE IS 2014, 24–26 September 2014, Warsaw, Poland, Volume 1: Mathematical Foundations, Theory, Analyses, Intelligent Systems 2014, Advances in Intelligent Systems and Computing, Vol. 322, pp. 331–340, Springer International Publishing Switzerland (2015)
5. Landowski, M.: RDM interval arithmetic for solving economic problem with uncertain variables. Logistyka **6**(2014), 13502–13508 (2014)
6. Liu, S., Lin, Y.: Grey Systems, Theory and Applications. Springer, Berlin, Heidelberg (2010)
7. Moore, R.E., Kearfott, R.B., Cloud, J.M.: Introduction to Interval Analysis. SIAM, Philadelphia (2009)
8. Moore, R.E.: Interval Analysis. Prentice Hall, Englewood Cliffs (1966)
9. Pedrycz, W., Skowron, A., Kreinovicz, V. (eds.): Handbook of Granular Computing. Wiley, Chichester (2008)
10. Petkovic, M.S., Petkovic, L.D.: Complex Interval Arithmetic and Its Applications, 1 edn. Mathematical Research, vol. 105. WILEY-VCH, Berlin (1998)
11. Piegat, A., Landowski, M.: Aggregation of Inconsistent Expert Opinions with Use of Horizontal Intuitionistic Membership Functions. In: Atanassov, K.T., Castillo, O., Kacprzyk, J., Krawczak, M., Melin, P., Sotirov, S., Sotirova, E., Szmidt, E., Tré, G.D., Zadrożny, S. (eds.) Novel Developments in Uncertainty Representation and Processing. AISC, vol. 401, pp. 215–223. Springer, Heidelberg (2016). doi:10.1007/978-3-319-26211-6_18
12. Piegat, A., Landowski, M.: Horizontal membership function and examples of its applications. Intl. J. Fuzzy Syst. **17**(1), 22–30 (2015)
13. Piegat, A., Landowski, M.: Two interpretations of multidimensional RDM interval arithmetic - multiplication and division. Intl. J. Fuzzy Syst. **15**(4), 486–496 (2013)
14. Roche, R.: Complex Interval Arithmetic with Some Applications. Lockheed Missiles & Space Company, Sunnyvale (1966)
15. Sengupta, A., Pal, T.K.: Fuzzy Preference Ordering of Interval Numbers in Decision Problems. Springer, Berlin, Heidelberg (2009)
16. Williamson, R.: Probabilistic arithmetic. Ph.D. thesis, Department of Electrical Engineering, University of Queensland (1989)

Homogeneous Ensemble Selection - Experimental Studies

Robert Burduk[✉] and Paulina Heda

Department of Systems and Computer Networks, Wroclaw University of Science and Technology, Wybrzeze Wyspianskiego 27, 50-370 Wroclaw, Poland
{robert.burduk,paulina.heda}@pwr.edu.pl

Abstract. The paper presents the dynamic ensemble selection method. Proposed method uses information from so-called decision profiles which are formed from the outputs of the base classifiers. In order to verify these algorithms, a number of experiments have been carried out on several public available data sets. The proposed dynamic ensemble selection is experimentally compared against all base classifiers and the ensemble classifiers based on the sum and decision profile methods. As a base classifiers we used the pool of homogeneous classifiers.

Keywords: Multiple classifier system · Decision profile · Ensemble selection

1 Introduction

In machine learning there are several groups of algorithms belonging to different areas of human activity. One of them is supervised machine learning that automatically induce patterns from data. In general, classification algorithm assign pre-defined class label to a new pattern based on the knowledge from so called training set wherein class labels are known. A number of method have been used in order to solve different classification task. The single supervised learning method may be called a base learning algorithm. The use of the multiple base classifier for a decision problem is known as an ensemble of classifiers (EoC) or as multiple classifiers systems (MCSs) [12,15].

The building of MCSs consists of three main phases: generation, selection and integration (fusion) [4]. In the generation phase a pool of the base classifiers is generated. The formation of this pool can be automated or can be heuristic choice. When injecting randomness into the learning algorithm or manipulating the training objects is done [3,11], then we are talking about homogeneous classifiers. Other approach it that the ensemble is composed of heterogeneous classifiers. It means, that some different learning algorithms to the same data set are apply.

In the second phase (selection), a single base classifier or a subset base classifiers having the best classifiers of the pool is selected [14]. Formally, if we choose one classifier then it is called the classifier selection. But if we choose a subset of

© Springer International Publishing AG 2017
S. Kobayashi et al. (eds.), *Hard and Soft Computing for Artificial Intelligence, Multimedia and Security*, Advances in Intelligent Systems and Computing 534, DOI 10.1007/978-3-319-48429-7_6

base classifiers from the pool then it is called the ensemble selection or ensemble pruning. Generally, in the ensemble selection, there are two approaches: the static ensemble selection and the dynamic ensemble selection [2,4]. In static approaches, the selection is performed during the training phase of the recognition system. The selected base classifier or a base classifier set is used for the classification of all new test samples. The dynamic ensemble selection approaches select a different base classifier or a different subsets of the base classifiers for each new test sample.

The last phase of building of MCSs presents also large number of fusion methods [7,16,21]. For example, in the third phase the simple majority voting scheme [19] is most popular. Generally, the final decision which is made in the third phase uses the prediction of the base classifiers and it is popular for its ability to fuse together multiple classification outputs for the better accuracy of classification. The fusion methods can be into selection-based, fusion-based and hybrid ones according to their functioning [6]. The fusion strategies can be also divided into fixed and trainable ones [7]. Another division distinguishes class-conscious and class-indifferent integration methods [16].

In this work we will consider the dynamic selection of classifiers ensemble. The proposed method is based on information from decision profiles. The decision scheme computed in this work generally not used information from wrong classified object belonging to training set. Generally, we presented two versions of dynamic selection algorithm. One of them uses the average, the second the median in the training phase.

The remainder of this paper is organized as follows. Section 2 presents the concept of the base classifier and ensemble of classifiers. Section 3 contains the proposed dynamic selection of classifiers ensemble. The experimental evaluation, discussion and conclusions from the experiments are presented in Sect. 4. The paper is concluded by a final discussion.

2 Supervised Classification

2.1 Base Classifiers

The aim of the supervised classification is to assign an object to a specific class label. The object is represented by a set of d features, or attributes, viewed as d-dimensional feature vector x. The recognition algorithm maps the feature space x to the set of class labels Ω according to the general formula:

$$\Psi : X \rightarrow \Omega. \tag{1}$$

The recognition algorithm defines the classifier, which in the complex classification task with multiple classifiers is called a base classifier.

The output of a base classifier can be divided into three types [16].

- The abstract level – the classifier ψ assigns the unique label j to a given input x.

- The rank level – in this case for each input (object) x, each classifier produces an integer rank array. Each element within this array corresponds to one of the defined class labels. The array is usually sorted and the label at the top being the first choice.
- The measurement level – the output of a classifier is represented by a confidence value (CV) that addresses the degree of assigning the class label to the given input x. An example of such a representation of the output is a posteriori probability returned by Bayes classifier. Generally, this level can provide richer information than the abstract and rank level.

In this work we consider the situation when each base classifier returns CVs. Additionally, before the final combination of base classifiers' outputs the CVs modification process is carried out.

2.2 Ensemble of Classifiers

Let us assume that $k \in \{1, 2, ..., K\}$ different classifiers $\Psi_1, \Psi_2, \ldots, \Psi_K$ are available to solve the classification task. The output information from all K component classifiers is applied to make the ultimate decision of MCSs. This decision is made based on the predictions of all the base classifiers.

One of the possible methods for integrating the output of the base classifier is the sum rule. In this method the score of MCSs is based on the application of the following sums:

$$s_\omega(x) = \sum_{k=1}^{K} p_k(\omega|x), \qquad \omega \in \Omega, \tag{2}$$

where $p_k(\omega|x)$ is CV for class label ω returned by classifier k.

The final decision of MCSs is made following the maximum rule:

$$\Psi_S(x) = \arg \max_\omega s_\omega(x). \tag{3}$$

In the presented method (3) CV obtained from the individual classifiers take an equal part in building MCSs. This is the simplest situation in which we do not need additional information on the testing process of the base classifiers except for the models of these classifiers. One of the possible methods in which weights of the base classifier are used is presented in [5].

Decision template (DT) is another approach to build the MCSs. DT was proposed in [17]. In this MCS model DT are calculated based on training set one per class label. In the operation phase the similarity between each DT and outputs of base classifiers for object x is computed. The class label with the closest DT is assigned to object x. In this paper algorithm with DT is labeled Ψ_{DT} and it is used as one of the reference classifier.

3   Dynamic Selection Algorithm

3.1   Training Phase

The proposed dynamic selection CV algorithm uses DPs. DP is a matrix containing CVs for each base classifier, i.e.:

$$DP(x) = \begin{bmatrix} p_1(0|x) & \cdots & p_1(\Omega|x) \\ \vdots & \cdots & \vdots \\ p_K(0|x) & \cdots & p_K(\Omega|x) \end{bmatrix}. \tag{4}$$

In the first step of the algorithm we remove CVs which relate to the misclassification on the training set. This set contains N labeled examples $\{(x_1, \overline{\omega}_1), ..., (x_N, \overline{\omega}_N)\}$, where $\overline{\omega}_i$ is true class label of the object described by feature vector x_i. CVs are removed according to the formula:

$$p'_k(\omega|x) = \begin{cases} p_k(\omega|x), & \text{if} \quad I(\Psi(x), \overline{\omega}) = 1 \\ 0, & \text{if} \quad I(\Psi(x), \overline{\omega}) = 0. \end{cases} \tag{5}$$

where $I(\Psi(x), \overline{\omega})$ is an indicator function having the value 1 in the case of the correct classification of the object described by feature vector x, i.e. when $\Psi(x) = \overline{\omega}$.

Selection with average. In the next step of our algorithm, the decision scheme (DS) is calculated according to the formula:

$$DS(\beta) = \begin{bmatrix} ds(\beta)_{10} & \cdots & ds(\beta)_{1\Omega} \\ \vdots & \cdots & \vdots \\ ds(\beta)_{K0} & \cdots & ds(\beta)_{K\Omega} \end{bmatrix}, \tag{6}$$

where

$$ds(\beta)_{k\omega} = \overline{ds}_{k\omega} + \beta \sqrt{\frac{\sum_{n=1}^{N} (p'_k(\omega_n|x_n) - \overline{ds}_{k\omega})^2}{N-1}} \tag{7}$$

and

$$\overline{ds}_{k\omega} = \frac{\sum_{n=1}^{N} p'_k(\omega_n|x_n)}{N}. \tag{8}$$

The parameter β in our algorithm determines how we compute DS elements. For example, if $\beta = 0$, then $ds_{k\omega}$ is the average of appropriate DFs received after the condition (5). The algorithm uses described above the method of calculation scheme is labeled as Ψ_{SA}.

Selection with median. The second version of the proposed algorithm uses median instead of the mean. DS is calculated in this case according to the formula:

$$DS_m(\beta) = \begin{bmatrix} ds_m(\beta)_{10} & \cdots & ds_m(\beta)_{1\Omega} \\ \vdots & \cdots & \vdots \\ ds_m(\beta)_{K0} & \cdots & ds_m(\beta)_{K\Omega} \end{bmatrix}, \tag{9}$$

where

$$ds_m(\beta)_{k\omega} = \text{median}(p'_k(\omega|x)) + \beta Q_1((p'_k(\omega|x)) \tag{10}$$

and Q_1 is first quartile. The described above algorithm is labeled as Ψ_{SM}.

3.2   Operation Phase

During the operation phase the selection of CVs is carried out using DSs calculated in the training phase. We present a selection method for $DS(\beta)$, selection for $DS_m(\beta)$ is analogous. For the new object being recognized x, the selection CVs is performed according to the formula:

$$p'_k(\omega|x) = \begin{cases} \overline{m} * p_k(\omega|x) & \text{if } p_k(\omega|x) \geq ds(\beta_1)_{k\omega} \\ m * p_k(\omega|x) & \text{if } ds(\beta_1)_{k\omega} < p_k(\omega|x) < ds(\beta_2)_{k\omega} \\ 0 & \text{if } p_k(\omega|x) \leq ds(\beta_2)_{k\omega} \end{cases} \tag{11}$$

In the selection we use parameters β_1, β_2, $\overline{m}$ and m. In particular, the parameter β_2, defines the values (from DS) that are used in the selection. The parameters $\overline{m}$ and m define changes in the values of CVs.

4   Experimental Studies

The aim of the experiments was to compare the quality of classifications of the proposed dynamic selection algorithms Ψ_{SM} and Ψ_{SA} with the base classifiers and their ensemble without selection Ψ_{DT} and Ψ_S. The ensemble classifier in our research is composed of homogeneous classifiers. We use ensemble classifier consists of different k-NN or SVM base classifiers. In the experiments the feature selection process [13,18] was not performed and we have used standard 10-fold-cross-validation method.

In the experiential we use 13 data sets. Nine of them come from the Keel Project [1] and four com from UCI Repository [9] (Blood, Breast Cancer Wisconsin, Indian Liver Patient and Mammographic Mass UCI). Table 1 shows some details of the used data sets. First group of four base classifiers work according to $k - NN$ rule, second group use the Support Vector Machines models. The base classifiers are labeled as $\Psi_1, ..., \Psi_4$ for the first group and $\Psi_5, ..., \Psi_8$ for the second group.

The parameters of selection algorithm were established on: case 1 – $\beta_1 = 1$, $\beta_2 = 0$, $\overline{m} = 1.5$ and $m = 1$; case 2 – $\beta_1 = 1$, $\beta_2 = 1$, $\overline{m} = 1.5$ and $m = 1$; case 3 – $\beta_1 = 1.5$, $\beta_2 = 1.5$, $\overline{m} = 1.5$ and $m = 1$; case 4 – $\beta_1 = 1.5$, $\beta_2 = 1.5$, $\overline{m} = 1.25$ and $m = 1$ and case 5 – $\beta_1 = 1$, $\beta_2 = 1$, $\overline{m} = 1.25$ and $m = 1$.

Tables 2 and 3 show the results of the classification for the proposed ensemble selection methods Ψ_{SM} and Ψ_{SA} for five given sets of parameters. Table 2 present the results for k-NN base classifiers and Table 3 for SVM base classifiers. We present the classification error and the mean ranks obtained by the Friedman test [20]. The results in Tables 2 and 3 demonstrate clearly that the value of the parameters is important in the proposed ensemble selection method.

Table 1. Description of data sets selected for the experiments

Data set	Example	Attribute	Ration (0/1)
Blood	748	5	3.2
Breast Cancer Wisconsin	699	10	1.9
Cylinder Bands	365	19	1.7
Haberman's Survival	306	3	2.8
Hepatitis	80	19	0.2
Indian Liver Patient	583	10	0.4
Mammographic Mass UCI	961	6	1.2
Mammographic Mass	830	5	1.1
Parkinson	197	23	0.3
Pima Indians Diabetes	768	8	1.9
South African Hearth	462	9	1.9
Spectf Heart	267	44	0.3
Statlog (Heart)	270	13	1.3

Table 2. Classification error and mean rank positions for the proposed selection algorithms Ψ_{SM} and Ψ_{SA} with the tested parameters – case of k-NN base classifiers

Data set	$\Psi_{SM}^{(1)}$	$\Psi_{SM}^{(2)}$	$\Psi_{SM}^{(3)}$	$\Psi_{SM}^{(4)}$	$\Psi_{SM}^{(5)}$	$\Psi_{SA}^{(1)}$	$\Psi_{SA}^{(2)}$	$\Psi_{SA}^{(3)}$	$\Psi_{SA}^{(4)}$	$\Psi_{SA}^{(5)}$
Blood	0.22	0.22	0.27	0.27	0.48	0.23	0.21	0.21	0.23	0.38
Cancer	0.04	0.04	0.04	0.04	0.68	0.03	0.04	0.04	0.03	0.04
Bands	0.39	0.39	0.40	0.40	0.45	0.38	0.38	0.38	0.39	0.45
Haber.	0.24	0.24	0.25	0.25	0.46	0.24	0.24	0.24	0.24	0.33
Hepat.	0.17	0.17	0.16	0.16	0.35	0.17	0.14	0.14	0.17	0.27
Liver	0.31	0.31	0.33	0.33	0.43	0.33	0.31	0.31	0.33	0.38
MamUCI	0.22	0.22	0.22	0.22	0.26	0.22	0.22	0.22	0.22	0.23
Mam.	0.18	0.18	0.18	0.18	0.24	0.18	0.18	0.18	0.18	0.19
Park.	0.13	0.13	0.13	0.13	0.41	0.18	0.15	0.15	0.18	0.23
Pima	0.26	0.26	0.26	0.26	0.32	0.28	0.27	0.27	0.28	0.29
Saheart	0.32	0.32	0.33	0.33	0.40	0.34	0.33	0.33	0.34	0.39
Spec.	0.20	0.20	0.20	0.20	0.32	0.20	0.21	0.21	0.21	0.31
Statlog	0.35	0.35	0.34	0.34	0.34	0.30	0.30	0.30	0.30	0.31
Avg. Rank	6.76	6.76	5.84	5.84	1.26	6.19	**7.07**	**7.07**	5.65	2.50

Table 3. Classification error and mean rank positions for the proposed selection algorithms Ψ_{SM} and Ψ_{SA} with the tested parameters – case of SVM base classifiers

Data set	$\Psi_{SM}^{(1)}$	$\Psi_{SM}^{(2)}$	$\Psi_{SM}^{(3)}$	$\Psi_{SM}^{(4)}$	$\Psi_{SM}^{(5)}$	$\Psi_{SA}^{(1)}$	$\Psi_{SA}^{(2)}$	$\Psi_{SA}^{(3)}$	$\Psi_{SA}^{(4)}$	$\Psi_{SA}^{(5)}$
Blood	0.21	0.21	0.21	0.21	0.40	0.26	0.25	0.25	0.26	0.34
Cancer	0.02	0.02	0.02	0.02	0.21	0.03	0.04	0.04	0.03	0.06
Bands	0.36	0.36	0.36	0.36	0.41	0.33	0.35	0.35	0.33	0.31
Haber.	0.27	0.27	0.27	0.27	0.48	0.27	0.27	0.28	0.29	0.34
Hepat.	0.16	0.16	0.16	0.16	0.14	0.16	0.16	0.16	0.16	0.16
Liver	0.27	0.27	0.27	0.27	0.35	0.30	0.28	0.28	0.30	0.35
MamUCI	0.23	0.23	0.23	0.23	0.26	0.22	0.22	0.22	0.22	0.22
Mam.	0.20	0.20	0.20	0.20	0.22	0.20	0.19	0.19	0.20	0.20
Park.	0.27	0.27	0.27	0.27	0.12	0.16	0.14	0.13	0.17	0.21
Pima	0.40	0.40	0.41	0.41	0.44	0.37	0.39	0.39	0.37	0.39
Saheart	0.30	0.30	0.31	0.31	0.51	0.34	0.33	0.33	0.34	0.36
Spec.	0.20	0.20	0.20	0.20	0.31	0.29	0.32	0.38	0.34	0.35
Statlog	0.44	0.44	0.44	0.44	0.24	0.3	0.3	0.31	0.31	0.27
Avg. Rank	5.96	5.96	5.65	5.65	3.42	6.15	**6.38**	5.80	5.34	4.65

Table 4. Classification error and mean rank positions for the base classifiers ($\Psi_1, ..., \Psi_4$), algorithms Ψ_{DT}, Ψ_S and the proposed selection algorithm Ψ_{SA} – case of k-NN base classifiers

Data set	Ψ_1	Ψ_2	Ψ_3	Ψ_4	Ψ_{DT}	Ψ_S	$\Psi_{SA}^{(2)}$
Blood	0.23	0.21	0.20	0.23	0.35	0.21	0.21
Cancer	0.04	0.04	0.04	0.05	0.04	0.04	0.04
Bands	0.38	0.44	0.39	0.37	0.42	0.38	0.38
Haber.	0.25	0.24	0.26	0.25	0.28	0.26	0.24
Hepat.	0.17	0.17	0.16	0.16	0.17	0.17	0.14
Liver	0.31	0.30	0.30	0.35	0.38	0.31	0.31
MamUCI	0.21	0.22	0.22	0.23	0.22	0.22	0.22
Mam.	0.19	0.19	0.19	0.19	0.18	0.18	0.18
Park.	0.14	0.18	0.20	0.14	0.17	0.13	0.15
Pima	0.27	0.27	0.26	0.28	0.28	0.27	0.27
Saheart	0.33	0.31	0.32	0.36	0.37	0.33	0.33
Spec.	0.20	0.22	0.23	0.20	0.24	0.21	0.21
Statlog	0.32	0.31	0.32	0.3	0.30	0.30	0.30
Avg. Rank	4.19	4.00	4.07	3.61	2.61	4.53	**4.96**

Table 5. Classification error and mean rank positions for the base classifiers $(\Psi_1, ..., \Psi_4)$, algorithms Ψ_{DT}, Ψ_S and the proposed selection algorithm Ψ_{SA} – case of SVM base classifiers

Data set	Ψ_5	Ψ_6	Ψ_7	Ψ_8	Ψ_{DT}	Ψ_S	$\Psi_{SA}^{(2)}$
Blood	0.20	0.26	0.21	0.23	0.23	0.21	0.25
Cancer	0.02	0.13	0.06	0.06	0.02	0.02	0.04
Bands	0.36	0.36	0.36	0.36	0.34	0.36	0.35
Haber.	0.26	0.31	0.25	0.29	0.30	0.26	0.27
Hepat.	0.14	0.16	0.16	0.16	0.15	0.16	0.16
Liver	0.27	0.72	0.27	0.26	0.26	0.27	0.28
MamUCI	0.22	0.26	0.21	0.24	0.22	0.21	0.22
Mam.	0.20	0.30	0.21	0.20	0.20	0.20	0.19
Park.	0.08	0.55	0.27	0.26	0.08	0.27	0.14
Pima	0.33	0.34	0.40	0.34	0.31	0.37	0.39
Saheart	0.31	0.34	0.33	0.34	0.33	0.33	0.33
Spec.	0.20	0.75	0.20	0.20	0.19	0.20	0.32
Statlog	0.21	0.44	0.44	0.44	0.2	0.44	0.3
Avg. Rank	5.46	1.73	3.73	3.46	**5.50**	4.23	3.88

In addition, for the method with average Ψ_{SA} is always achieved better results as compared to the method with median Ψ_{SM}. The algorithm $\Psi_{SA}^{(2)}$ was selected for the comparison with other classifiers, i.e. with the base classifiers and the algorithms without the selection.

Table 4 presents the classification error and the average ranks obtained by the Friedman test for the case when we use k-NN as a base classifier. Table 5 summarizes the results of experiments for the case when we use the SVM model as a base classifier.

Generally, ensembles of classifiers achieve better results than the base classifier. The proposed in the paper dynamic classifier selection method achieves better results than the algorithm with k-NN rule as a base classifier. For the case when we use the SVM model as a base classifier the best is DT method. The critical difference (CD) for Friedman test test at $p = 0.05$ is equal to $CD = 2.49$ – for 7 methods used for classification and 13 data sets. We can conclude that the post-hoc Nemenyi test detects significant differences only between Ψ_{DT} and Ψ_6 algorithms.

5   Conclusion

This paper presents dynamic ensemble classifier selection algorithm with uses average or median in the training phase. The proposed algorithm uses information from so-called decision profiles which are formed from the outputs of the base classifiers. In the experiments we examined quality of classification in the cases of homogeneous ensemble of base classifiers. The proposed algorithm has been compared with all base classifiers and the ensemble classifiers based on the sum and decision profile methods. The experiments have been carried out on 13 benchmark data sets.

Future work might involve the application of the proposed methods for various practical tasks [8,10,22] in which base classifiers and their ensemble can be used. Additionally, in future work should examine the proposed dynamic ensemble selection method on a larger set of base classifiers, other base classifiers as well as on heterogeneous pool of base classifiers.

Acknowledgments. This work was supported by the Polish National Science Center under the grant no. DEC-2013/09/B/ST6/02264 and by the statutory funds of the Department of Systems and Computer Networks, Wroclaw University of Technology.

References

1. Alcalá, J., Fernández, A., Luengo, J., Derrac, J., García, S., Sánchez, L., Herrera, F.: Keel data-mining software tool: data set repository, integration of algorithms and experimental analysis framework. J. Multiple Valued Logic Soft Comput. **17**(255–287), 11 (2010)
2. Baczyńska, P., Burduk, R.: Ensemble selection based on discriminant functions in binary classification task. In: Jackowski, K., Burduk, R., Walkowiak, K., Woźniak, M., Yin, H. (eds.) IDEAL 2015. LNCS, vol. 9375, pp. 61–68. Springer, Heidelberg (2015). doi:10.1007/978-3-319-24834-9_8
3. Breiman, L.: Randomizing outputs to increase prediction accuracy. Mach. Learn. **40**(3), 229–242 (2000)
4. Britto, A.S., Sabourin, R., Oliveira, L.E.: Dynamic selection of classifiers-a comprehensive review. Pattern Recogn. **47**(11), 3665–3680 (2014)
5. Burduk, R.: Classifier fusion with interval-valued weights. Pattern Recogn. Lett. **34**(14), 1623–1629 (2013)
6. Canuto, A.M., Abreu, M.C., de Melo Oliveira, L., Xavier, J.C., Santos, A.D.M.: Investigating the influence of the choice of the ensemble members in accuracy and diversity of selection-based and fusion-based methods for ensembles. Pattern Recogn. Lett. **28**(4), 472–486 (2007)
7. Duin, R.P.: The combining classifier: to train or not to train? In: Proceedings of the 16th International Conference on Pattern Recognition, vol. 2, pp. 765–770. IEEE (2002)
8. Forczmański, P., Łabędź, P.: Recognition of occluded faces based on multi-subspace classification. In: Saeed, K., Chaki, R., Cortesi, A., Wierzchoń, S. (eds.) CISIM 2013. LNCS, vol. 8104, pp. 148–157. Springer, Heidelberg (2013). doi:10.1007/978-3-642-40925-7_15
9. Frank, A., Asuncion, A.: UCI machine learning repository (2010)
10. Frejlichowski, D.: An algorithm for the automatic analysis of characters located on car license plates. In: Kamel, M., Campilho, A. (eds.) ICIAR 2013. LNCS, vol. 7950, pp. 774–781. Springer, Heidelberg (2013). doi:10.1007/978-3-642-39094-4_89
11. Freund, Y., Schapire, R.E., et al.: Experiments with a new boosting algorithm. In: ICML, vol. 96, pp. 148–156 (1996)

12. Giacinto, G., Roli, F.: An approach to the automatic design of multiple classifier systems. Pattern Recogn. Lett. **22**, 25–33 (2001)
13. Inbarani, H.H., Azar, A.T., Jothi, G.: Supervised hybrid feature selection based on pso and rough sets for medical diagnosis. Comput. Methods Programs Biomed. **113**(1), 175–185 (2014)
14. Jackowski, K., Krawczyk, B., Woźniak, M.: Improved adaptive splitting and selection: the hybrid training method of a classifier based on a feature space partitioning. Int. J. Neural Syst. **24**(3), 1430007 (2014)
15. Korytkowski, M., Rutkowski, L., Scherer, R.: From ensemble of fuzzy classifiers to single fuzzy rule base classifier. In: Rutkowski, L., Tadeusiewicz, R., Zadeh, L.A., Zurada, J.M. (eds.) ICAISC 2008. LNCS (LNAI), vol. 5097, pp. 265–272. Springer, Heidelberg (2008). doi:10.1007/978-3-540-69731-2_26
16. Kuncheva, L.I.: Combining Pattern Classifiers: Methods and Algorithms. John Wiley & Sons, Hoboken (2004)
17. Kuncheva, L.I., Bezdek, J.C., Duin, R.P.: Decision templates for multiple classifier fusion: an experimental comparison. Pattern Recogn. **34**(2), 299–314 (2001)
18. Rejer, I.: Genetic algorithm with aggressive mutation for feature selection in bci feature space. Pattern Anal. Appl. **18**(3), 485–492 (2015)
19. Ruta, D., Gabrys, B.: Classifier selection for majority voting. Inf. Fusion **6**(1), 63–81 (2005)
20. Trawiński, B., Smętek, M., Telec, Z., Lasota, T.: Nonparametric statistical analysis for multiple comparison of machine learning regression algorithms. Int. J. Appl. Math. Comput. Sci. **22**(4), 867–881 (2012)
21. Xu, L., Krzyżak, A., Suen, C.Y.: Methods of combining multiple classifiers and their applications to handwriting recognition. IEEE Trans. Syst. Man Cybern. **22**(3), 418–435 (1992)
22. Zdunek, R., Nowak, M., Pliński, E.: Statistical classification of soft solder alloys by laser-induced breakdown spectroscopy: review of methods. J. Eur. Opt. Soc. Rapid Publ. **11**(16006), 1–20 (2016)

Deterministic Method for the Prediction of Time Series

Walery Rogoza[⊠]

Faculty of Computer Science and Information Technology, West Pomeranian
University of Technology, Zolnierska 49, 71-210 Szczecin, Poland
wrogoza@wi.zut.edu.pl

Abstract. The business analyst is frequently forced to make decision using data
on a certain business process obtained within a short time interval. Under these
conditions, the analyst is not in a position to use traditional statistical methods
and should be satisfied with experimental samples that are few in number. The
paper deals with the new method for the prediction of time series based on the
concepts of system identification. The suggested method shows a reasonable
flexibility and accuracy when the analyzed process possesses a certain regularity
property, and it is usable for the prediction of time series in a short-term
perspective.

Keywords: Time series · Prediction · Deterministic method · Short-term
intervals

1 Introduction

The business analyst is frequently dealt with data that represent the dynamics of a
process in time. Data of this kind are classified as time series. An investigation of
regularities exhibited by time series is of great importance in the establishment of
relations between changes in parameters, which describe the process under consideration, as well as in the control over the process and in the prediction of parameter
values in future time instances. If parameter variations are subject to certain regularities, one can find algebraic relations between them due to which it is possible to build
mathematical models which facilitate the solution of the above problems due to use of
computerized algorithms.

A typical strategy of model construction in the solution of the problem of time
series prediction provides the use of the set of experimental samples Psam with the
known values of parameters describing the analyzed process. Experimental samples are
usually obtained at a number of discrete time instances and are used to build equations
which express functional dependences between parameters. Traditionally, this procedure is divided into two stages: (a) model synthesis (sometimes it is called model
learning) and (b) model testing [1]. To realize the procedure, the set of samples Psam is
divided into two subsets – the subset which is used for model learning Plearn and the
subset Ptest for testing the obtained models: $P_{sam} = P_{train} \cup P_{test}$. Time series are
usually studied with the use of statistical methods, which require a rather great number
of samples (as a rule, several tens samples). Statistical methods have their greatest

© Springer International Publishing AG 2017
S. Kobayashi et al. (eds.), *Hard and Soft Computing for Artificial Intelligence, Multimedia and
Security*, Advances in Intelligent Systems and Computing 534, DOI 10.1007/978-3-319-48429-7_7

impact when the studied processes adhere to certain statistical laws [1]. For example, statistical methods are well suited to the simulation of ergodic and stationary processes.

On occasion, researchers are up against the cases when they have a little number of experimental samples (say, 7 – 10) at their disposal, which do not allow them to find statistical characteristics (expectation, deviation, and so on), which are required to establish statistical regularities. Moreover, in such situations, there is no possibility to estimate, whether the mentioned regularities exist for the given process at all. It is obvious that statistical methods fail in such cases, and in addition, no possibility exists to estimate the accuracy of created models for lack of a sufficient number of samples.

Under the above circumstances, deterministic methods based on the principals of system identification might be considered as an alternative approach, when the researcher has a small set of experimental data at hand and is forced to adopt the results of simulation bypassing the stage of model testing. It is evident that the researcher cannot rely on a high accuracy and long-term prediction of the process, whose analysis is based on the small set of experimental data, but sometimes it is the only way to initiate the analysis of a process at early stages of investigation.

This paper discusses a deterministic method dedicated to the prediction of time series for the short time interval in future with small number of experimental samples.

2 Theoretical Grounding in the Suggested Method

Assume that we have in our disposition an object, whose states are described by the n-vector of time dependent real variables $X(t) = (x_1(t), x_2(t), \ldots, x_n(t))^T$, where t is time and T is the transposition symbol. Assume also that components of vector $X(t)$ have been determined experimentally at discrete time instances $t_0 > t_1 > t_2 > , \ldots, > t_N$, which yields the set of vectors with known components $X(t_0), X(t_1), \ldots, X(t_N) = X_0, X_1, \ldots, X_N$, where 0, 1, ..., N are numbers of samples. It is desired to predict the values of the vector $X(t)$ components in future time instances $t_{N+1} > t_{N+2} > , \ldots, > t_{N+K}$. In other words, using vectors $X_0, X_1, \ldots, X_N$ of known values we would like to predict values of variables consisting in vectors $X_{N+1}, X_{N+2}, \ldots, X_{N+K}$. The limiting time point t_{N+K}, for which the created predicting models possess satisfactory accuracy, depends on the regularity of the process under consideration. Eventually the accuracy becomes unsatisfactory, and the created models should be formed again by using the newest experimental samples.

We can formalize the description of the studied problem using concepts of the system identification theory. The analyzed object can be considered as a "black box", known variables as input signals (input variables) of the "black box", that is, variables, whose values are known from experimental sample, and variables to be computed using known variables as output signals (output variables), that is, variables, whose values are to be predicted. The discussed problem can be interpreted as that one which consists in the formation of mathematical models allowing us to compute the values of output variables as functions of input variables. For the sake of convenience, let us denote input variables by small letters $x_1, x_2, \ldots, x_n$ and output variables by capital letters $Y_1, Y_2, \ldots, Y_n$.

Figure 1 demonstrates pictorially this idea for the case, when we deal with n input variables determined at the instance of time t_N and n output variables, that is, the same variables, whose values should be computed at the next instance of time t_{N+1}. In the general case, each output variable should be considered as a function of all the input variables, that is, $Y_i(t_{N+1}) \equiv x_i(t_{N+1}) = f_i(x_1(t_N), \ldots, x_n(t_N))$, where $f_i()$ is a specific function for each Y_i. Attempts at the construction of the above generalized functions lead to the construction of too cumbersome equations and provide the use of a lot of experimental samples. This may be attributed to a failure of the above assumption on a small number of the experimental samples being in the researcher's disposition, which sends us in search of another approach, which is usable to our specific conditions.

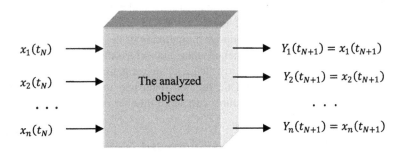

Fig. 1. Pictorial representation of the analyzed object as a "black box"

With this aim, we will build the desired models in the form of Kolmogorov-Gabor second-order polynomials (KGP), which are polynomials of the least possible order that are sufficient to take into account mutual relationships between the parameters of the process under consideration. We would like to remark that KGP of different orders form a basis for traditional methods of building regression equations [3]. However, in contrast to the mentioned traditional approach, we will use KGP of the second order to establish deterministic models based on the ideas of system identification.

For example, if we desire to express the value of variable $Y_i \equiv Y_i(t_{N+1})$ at instance t_{N+1}, as a function of variables $x_i \equiv x_i(t_N)$ and $x_j \equiv x_j(t_N)$, where $i \neq j$, determined at the instance t_N, we can represent the mentioned second order KGP as follows:

$$Y_i = a_0 + a_1 x_i + a_2 x_j + a_3 x_i^2 + a_4 x_j^2 + a_5 x_i x_j, \tag{1a}$$

where $a_0, a_1, \ldots, a_5$ are real coefficients, to be computed using data containing in experimental samples. Such a model implies that we adopt the hypothetical condition, according to which only two input variables x_i and x_j change at instance t_N and the rest of input variables are constant. It is easy to estimate that if the process is described by n variables, we can form $(n - 1)$ equations for the prediction of the Y_i variable, which include variable x_i and one of the rest variables $x_1, x_2, \ldots, x_{i-1}, x_{i+1}, \ldots, x_n$ as input variables. For this reason we call equations of the form (1a) the particular equations (models). For every particular model there is a dual particular model, which is determined for the same conditions. For example, equation

$$Y_j = b_0 + b_1 x_j + b_2 x_i + b_3 x_j^2 + b_4 x_i^2 + b_5 x_j x_i \tag{1b}$$

is dual with respect to (1a), because it implies that we compute $Y_j \equiv Y_j(t_{N+1})$ at instance t_{N+1} as a function of variables $x_j \equiv x_j(t_N)$ and $x_i \equiv x_i(t_N)$, where $i \neq j$, determined at instance t_N, under the condition that the rest of input variables are constant.

As was mentioned above, in order to compute polynomial coefficients of a particular model, we can form and solve the set of six equations with known values of output and input variables. For example, coefficients of model (1a) can be find by the solution of the following set of equations:

$$
\begin{aligned}
Y_i(t_N) &= a_0 + a_1 x_i(t_{N-1}) + a_2 x_j(t_{N-1}) + a_3 x_i^2(t_{N-1}) + a_4 x_j^2(t_{N-1}) \\
&\quad + a_5 x_i(t_{N-1}) x_j(t_{N-1}); \\
Y_i(t_{N-1}) &= a_0 + a_1 x_i(t_{N-2}) + a_2 x_j(t_{N-2}) + a_3 x_i^2(t_{N-2}) + a_4 x_j^2(t_{N-2}) \\
&\quad + a_5 x_i(t_{N-2}) x_j(t_{N-2}); \\
&\quad \cdots \\
Y_i(t_{N-5}) &= a_0 + a_1 x_i(t_{N-6}) + a_2 x_j(t_{N-6}) + a_3 x_i^2(t_{N-6}) + a_4 x_j^2(t_{N-6}) \\
&\quad + a_5 x_i(t_{N-6}) x_j(t_{N-6}),
\end{aligned}
\tag{2}
$$

where the values of variables $y_i(t_N), \ldots, y_i(t_{N-6})$ and $x_i(t_{N-1}), \ldots, x_i(t_{N-5})$ are obtained from experimental samples $P(N), \ldots, P(N-6)$.

Knowing coefficients $a_0, a_1, \ldots, a_5$, we can then form model (1a) for the prediction of the $Y_i(t_{N+1})$ variable at the instance t_{N+1}. If process under consideration possesses certain regularity, we can use the same model to predict $Y_i(t_{N+1}), Y_i(t_{N+2}), \ldots, Y_i(t_{N+K})$. It should be noted that the traditional approach to prediction of time series is based on the strategy of increasing orders of KGPs at the consequent time instances, which as a rule requires a great number of experimental samples [3, 4]. In distinction to the traditional approach, the suggested method is based on the use the second order KGP, only, which weakens the requirements on the number of experimental samples. And the final predicted value $Y_i(t_{N+1})$ can be computed as the arithmetic mean of all the predicted values obtained using particular equation for the variable $Y_i(t_{N+1})$.

It is obvious that the contribution of changes in values of different input variables to changes of output variables may be different in different particular models. The information about the extent to which changes of some variables are affected by the changes of other variables may be of fundamental importance in the selection of principal particular models as well as in the selection of a principal variable in each particular model. Following a classical system identification theory, we can estimate such impacts using sensitivity functions. For any output function y of n input variables $y = f(x_1, \ldots, x_i, \ldots, x_n)$, the classical definitions of the first order sensitivity functions of y with respect to variable x_i have the forms [5]:

- relative sensitivity function

$$S_{x_i}^f = \frac{\partial f / f}{\partial x_i / x_i},$$ (3)

- semirelative sensitivity function

$$S_{x_i}^f = \frac{\partial f}{\partial x_i / x_i},$$ (4)

- absolute sensitivity function

$$S_{x_i}^f = \frac{\partial f}{\partial x_i}.$$ (5)

Each definition is usable under the condition that there are derivatives of function f with respect to variable x_i at each instance of time. The special convenience of definition (3) is that it provides the difference quotient of function f and parameter x_i, which is independent of absolute values of variables presented in the equation (it is sometimes important because changes in variables x_i possessing small absolute values may have great impact on changes in the resulting value of function f). However (3) is usable if both function f and variable x_i are nonzero. Equation (4) is usable if parameter x_i is nonzero. For the sake of definiteness, we use (4) in our following discussion.

By analogy, we can define the second order sensitivity functions as follows:

- semirelative sensitivity function with respect to variable x_i:

$$S_{x_i^2}^f = \frac{\partial^2 f}{\partial x_i^2} \cdot x_i^2,$$ (6)

- semirelative sensitivity function with respect to variables x_i and x_j:

$$S_{x_i x_j}^f = \frac{\partial^2 f}{\partial x_i \partial x_j} \cdot x_i \cdot x_j.$$ (7)

Substituting (1a) into (4, 6, 7), we can obtain equations for the first and second order sensitivities of variable Y_i with respect to parameters x_i and x_j in terms of coefficients of the particular model (1a):

$$S_{x_i}^{Y_i} = \frac{\partial Y_i}{\partial x_i} \cdot x_i = \left(a_1 + 2a_3 x_i + a_5 x_j\right) \cdot x_i,$$ (8)

$$S_{x_j}^{Y_i} = \frac{\partial Y_i}{\partial x_j} \cdot x_j = \left(a_2 + 2a_4 x_j + a_5 x_i\right) \cdot x_j,$$ (9)

$$S_{x_i}^{Y_i} = \frac{\partial^2 Y_i}{\partial x_i^2} \cdot x_i^2 = 2a_3 x_i^2, \tag{10}$$

$$S_{x_j^2}^{Y_i} = \frac{\partial^2 Y_i}{\partial x_j^2} \cdot x_j^2 = 2a_4 x_j^2, \tag{11}$$

$$S_{x_i x_j}^{Y_i} = \frac{\partial^2 Y_i}{\partial x_i x_j} \cdot x_i \cdot x_j = a_5 \cdot x_i \cdot x_j. \tag{12}$$

The similar equations can be obtained for the dual model (1b) as well as for any other particular model.

There is no difficulty to show that using the semirelative sensitivity functions (4), (6), (7), we can develop particular models in terms of semirelative sensitivity functions and increments of variables x_i and x_j. Indeed, the expansion of function $Y_i(t_{N+1})$ of the form (1a) in Taylor's series at point t_N yields:

$$
\begin{aligned}
Y_i(t_{N+1}) &= Y_i(t_N) + \frac{\partial Y_i(t_N)}{\partial x_i} \cdot (\Delta x_i)|_{P(N)} + \frac{\partial Y_i(t_N)}{\partial x_j} \cdot (\Delta x_j)|_{P(N)} \\
&+ \frac{1}{2} \cdot \frac{\partial^2 Y_i(t_N)}{\partial x_i^2} \cdot (\Delta x_i)^2|_{P(N)} + \frac{1}{2} \cdot \frac{\partial^2 Y_i(t_N)}{\partial x_j^2} \cdot (\Delta x_j)^2|_{P(N)} + \frac{\partial^2 Y_i(t_N)}{\partial x_i \cdot \partial x_j} \cdot (\Delta x_i) \cdot (\Delta x_j)|_{P(N)} + R_2 \\
&= Y_i(t_N) + \frac{\partial Y_i(t_N)}{\partial x_i} \cdot x_i(t_N) \cdot \frac{\Delta x_i}{x_i}|_{P(N)} + \frac{\partial Y_i(t_N)}{\partial x_j} \cdot x_j(t_N) \cdot \frac{\Delta x_j}{x_j}|_{P(N)} \\
&+ \frac{1}{2} \cdot \frac{\partial^2 Y_i(t_N)}{\partial x_i^2} \cdot x_i^2(t_N) \cdot \left(\frac{\Delta x_i}{x_i}\right)^2|_{P(N)} + \frac{1}{2} \cdot \frac{\partial^2 Y_i(t_N)}{\partial x_j^2} \cdot x_j^2(t_N) \cdot \left(\frac{\Delta x_j}{x_j}\right)^2|_{P(N)} \\
&+ \frac{\partial^2 Y_i(t_N)}{\partial x_i \cdot \partial x_j} \cdot x_i(t_N) \cdot x_j(t_N) \cdot \frac{\Delta x_i}{x_i} \cdot \frac{\Delta x_j}{x_j}|_{P(N)} + R_2 \\
&= x_i(t_N) + S_{x_i}^{Y_i} \cdot \frac{\Delta x_i}{x_i}|_{P(N)} + S_{x_j}^{Y_i} \cdot \frac{\Delta x_j}{x_j}|_{P(N)} \\
&+ \frac{1}{2} \cdot S_{x_i^2}^{Y_i} \cdot \left(\frac{\Delta x_i}{x_i}\right)^2|_{P(N)} + \frac{1}{2} \cdot S_{x_j^2}^{Y_i} \cdot \left(\frac{\Delta x_j}{x_j}\right)^2|_{P(N)} + S_{x_i x_j}^{Y_i} \cdot \frac{\Delta x_i}{x_i} \cdot \frac{\Delta x_j}{x_j}|_{P(N)} + R_2,
\end{aligned}
\tag{13}
$$

where R_2 is the remainder term of Taylor's series. We take into account that $Y_i(t_N)$ and $x_i(t_N)$. are different notations of the same variable.

Substituting Eqs. (8)–(12) for sensitivities into (13), we can develop the predicting equation in terms of increments of input variables Δx_i and Δx_j and polynomial coefficients $a_0, a_1, \ldots, a_5$ of KGP (1a):

$$
\begin{aligned}
Y_i(t_{N+1}) &\approx x_i(t_N) + \left[a_1 + 2a_3 x_i(t_N) + a_5 x_j(t_N)\right] \cdot \Delta x_i \\
&+ \left[a_2 + 2a_4 x_j(t_N) + a_5 x_i(t_N)\right] \Delta x_j + a_3 \cdot (\Delta x_i)^2 + a_4 \cdot (\Delta x_j)^2 + a_5 \cdot (\Delta x_i) \cdot (\Delta x_j)
\end{aligned}
\tag{14}
$$

Two remarks should be made on the above increments. First, to gain a better understanding of the sense of increments Δx_i and Δx_j, let us take into account that the value of variable $x_i(t_N)$ just as the variable $x_j(t_N)$ at the instance t_N can be presented in the form of the sum of that variable at the instance t_{N-1} and a certain increment, that is, $x_{i,j}(t_N) = x_{i,j}(t_{N-1}) + \Delta x_{i,j}(t_{N-1})$, where $\Delta x_{i,j}(t_{N-1}) = x_{i,j}(t_N) - x_{i,j}(t_{N-1})$ is the backward difference of the $x_i(t_N)$ variable (the $x_j(t_N)$ variable, respectively).

Rewriting (1a) with the use of the above increments, we can write the equation:

$$
\begin{aligned}
Y_i(t_{N+1}) &= a_0 + a_1 x_i(t_N) + a_2 x_j(t_N) + a_3 x_i^2(t_N) + a_4 x_j^2(t_N) + a_5 x_i(t_N) x_j(t_N) \\
&= a_0 + a_1 [x_i(t_{N-1}) + \Delta x_i(t_{N-1})] + a_2 [x_j(t_{N-1}) + \Delta x_j(t_{N-1})] \\
&\quad + a_3 [x_i(t_{N-1}) + \Delta x_i(t_{N-1})]^2 + a_4 [x_j(t_{N-1}) + \Delta x_j(t_{N-1})]^2 \\
&\quad + a_5 [x_i(t_{N-1}) + \Delta x_i(t_{N-1})][x_j(t_{N-1}) + \Delta x_j(t_{N-1})].
\end{aligned}
$$

And finally removing the parentheses and considering that

$$
\begin{aligned}
Y_i(t_N) &\equiv x_i(t_N) = a_0 + a_1 x_i(t_{N-1}) + a_2 x_j(t_{N-1}) + a_3 x_i^2(t_{N-1}) + a_4 x_j^2(t_{N-1}) \\
&\quad + a_5 x_i(t_{N-1}) x_j(t_{N-1}),
\end{aligned}
$$

we can obtain the same equation as (14). It follows that increments $\Delta x_i(t_{N-1})$ and $\Delta x_j(t_{N-1})$ in (14) are backward differences of variables $x_i(t_N)$ and $x_j(t_N)$, and $\frac{\Delta x_{i,j}}{x_{i,j}}\big|_{P(N)}$ in (13) are relative backward differences of variables $x_i(t_N)$ and $x_j(t_N)$, which we denote as follows: $\frac{\Delta x_{i,j}(t_{N-1})}{x_{i,j}(t_{N-1})} \equiv ddx_{i,j}(t_{N-1})$.

Thus, in order to compute, say, value $Y_i(t_{N+1})$ using Eq. (13), we should substitute values of input variables $x_i(t_N)$ and $x_j(t_N)$ obtained at the t_N instance and relative backward differences $ddx_i(t_{N-1})$ and $ddx_j(t_{N-1})$ obtained at the t_{N-1} instance. In this connection, it is important to emphasize that it follows from the above consideration that the considered particular models are functions with lagging arguments.

And the second remark is related to the way, in which the mentioned relative backward differences can be computed. Recall that particular models in any form (1a), (13) or (14) are constructed on the hypothetical assumption that changes in $Y_i(t_{N+1})$ depend upon changes of variables $x_i(t_N)$ and $x_j(t_N)$, only, whereas the rest of input variables are assumed to be constant. It means that using particular models, we deal with the projections of the n–dimensional vectors of variables $X_i(t)$ on the two–dimensional (x_i, x_j)-subspaces, where $i, j = 1, \ldots, n$. Because of this, the above relative backward differences should be interpreted as the projections of actual increments of variables x_i and x_j on the (x_i, x_j)-subspace, either. Therefore ddx_i and ddx_j are different of actual relative increments of like variables x_i and x_j. Thus we should find a special procedure to compute values of ddx_i and ddx_j for $i, j = 1, \ldots, n$, at each sequential instance of time.

The desired procedure is readily apparent from the fact that any two dual particular models Y_i and Y_j include the same relative backward differences ddx_i and ddx_j and are valid on the same assumptions. Therefore we can compute relative backward differences, say, $ddx_i\big|_{P(N-1)}$ and $ddx_j\big|_{P(N-1)}$, through the formation and solution of the pair of dual predicting equations in increments for each travelled instance of time:

$$Y_i(t_N) = x_i(t_{N-1}) + S_{x_i}^{Y_i} \cdot ddx_i|_{P(N-1)} + S_{x_j}^{Y_i} \cdot ddx_j|_{P(N-1)}$$

$$+ \frac{1}{2} \cdot S_{x_i^2}^{Y_i} \cdot (ddx_i)^2|_{P(N-1)} + \frac{1}{2} \cdot S_{x_j^2}^{Y_i} \cdot (ddx_j)^2|_{P(N-1)} + S_{x_i x_j}^{Y_i} \cdot ddx_i \cdot ddx_j|_{P(N-1)},$$

$$Y_j(t_N) = x_j(t_{N-1}) + S_{x_j}^{Y_j} \cdot ddx_j|_{P(N-1)} + S_{x_i}^{Y_j} \cdot ddx_i|_{P(N-1)}$$

$$+ \frac{1}{2} \cdot S_{x_j^2}^{Y_j} \cdot (ddx_j)^2|_{P(N-1)} + \frac{1}{2} \cdot S_{x_i^2}^{Y_j} \cdot (ddx_i)^2|_{P(N-1)} + S_{x_j x_i}^{Y_j} \cdot ddx_j \cdot ddx_i|_{P(N-1)},$$

$$(15)$$

where all the components except $ddx_i|_{P(N-1)}$ and $ddx_j|_{P(N-1)}$ are known values or can be computed using known values of polynomial coefficients from (1a) and (1b). In other words, we compute relative backward differences $ddx_i|_{P(N-1)}$ and $ddx_j|_{P(N-1)}$ with one time point step lag. It should be remarked that having in our disposition a series of relative backward differences obtained for the series instances of time in the past, we could compute the next values of relative backward differences using an extrapolation method (for example, Lagrange's method). But we follow the method of the direct computation of the above relative backward differences step by step.

Thus our consideration gives us insight into what techniques can be applied to provide the solution of the above prediction problem. Two approaches may be proposed. The first approach is based on the direct use of particular models of the form (1a) and (1b). And the second approach provides the use of particular models in increments (13). It is evident that the first approach is simpler, but it imposes more severe limits on the conditions of regularity of the analyzed process. The second approach exhibits better flexibility with some sacrifice in the computational complexity. These statements will be illustrated in the example below.

The following sequence of operations can be suggested in the solution of the problem of the short-term prediction of time series.

Step 1. Selection of the newest seven sequential experimental samples, formation and solution of the sets of equations of the form (2) with respect to unknown polynomial coefficients, and formation of particular models for every predicted variable.

Step 2. Following the computation of sensitivity functions (8)–(12), formation and solution of pairs of dual equations of form (15) with respect to relative backward differences ddx_i and ddx_j, $i,j = 1,\ldots,n$ at the time point t_{N-1} preceding the newest time point t_N.

Step 3. Solution of the problem of prediction of values $Y_i(t_{N+1})$ and $Y_j(t_{N+1})$, $i,j = 1,\ldots,n$ for every pair of dual equations using equations of form (15).

Step 4. Computation of the resulting predicted values $Y_i(t_{N+1}), i = 1,\ldots,n$ through the computation of the arithmetic means of the predicted values obtained by the use of all the particular models for every variable separately.

Step 5. The estimation of errors of prediction for each predicted variable, which can be realized by the comparison of the predicted values with those given from the received new experimental samples. If the accuracy of the existing models becomes unsatisfactory, the development of new particular models should be repeated with the use of the newest experimental samples.

3 Example

To illustrate the usage of the above method, let us consider a simple example. Assume that we are at the beginning of the computation process and have data on a process presented in Table 1.

Table 1. Data obtained experimentally for nine samples

Samples	x	y	z
$P(0)$	0.2366	0.2068	0.2496
$P(1)$	0.2403	0.2462	0.2731
$P(2)$	0.3033	0.2406	0.2856
$P(3)$	0.2367	0.2688	0.2966
$P(4)$	0.2679	0.2607	0.2951
$P(5)$	0.3308	0.2484	0.2823
$P(6)$	0.2410	0.3136	0.3370
$P(7)$	0.3151	0.2736	0.3713
$P(8)$	0.4963	0.2788	0.3282

To simplify the notation and to facilitate further interpretation, let us denote the predicted variables (output variables, in the sense of our "black box" model) by capital letters X, Y, and Z, and the variables which are considered as known input variables by small letters x, y, and z.

The values of variables x, y, and z extracted from samples $P(0), P(1), \ldots, P(6)$, which arrive at instances $t_0, t_1, \ldots, t_6$, respectively, we can use to form particular models. The mentioned models then can be used to predict the values of the above variables at the next instances of time t_{N+7} and t_{N+8}. The estimation of the accuracy of the developed predicting models is possible in our example through the comparison of the predicted values with those obtained experimentally. For reasons of space, let us restrict our consideration to the prediction of two variables x and y only.

Step 1. Note that six particular models are to be formed in this problem, which reflect functional relationships between the variables X, Y, and Z and variables x, y, and z. The mentioned particular models can be divided into three groups of dual models: $X(x, y)$ and $Y(y, x)$, $X(x, z)$ and $Z(z, x)$, and $Y(y, z)$ and $Z(z, y)$.

Using data presenting in samples $P(0), P(1), \ldots, P(6)$, we can form the sets of equations of the form (2) to compute polynomial coefficients of all the particular models. For example, to find polynomial coefficients $a_0, a_1, \ldots, a_5$ of model $X(x, y)$, we can form the following set of equations:

$$0.2403 = a_0 + 0.2366a_1 + 0.2068a_2 + 0.2366^2 a_3 + 0.2068^2 a_4 + 0.2366 \cdot 0.2068a_5;$$
$$0.3033 = a_0 + 0.2403a_1 + 0.2462a_2 + 0.2403^2 a_3 + 0.2462^2 a_4 + 0.2403 \cdot 0.2462a_5;$$
$$\ldots$$
$$0.2410 = a_0 + 0.3308a_1 + 0.2484a_2 + 0.3308^2 a_3 + 0.2484^2 a_4 + 0.3308 \cdot 0.2484a_5,$$

whose solution is as follows: $a_0 = 2.0455$, $a_1 = -16.2210$, $a_2 = 1.3224$, $a_3 = -14.3273$, $a_4 = -49.3550$, $a_5 = 95.4847$.

Thus the desired particular model $X(x, y)$ takes the form:

$$X(x,y) = 2.0455 - 16.2210x + 1.3324y - 14.3273x^2 - 49.3550y^2 + 95.4847xy$$
(16)

To find polynomial coefficients $b_0, b_1, \ldots, b_5$, of the dual model $Y(y, x)$, we proceed in a similar way to obtain the following particular model:

$$Y(y,x) = 1.5550 - 5.6708y - 5.252x + 14.9988y^2 + 12.8445x^2 - 5.1894yx.$$
(17)

Just in the same way we can form other particular models:

$$X(x,z) = 39.3362 - 250.489x - 54.623z + 96.1379x^2 - 197.77z^2 + 689.9113xz;$$

$$Z(z,x) = 1.9487 + 2.0640z - 15.5134x - 11.7447z^2 + 18.0084x^2 + 20.3766zx;$$

$$Y(y,z) = -25.29022 - 179.125x + 340.38z - 327.215x^2 - 1131.1z^2 + 1202.8xz;$$

$$Z(z,y) = -20.3054 + 268.05z - 136.966y - 875.559z^2 - 243.37y^2 + 909.63zy.$$

Before we pass to the next steps of the proposed method, let us try to predict the values of variables x and y at instants t_7 and t_8 using the above particular models $X(x, y)$ and $Y(y, x)$ as regression equations. For this purpose we substitute values $x(t_6) = 0.2410$ and $y(t_6) = 0.3136$ into (16) and (17) to give the following predicted values: $x(t_7)_{pr} = X_7(x, y) = 0.0815$, $y(t_7)_{pr} = Y_7(y, x) = 0.3398$, whereas actual values of the mentioned variables in sample $P(7)$ (see) $x(t_7)_{act} = 0.3151$, $y(t_7)_{act} = 0.2736$. Consequently, the errors of prediction become substantial even at the first time step ahead: $\delta_{x_7} = \frac{|x(t_7)_{pr} - x(t_7)_{act}|}{x(t_7)_{act}} = 0.741$ (that is, about 74 %), $\delta_{y_7} = \frac{|y(t_7)_{pr} - y(t_7)_{act}|}{y(t_7)_{act}} = 0.242$ (that is, about 24.2 %). If we then substitute predicted values $x(t_7)_{pr}$ and $y(t_7)_{pr}$ into models (16) and (17) to predict values of these variables at the next instance t_8, we can obtain $x(t_8)_{8,pr} = -1.9756$ ($\delta_{x_8} = 4.98$, that is, the error is about 498 %) and $y_{8,pr} = 0.873$ ($\delta_{y_8} = 2.131$, that is, the error is about 213%).

Hence, the presented results show that the computation process diverges and we deal with a rather stiff problem, whose solution cannot be obtained using traditional regression equations.

Now let us continue the solution of the problem using the method proposed.

Step 2. Compute sensitivity functions (8)–(12). For better visualization of impacts of different input variables on the output variables, Tables 2 and 3 present sensitivity values for particular models $X(x, y)$ and $Y(y, x)$.

As can be seen, sensitivities vary and are substantially different for different instances of time and different variables. Next, we can form and solve the sets of dual equations of the form (15), using experimental data presenting in two neighboring

Table 2. Values of sensitivity functions obtained for model $X(x,y)$

Sample	$S_x^{X(x,y)}$	$S_y^{X(x,y)}$	$S_{x^2}^{X(x,y)}$	$S_{y^2}^{X(x,y)}$	$S_{xy}^{X(x,y)}$
$P(0)$	−0.77	0.724	−1.604	−4.2214	4.672
$P(1)$	0.0965	−0.0086	−1.655	−5.9832	5.649
$P(2)$	−0.588	1.572	−2.636	−5.714	6.968
$P(3)$	0.6303	−0.7015	−1.605	−7.132	6.0752
$P(4)$	0.2666	0.305	−2.0565	−6.709	6.669
$P(5)$	−0.6555	2.0839	−3.136	−6.09	7.846
$P(6)$	1.643	−2.076	−1.664	−9.708	7.2165
$P(7)$	0.2757	1.2045	−2.845	−7.389	8.2319

Table 3. Values of sensitivity functions obtained for model $Y(y,x)$

Sample	$S_y^{Y(y,x)}$	$S_x^{Y(y,x)}$	$S_{y^2}^{Y(y,x)}$	$S_{x^2}^{Y(y,x)}$	$S_{yx}^{Y(y,x)}$
$P(0)$	−0.1438	−0.0585	1.283	1.438	−0.254
$P(1)$	0.1151	−0.0857	1.818	1.4834	−0.307
$P(2)$	−0.0066	0.3915	1.7365	2.363	−0.3787
$P(3)$	0.3129	−0.134	2.167	1.439	−0.3301
$P(4)$	0.1979	0.0743	2.039	1.844	−0.362
$P(5)$	0.0159	0.647	1.851	2.811	−0.4264
$P(6)$	0.7795	−0.1659	2.9501	1.492	−.0392
$P(7)$	0.2466	0.4483	2.2455	2.5506	−0.447

samples and the computed sensitivity values. For example, using data presented in samples $P(4)$ and $P(5)$ (see Table 1) and the values of sensitivities for sample $P(5)$ given in Tables 2 and 3, we can form two equations in increments as follows:

$$0.2410 = 0.3308 - 0.6555 \cdot ddx(t_5) + 2.0839 \cdot ddy(t_5)$$
$$- 0.5 \cdot 3.136 \cdot (ddx(t_5))^2 - 0.5 \cdot 6.09 \cdot (ddy(t_5))^2 + 7.846 \cdot ddx(t_5) \cdot ddy(t_5);$$

$$0.3136 = 0.2484 - 0.0159 \cdot ddy(t_5) + 0.6470 \cdot ddx(t_5)$$
$$+ 0.5 \cdot 1.8510 \cdot (ddy(t_5))^2 - 0.5 \cdot 2.811 \cdot (ddx(t_5))^2 - 0.4264 \cdot ddy(t_5) \cdot ddx(t_5).$$

The solution of this set of equations is as follows: $ddx(t_4) = 0.0848$, $ddy(t_4) = -0.0083$.

In the same manner, we can obtain relative backward increments ddx and ddy for particular models $X(x, y)$ and $Y(y, x)$ for every experimental sample $P(0), P(2), \ldots$. Table 4 represents values of relative backward differences for the samples, which we deal with.

Using data presented in Table 4, we can obtain relative backward differences for the next samples using extrapolation methods, say, Lagrange's polynomials. As an example, we computed values ddx and ddy for $P(7)$ (bold numbers in Table 3).

Table 4. Relative backward increments ddx_i and ddy_i for the dual particular models $X(x, y)$ and $Y(y, x)$

Sample	ddx_i	ddy_i
$P(0)$	−0.1024	−0.129
$P(1)$	0.0799	0.0599
$P(2)$	0.0598	−0.0132
$P(3)$	0.0443	−0.0126
$P(4)$	0.05214	−0.01723
$P(5)$	0.0848	−0.0083
$P(6)$	−0.0383	−0.0684
$P(7)$	**−0.0814**	**0.1690**

However, the suggestion of the proposed method is to use the current values of variables at sample $P(N)$ and the relative backward differences obtained at the previous sample $P(N-1)$ to predict the values of the mentioned variables at next sample $P(N+1)$. For example, trying to predict $X(t_7)$ and $Y(t_7)$, we can use the current values of these variables $x(t_6)$ and $y(t_6)$ and relative backward differences $ddx(t_5)$ and $ddy(t_5)$. This idea was proven above and is illustrated in the computation of the predicted value of variable $X(t_7)$ as is shown at the next step.

Step 3. The use of (15) gives:

$$X(t_7) = 0.241 + 1.643 \cdot 0.0848 - 2.076 \cdot (-0.0083) - 0.5 \cdot 1.664 \cdot (0.0848)^2$$
$$- 0.5 \cdot 9.708 \cdot (-0.0083)^2 + 7.2165 \cdot 0.0848 \cdot (-0.0083) = 0.3862.$$

The exact value of variable $x(t_7)$ (see Table 1) is 0.3151. Thus the relative error of prediction is $\delta_{x7} = \frac{|X(t_7) - x(t_7)|}{|x(t_7)|} = |0.3862 - 0.3151|/0.3151 = 0.225$, that is, about 22.5 %, which is much less than was obtained above using the regression equation, which gives 77 % error.

By way of illustration we computed predicted values of variables $x(t)$ and $y(t)$ for several samples using the same technique and then we compared the obtained results with the known data presented in Table 1 (see Table 5). The values of variables presented in Table 1, we arbitrarily call the actual data and denote them by x_{act} and y_{act}, and those obtained with the use of the suggested method we call the predicted data and denote them by x_{pr} and y_{pr}.

Table 5. The exact and predicted values of variables $x(t)$ and $y(t)$

Sample	x_{act}	x_{pr}	Relative error δ_x	y_{act}	y_{pr}	Relative error δ_y
$P(4)$	0.2679	0.2384	0.11	0.2607	0.2315	0.11
$P(5)$	0.3308	0.2696	0.18	0.2484	0.2637	0.06
$P(6)$	0.2410	0.2485	0.03	0.3136	0.2860	0.09
$P(7)$	0.3151	0.3862	0.22	0.2736	0.2878	0.05
$P(8)$	0.4963	0.2243	0.55	0.2788	0.2455	0.12

It is evident that the proposed method yields a satisfactory accuracy in every experimental sample and it does not diverge at least at two samples $P(7)$ and $P(8)$, as distinguished from the classical regression method mentioned above. We drop out steps 4 and 5 in this example for they are quite apparent.

4 Conclusion

The method suggested in the paper we can call the deterministic method for the prediction of time series (DPTS). It is dedicated to the use in cases, when we deal with small sets of experimental samples (at least 7 samples) and solve the problem of the prediction of values of parameters at several future instances of time.

The method is based on the ideas of the theory of system identification, and it does not provide the estimation of statistical characteristics of the process under consideration. On the one hand, the suggested method is free of the necessity of establishing statistical relations between the data, but on the other hand, the efficiency of the method is significantly dependent on the regularity of the process analyzed.

Investigations of the method show that it is a flexible tool for the prompt estimation of available states of complex objects and quick responding to changes in the values of parameters describing the dynamic processes under consideration.

References

1. Witten, I.H., Frank, E.: Data Mining: Practical Machine Learning Tools and Techniques. Morgan Kaufmann, San Francisco (2005)
2. Rogoza, W.: Adaptive simulation of separable dynamical systems in the neural network basis. In: Pejas, J., Piegat, A. (eds.) Enhanced Methods in Computer Security, Biometric and Artificial Intelligence Systems, pp. 371–386. Springer, US (2005)
3. Madala, H.R., Ivakhnenko, A.G.: Inductive Learning Algorithms for Complex Systems Modeling. CRC Press, Boca Raton (1994)
4. Freedman, D.A.: Statistical Models: Theory and Practice. Cambridge University Press, Cambridge (2005)
5. Vlach, J.: Linear Circuit Theory: Matrices in Computer Applications. Apple Academic Press (2014)

A Study on Directionality in the Ulam Square with the Use of the Hough Transform

Leszek J. Chmielewski$^{(\boxtimes)}$, Arkadiusz Orłowski, and Maciej Janowicz

Faculty of Applied Informatics and Mathematics (WZIM),
Warsaw University of Life Sciences (SGGW),
ul. Nowoursynowska 159, 02-775 Warsaw, Poland
{leszek_chmielewski,arkadiusz_orlowski,maciej_janowicz}@sggw.pl
http://www.wzim.sggw.pl

Abstract. A version of the Hough transform in which the direction of
the line is represented by a pair of co-prime numbers has been used to
investigate the directional properties of the Ulam spiral. The method
reveals the detailed information on the intensities of the lines which can
be found in the square and on the numbers of primes contained in these
lines. This makes it possible to make quantitative assessments related to
the lines. The analysis, among others, confirms the known observation
that one of the diagonal directions is more populated with lines than
the other one. The results are compared to those made for a square
containing randomly located points with a density close to that for the
Ulam square of a corresponding size. Besides its randomness, such square
also has a directional structure resulting from the square shape of the
pixel lattice. This structure does not depend significantly on the size
of the square. The analysis reveals that the directional structure of the
Ulam square is both quantitatively and qualitatively different from that
of a random square. Larger density of lines in the Ulam square along one
of the diagonal directions in comparison to the other one is confirmed.

Keywords: Ulam · Square · Spiral · Directionality · Random · Hough
transform

1 Introduction

Since its discovery in 1963 [1], the Ulam spiral [2] called also prime spiral [3] gains
much attention as the visual way to get insight into the domain of prime numbers.
It seems that the geometric structures which have an important mathematical
meaning are, among others, the diagonal lines. One of the observations was that
there are more primes on one diagonal than the other. It seems that there are
still discoveries to be made about the Ulam spiral (cf. [4], Section *Why It's
Interesting*).

A proper tool for detection and analysis of line segments in an image is the
Hough transform (HT) in its version for lines. The Ulam square is not a typical image in which real-world objects are represented by means of a projection onto the camera sensor, which is an approximation in itself. Rather, it is

© Springer International Publishing AG 2017
S. Kobayashi et al. (eds.), *Hard and Soft Computing for Artificial Intelligence, Multimedia and
Security*, Advances in Intelligent Systems and Computing 534, DOI 10.1007/978-3-319-48429-7_8

a strictly defined mathematical object. The analysis of lines treated as mathematical objects with the HT was considered by Cyganski, Noel and Orr [7,8]. They paid attention to the problem of all the digital straight lines possible to be represented in an image, being a digitization of a mathematical straight line crossing the image. Kiryati, Lindenbaum and Bruckstein [6] compared the *digital* and *analog* Hough transforms. They discussed the relation between the digital HT according to [7] and the conventional HTs which are analog in nature.

The lines present in the Ulam square are not the approximations of any actual lines. These lines are, however, the sequences of points having strictly specified ratios of coordinate increments. Therefore, in the case of our interest, the concept of the HT described in [7] is too complex. In the present paper we shall use the version of the Hough transform according to [5] in which the ratios of coordinate increments along the object to be detected have been directly used in the process of accumulating the evidence for the existence of lines.

The HT according to [5] has already been used in [9] to find long contiguous sequences of points which form segments of straight lines. In the present paper, however, we shall investigate the directional structure of the Ulam spiral. We shall make an attempt to answer two questions. First, to what extent in some of the directions there are more linear objects in the Ulam square than in the other ones. Second, what is the difference between the directional structure of the Ulam square and the square of corresponding size with randomly displaced points. The use of the method proposed will enable us to give quantitative answers to these questions.

2   Method

Let us recall some basic ideas concerning the Hough transform version we use in this paper, according to [5]. The Ulam square [2] is the $U \times U$ square, where U is and odd integer, with the coordinate system Opq, $p, q \in [-(U-1)/2, (U-1)/2]$, having its origin in the middle element occupied by the number 1, as shown in Fig. 1. A line in the Ulam square is considered as a sequence of pixels, in which the increments $\Delta p, \Delta q$ of the coordinates p, q between the subsequent pixels fulfill the condition $\Delta p/\Delta q = n_1/n_2$, where n_1, n_2 are small integers. Therefore, the slope of the line can be represented by two small integer numbers which form an array D_{ij}, where $i = n_1$ and $j = n_2$ (Fig. 2). Thus, i/j is the reduced fraction $\Delta p/\Delta q$. For the sake of uniqueness it is assumed $i \geq 0$. The dimensions of D are restricted to $[0, N] \times [-N, N]$. From all the elements of D only those are used which correspond to reduced fractions.

To accumulate the evidence of the existence of lines in the square, pairs of prime numbers are considered. Each pair is a voting set and it votes for a line having the slope determined by $\Delta p/\Delta q$. If the slope corresponds to an element of D, the pair is stored in D_{ij}, where i/j is the reduced fraction $\Delta p/\Delta q$; otherwise it is neglected. The votes are stored along the third dimension k of the accumulator, not shown in Fig. 2. For each such vote, the line offset is also stored, defined as the intercept with the axis Op for horizontal lines and with Oq for the remaining ones.

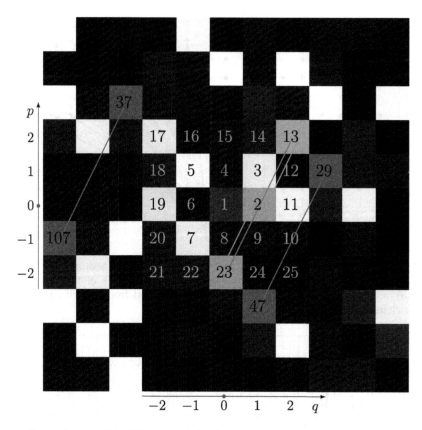

Fig. 1. Central part of the Ulam spiral for dimensions 11×11 with some of the voting pairs. Notations: (p, q) – coordinates with origin in the first number of the spiral; primes: black on white, green or red background; other numbers: grey on black or blue. Selected voting pairs for direction $(\Delta p, \Delta q) = (2, 1)$ are shown in colors. Two green pairs for primes 23, 2 and 2, 13 represent an increment $(\Delta p, \Delta q) = (2, 1)$; two red pairs 47, 29 and 107, 37 the pair 23, 13 marked by a darker green line represent an increment $(4, 2)$ which also resolves to $(2, 1)$. (Color figure online)

The relation of the coordinates in the direction table and the angle α can be seen in Fig. 3. Only some specific angles can be represented. This is in conformity with our interest in sequences of points for which the increments of coordinates are expressed by small integers.

After the accumulation process, in each element of the accumulator, which corresponds to a specific line characterized with the slope and the offset, as previously defined, the following data are stored: the number of voting pairs and the number of primes in this line, and the list of primes which lie on this line. For each prime in this list the following data are stored: its value, its index in the spiral, and its coordinates p, q. Each list can be easily sorted according to the chosen item in these data.

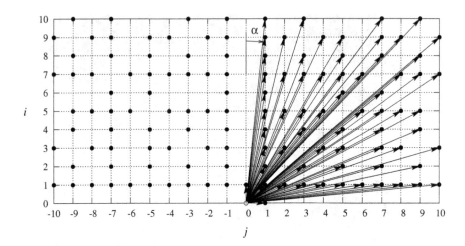

Fig. 2. Directions represented as a table D_{ij}. The directional vector has the initial point at the empty circle $(0,0)$ and the terminal point in one of the black circles $(\Delta p, \Delta q) = (i,j)$; $i \neq 0 \vee j \neq 0$, and $i \geq 0$ (reproduced from [5] with permission, slightly changed).

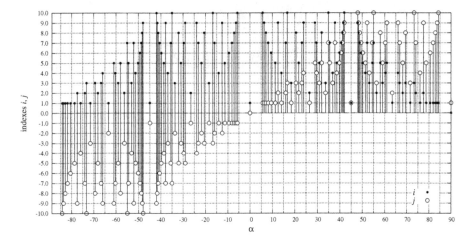

Fig. 3. Relation of the indexes in the direction table shown in Fig. 2 and the angle α.

The accumulator can be analyzed with respect to such aspects like, for example, the directional structure of the Ulam spiral, the existence of lines with large numbers of primes, the contiguous strings of points representing primes in the square, etc.

3   Directionality – Selected Results

Before we go on to the results, one more issue needs consideration. In any of our experiments we consider an Ulam square having a specific size, limited by the memory of the computer used and the available time for calculations. Both these requirements do not seem to be crucial, due to that on the one hand very large memories are available now and on the other hand we have to make the calculations only once. However, at this stage we have tried to see if there is a variability in the results with the growing size of the square, or if the results have a tendency to stabilize above some size. As it will be seen further, for some of the characteristics of the directionality, the second case holds. Therefore, we have come to a conclusion that, for the problem of our interest, we can preform the calculations for the square of quite a moderate size of 1001×1001 points. It contains 78 650 primes, where the largest one is 1 001 989.

3.1   Ulam Square

Let us consider several ways in which the directionality can be presented. In general, it is related to the intensity of lines having a specified direction. However, let us start from looking at the number of lines having a specific direction, shown in Fig. 4. The number of lines have been normalized with the maximum number for the given square size, which appeared at a different angle for each square size. No tendency of stabilization of the numbers with the growing size of the square can be seen. This characteristic can not be considered suitable for the analysis.

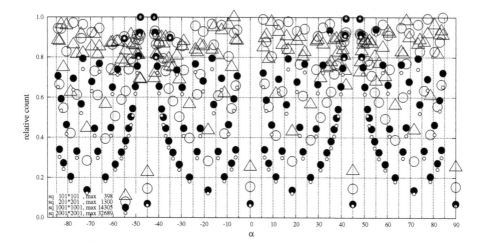

Fig. 4. Normalized numbers of lines in the given direction, for the squares of sizes 101×101, 201×201, 1001×1001 and 2001×2001. No tendency of stabilization of the numbers with the growing size of the square can be seen.

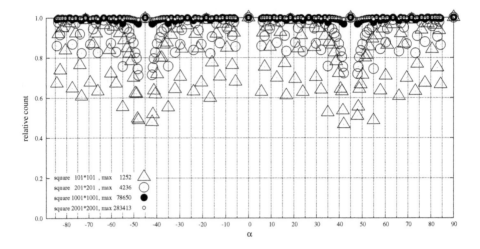

Fig. 5. Normalized numbers of primes on all the lines in the given direction, for the squares of selected sizes from 101×101 to 2001×2001. The numbers stabilize at a nearly uniform value.

Let us check if the number of primes on lines in each direction reveals some information on the directionality. This is shown in Fig. 5. It can be seen that the directional structure of the square is not revealed in this plot.

The basic parameter of line intensity in the Hough transform is the number of votes given to the specific line in the accumulation process, that is, the number of pairs of primes in a line. Let us then look at the number of voting pairs in the lines having a specific direction, shown in Fig. 6.

Let us also check if the number of voting pairs of primes per line, in the given direction, is useful for investigating the directionality of the Ulam spiral. This is shown in Fig. 7. This measure also has the tendency to stabilize as the size of the square grows, and exhibits the ability to reveal its directional structure.

In the plots of Figs. 6 and 7 is is clearly see that the horizontal, vertical and diagonal lines are the most populated in the Ulam square. It can also be seen that one of the diagonals has more representation that the other in both graphs. This is in conformity with the observations found in literature [1]. These two plots will be used in the further analysis.

3.2 Random Square

The question of directionality of the Ulam spiral can be well assessed in relation to that of a square with random dots of similar density. The tendency of the human visual system of seing regularities where they do not exist is known (cf. the famous dispute on the *canalli* on Mars, see e.g. [10]). The other side of the problem is that the image of our interest is formed by square pixels, which can be of importance as far as the directions are concerned, due to that vertical, horizontal and diagonal lines are the most naturally represented in the square

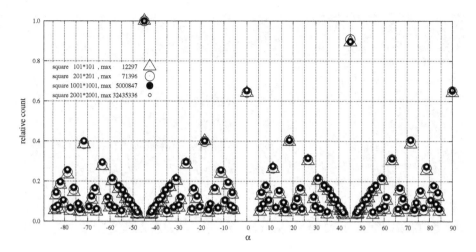

Fig. 6. Normalized numbers of votes (pairs of primes) on all the lines in the given direction, for the squares of various sizes. A tendency of stabilization of the points in the graphs with the growing size of the square can be seen.

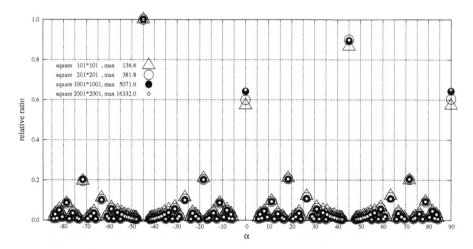

Fig. 7. Normalized numbers of pairs of primes per line in the given direction, for the squares of various sizes. Tendency of stabilization of the points in the graphs with the growing number of primes can be seen.

grid. Therefore, in the two random dot images of Fig. 8 the human eye tends to see some regularities, but they are different in each one.

In Figs. 9 and 10 the results for ten different realizations of the random square are compared to those of the Ulam square. Most of these realizations were taken for 1001×1001 pixels, as this size was considered reasonable for the case of the primes square. However, to see whether the properties depend on the size of the

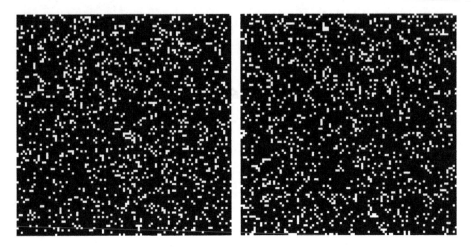

Fig. 8. Two random dot images of 101 × 101 pixels with the density of points close to that of the Ulam spiral.

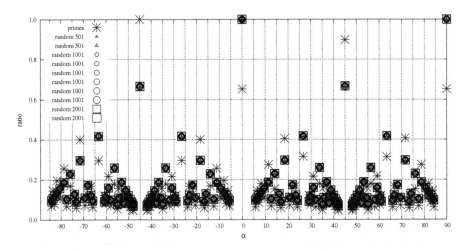

Fig. 9. Numbers of pairs of primes on all the lines in the given direction for 10 realizations of the random square compared to those for the Ulam square. Marks corresponding to the realizations are nearly in the same locations, with no significant dependence on the square size.

random square, two realizations were taken for size 501 × 501 and two for size 2001 × 2001 pixels. For each of the ten realizations, the results are very similar. This means that the directional structure of the random square is stable, besides that any particular realization differs in its details and that the size of the image changes.

In the plots of Figs. 9 and 10 it can be seen that the random squares exhibit the directional structure in which the horizontal, vertical and both diagonal lines

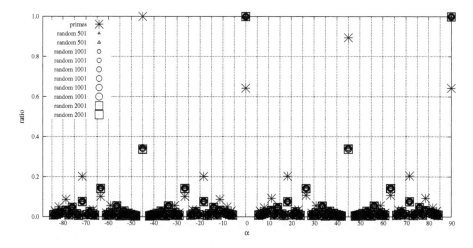

Fig. 10. Numbers of pairs of primes per line in the given direction for random squares compared to those in the Ulam square.

are dominating above the lines in other directions and that the directions are represented by different numbers of voting pairs. This phenomenon does not significantly depend on the realization of the random process which gave rise to a particular image. As it could be expected, the horizontal direction is not favored versus the vertical one; similarly, one diagonal direction is not favored versus the other one.

On the contrary, in the Ulam square one diagonal direction is dominating over the other one. This is the most conspicuous difference. The pattern of other points for subsequent directions is also different from that for the random square.

4 Summary and Prospect

A version of the Hough transform proposed previously has been used to investigate the directional properties of the Ulam spiral. The method makes it possible to assess the difference in the intensities of lines related to subsequent directions in the quantitative way, due to that the data on the points related to the angles are available.

The measures of line intensities used here allowed us to show that one of the diagonal directions is more populated with lines than the other one. Such a phenomenon is absent in the random square, although in that square also some directionality is observed.

The proposed method gives the possibility to study other quantitative characteristics of the Ulam square [11].

References

1. Wikipedia: Ulam spiral — Wikipedia, The Free Encyclopedia (2016). https:// en.wikipedia.org/w/index.php?title=Ulam_spiral&oldid=726734495. Accessed 20 June 2016
2. Stein, M., Ulam, S., Wells, M.: A visual display of some properties of the distribution of primes. Am. Math. Mon. **71**(5), 516–520 (1964). doi:10.2307/2312588
3. Weisstein, E.W.: Prime spiral. From MathWorld–A Wolfram Web Resource (2016). http://mathworld.wolfram.com/PrimeSpiral.html. Accessed 20 June 2016
4. Authors of the *Prime Spiral* chapter: Prime spiral (Ulam spiral). The Math Images Project (2016). http://mathforum.org/mathimages/index.php/Prime_spiral_(Ulam_spiral). Accessed 20 June 2016
5. Chmielewski, L.J., Orłowski, A.: Hough transform for lines with slope defined by a pair of co-primes. Mach. Graph. Vis. **22**(1/4), 17–25 (2013)
6. Kiryati, N., Lindenbaum, M., Bruckstein, A.: Digital or analog Hough transform? Pattern Recogn. Lett. **12**(5), 291–297 (1991). doi:10.1016/0167-8655(91)90412-F
7. Cyganski, D., Noel, W.F., Orr, J.A.: Analytic Hough transform. In: Proceedings of SPIE. Sensing and Reconstruction of Three-Dimensional Objects and Scenes, vol. 1260, pp. 148–159 (1990). doi:10.1117/12.20013
8. Liu, Y., Cyganski, D., Vaz, R.F.: Efficient implementation of the analytic Hough transform for exact linear feature extraction. In: Proceedings of SPIE. Intelligent Robots and Computer Vision X: Algorithms and Techniques, vol. 1607, pp. 298–309 (1992). doi:10.1117/12.57109
9. Chmielewski, L.J., Orłowski, A.: Finding line segments in the Ulam square with the Hough transform. In: Chmielewski, L.J., Datta, A., Kozera, R., Wojciechowski, K. (eds.) ICCVG 2016. LNCS, vol. 9972, pp. 617–626. Springer, Heidelberg (2016). doi:10.1007/978-3-319-46418-3_55
10. Evans, J.E., Maunder, E.W.: Experiments as to the actuality of the "canals" observed on Mars. Mon. Not. R. Astron. Soc. **63**, 488–499 (1903). doi:10.1093/mnras/63.8.488
11. Chmielewski, L.J., Orwski, A.: Prime numbers in the Ulam dsquare (2016). www.lchmiel.pl/primes. Accessed 01 Aug 2016

Design of Information and Security Systems

Ontological Approach Towards Semantic Data Filtering in the Interface Design Applied to the Interface Design and Dialogue Creation for the "Robot-Aircraft Designer" Informational System

Nikolay Borgest$^{(\boxtimes)}$ and Maksim Korovin

Samara University, Samara, Russia
borgest@yandex.ru, maks.korovin@gmail.com

Abstract. The article discusses an approach towards data compression in the "Robot- airplane designer" system – a tool for automated airplane design. The proposed approach has been tested on a prototype of the "Robot- airplane designer" system that was specially created as a demo unit. The demo unit should be capable of solving the task within a given amount of time, as the exhibition format has its limitations. On the other hand, it is important to keep the fidelity of the project as high as possible. It is important to provide the user with just enough information to help him make a correct decision, without overloading him with data that is, perhaps, valid, but is not essential for the current task. The proposed solution is an ontological approach, where the required bits of information are extracted from the knowledge base with regard to the user's level of competence and personal preferences. The project data, including statistical information, design decision making strategies and values for the target values is stored in a number of databases, interconnected using a thesaurus.

Keywords: Design · User interface · Aircraft · Automation

1 Introduction

Real datasets, used in aircraft design, are often large enough to necessitate data compression. However, decreasing the amount of information, presented to the user, also leads to the decrease of the overall fidelity of the project. The tradeoff in such cases is usually between the ease of comprehension and the time effectiveness of the process. In this regard, the use of semantic compression is of considerable interest and can be efficiently used to streamline the design process.

Sources in modern communication networks, humans, computers, or smart devices, aim at sharing meaningful and useful information, instead of merely delivering any information and maximizing the throughput [1]. The communication problem discussed in this paper is the two-way communication between the user and the interface of an airplane designing system "Robot-aircraft designer" [2], aimed to provide the user

S. Kobayashi et al. (eds.), *Hard and Soft Computing for Artificial Intelligence, Multimedia and Security*, Advances in Intelligent Systems and Computing 534, DOI 10.1007/978-3-319-48429-7_9

with information that leads to useful conclusions, without overloading him with data, irrelevant for solving his current task.

It is important to note, that the aim of the discussed approach is to compress only the data that is displayed to the user of the system, and not the data that he system is using itself. The project data is stored in the form of numerous datasets, attributed using a thesaurus. Displayed data is filtered based on declared qualification and goals of the user (i.e. if the user states that he is particularly interested in the structural design of the wing, the system will reallocate the resources of the interface in order to provide him with more data about the decisions, made during the wing design by the cost of reducing the amount of displayed information regarding less desired subjects). Since the approach was developed for a demo unit, the time constraint was considered the most important.

2 Data Compression Techniques

Data compression techniques can be divided into two big groups – lossless and lossy. A lossless compression means that the restored data file is identical to the original. In comparison, data files that represent images and other acquired signals do not have to be keep in perfect condition for storage or transmission. Compression techniques that allow this type of degradation are called lossy. This distinction is important because lossy techniques are much more effective at compression than lossless methods [3]. When it comes to displaying the information, since we are only interfering with the amount of data displayed and the means of its display, we can assume that the data compression is lossless, as the initial data is still present in the database and can be retrieved at any given moment. However, since we are aiming to reduce the amount of time that I takes for an average user to process the given information, it is important to deliver information in the most comprehensive form. For example, diagrams are more convenient for showing statistical data in comparison to tables. Tables generally are harder to process and take more screen space than diagrams. Therefore, it is more beneficial to filter relevant information from the database and generate dynamic diagrams, rather than just deliver raw numbers to the user. In certain cases, the user does not require detailed information about a particular value. For example, if the demo unit is used by a student to create a project of an airplane, the very basic definition of a chosen fuselage cross-section shape (i.e. elliptical or rectangular with rounded corners) might just be sufficient. However, if a more experienced person uses the unit, the dimensions of the cabin might be of importance (for example if the user has a particular payload in mind and wants to ensure that the cabin size is sufficient). In order to preserve relative simplicity of the interface, the default value representations are limited to the integral parameters (such as the fuselage cross-section shape), were corresponding detailed data is available on demand. The compression of displayed information is conveyed based on the semantics of the aircraft description.

Although semantic compression has several advantages over syntactic compression, the two types of compression are not mutually exclusive [4].

Since the semantics of the data are taken into consideration, complex correlation and data dependency between the data attributes can be exploited in case of semantic compression of data. Further, the exploratory nature of many data analysis applications implies that exact answers are usually not needed and analysts may prefer a fast approximate answer with an upper bound on the error of approximation. By taking into consideration the error tolerance that is acceptable in each attribute, semantic compression can be used to perform lossy compression to enhance the compression ratio. (The benefits can be substantial even when the level of error tolerance is low) [5].

Semantic compression allows creation of generalized descriptions presented in a human-readable form and in a way that correlates with the user's preferences and his/her knowledge of a given domain. However, it is important to control the level of generalization, as if the generalization is too broad, the information loss can be so considerable, that the user might not be able to access decompressed data, since there will be no hints of how the initial dataset looks like and what parameters are actually being changed. When semantic compression is used for a user that is not qualified in aircraft this is not acceptable.

Compression of data is achieved by utilizing a semantic network and data on term relevancies (in the form of a relevancy dictionary). The least relevant terms are treated as unnecessary and are replaced by more general interpretations (their variations are stored in a semantic network). As a result, the reduced number of definitions can be used to represent a dataset without significant information loss and while preserving the interconnections between variables, which is important from the perspective of processing. The connections between terms are stored in the airplane thesaurus (Fig. 1).

Domain thesaurus is a subclass of ontology and thus should be developed as one. The process of creation of thesaurus starts with the localization of a subject domain.

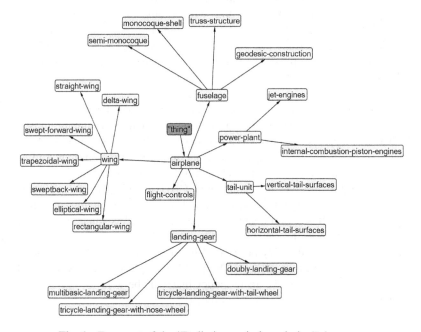

Fig. 1. Fragment of the "Preliminary airplane design" thesaurus

In case of the robot-designers' thesaurus, global subject domain is aircraft design, local subject domain - preliminary airplane design. Preliminary airplane design is a process of creation of a prototype project for a designed airplane. In order to create a thesaurus it is necessary to conduct a system analysis of the subject domain - to determine interconnections between elements of the studied system, to perform ontological analysis of the domain.

Basic terminology of the subject domain is taken from the FAR 25, student's books, terminological dictionaries, encyclopedias, monographs and papers in the studied domain. Ontology in the form of thesaurus serves to present information, necessary for understanding of the place of a particular term in the terminological environment [6].

3   Robot-Aircraft Designer

Modern CAD systems are gradually making a transition from being a tool towards serving as an "intelligent partner" for the designer. High level of formalization of the common design procedures allows CAD systems to perform the ever-increasing amount of tasks without the need of the human designer's involvement. CAD knowledge is an essential quality for a modern engineer. However, modern heavy CAD systems have more and more complex interfaces that make the task of familiarization quite challenging. Even for the engineers, familiar with the CAD interface within their daily tasks range, the task of identifying the right instruments for an un-familiar task may present a challenge.

Provided that the process of designing is fully formalized within a certain set of tasks it is possible to create a system that will be able to perform parametric optimization (including structural) of a predefined parameterized model. During the optimization of design parameters, an instance of the parameterized meta-model, that potentially includes all the possible solutions, is generated based on the input data to satisfy the design conditions [7, 8]. Robot-designer's operation scheme is presented on Fig. 2.

It is well known that designing is a field, where many heuristic solutions are implemented. Nevertheless, in the field of traditional structures designing, most of the initially heuristic approaches have been formalized, which makes it possible for an automated system to use them successfully, obtaining the first appearance of an airplane and then refining it as more project data becomes defined (Fig. 3).

However, despite the fact that automated CAD models do, in fact, somewhat simplify the process of designing an artifact, the task of their tuning usually presents a challenge of its own. Based on the authors experience, it takes some time (depending on the users qualification) to get acquainted with the meta-model to fully utilize its potential, that contradicts the very purpose of the meta-model implication – to reduce the time required to design an artifact. The problem might be addressed with developing a special interface for the meta-model that will be able to "guide" the user through the process of the model description [9].

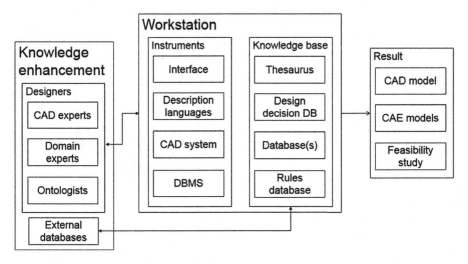

Fig. 2. Robot-designer's operation scheme

Fig. 3. Airplane model's evolution

4 Robot-Designer's Interface

The interface, described in the paper is intended for a demo unit of an aircraft preliminary design system based on a meta-model of the aircraft (created in the CATIA CAD system) [10]. The overview of the systems interface is presented on Fig. 4, the unit itself is shown on Fig. 5.

The interface consists of several major parts: timeline (1), model preview window (2), collapsible general information window (3) and the interactive section (4).

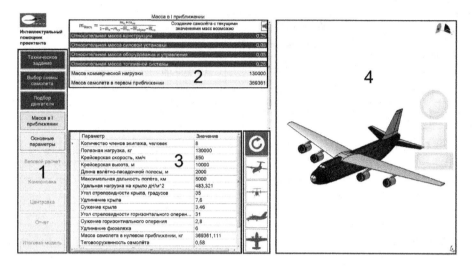

Fig. 4. The overview of the systems interface

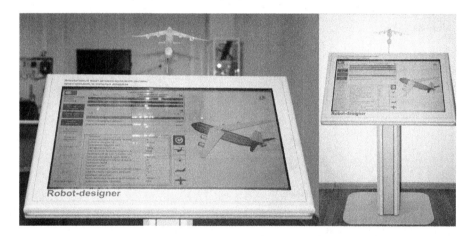

Fig. 5. Robot-aircraft designers demo unit

The timeline represents the scenario of creating an aircraft on a high level of abstraction. Therefore, subsections of the timeline represent business processes of a lower level of abstraction.

For example, for the wing design module the processes are:

- defining the planform of the wing;
- defining the aspect ratio of the wing;
- defining the tapering of the wing;
- defining the aerodynamic profile of the wing;
- defining the parameters of roll control;
- defining the mechanization of the wing.

Each of the processes has its own corresponding module that loads automatically based on a design scenario preset. The design scenario defines the level of the users involvement into the designing process. Implication of the design scenarios allows for tuning of the interface towards individual needs of users within a single design environment utilizing the same interface rules and allowing, if needed, natural translation between the design scenarios [11].

Every basic module belongs to a certain type that is defined by its nature (for example weather some numerical variable must be defined, or a particular type of airfoil must be chosen). The module gives a brief overview of the corresponding design process and gives the user hints on how changing a certain parameter will affect the project as a whole.

The properties of each module also include the average time t it takes the user to interact with the module. The t parameter usually depends on the module's fidelity and serves to allocate the desired modules on screen.

When the user starts interacting with the unit, he is asked about the available amount of time and topics of particular interest (i.e. 5 min, structural design of the fuselage). Using the term relations, stored in the semantic network, the system forms a design scenario, picking design modules that are important for the task and therefore require maximum fidelity. The modules with the highest fidelity (and the biggest t parameters) would be the ones that correspond with the topic of interest. Other design operations will be presented with quicker, lower-fidelity modules, in order to fulfill the time constraint. Interface optimization is performed under assumption that within the stated time constraints the user wants to get as much meaningful data regarding the project as possible.

In order to evaluate the effectiveness of the approach, the basic procedure of adjustment of sweep angle X of the wing was considered. The planform of the model's wing is defined by four airfoils, as shown on Fig. 6. In the simplest scenario, the sweep angle of the model can be altered by conjoint displacement of the airfoils. The values for displacement can be obtained from the distance between the airfoils and the increment of the sweep angle.

In order to manually adjust the sweep angle of the model, the user must perform a series of actions. First, he must measure the distance between the airfoils, then use the measurements to calculate absolute displacements. After obtaining the values, the user must modify sketches of all the airfoils except the first one. Based on the method, described in [12], average interaction times were calculated. Manual alteration of the sweep angle of the wing for the model, shown on Fig. 6, takes approximately 63 s, provided the user does know all the required steps in advance and is familiar with the model. Alteration of a predefined design value on a parameterized model from the model tree takes about 8 s (for people, inexperienced in the field of CAD, the actual interaction times are considerably longer). Usage of the developed model interface decreases the needed interaction time to approximately 4 s and is less dependent on the user's CAD skills, as the interface uses natural language definitions.

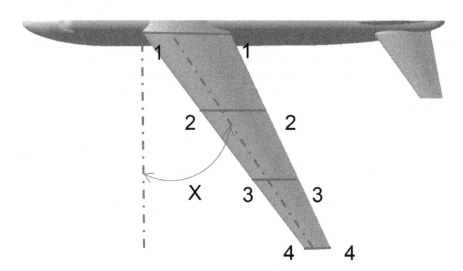

Fig. 6. Definition of the wing's sweep angle

5   Conclusion

The described approach towards data compression in the demo unit of the "Robot-airplane designer" system allows for automated time-optimization of the airplane design process description while maintaining a reasonable level of fidelity in the user's areas of interest. Each of the design processes has its own corresponding module that loads automatically based on a generated preference-sensitive design scenario. The design scenario defines the level of the user's involvement into the designing process. Implication of the design scenarios allows for tuning of the interface towards individual needs of users within a single design environment utilizing the same interface rules and allowing, if needed, natural translation between the design scenarios. Thesaurus is used to attribute project's parameters and simultaneously serves as the prime source of terms and definitions for the interface generation and a means of linking of terms and parameters from databases, containing information about past projects.

References

1. Basu, P., Bao, J., Dean, M., Hendler, J.A.: Preserving quality of information by using semantic relationships. In: PerCom Workshops, pp. 58–63 (2012)
2. Borgest, N., Gromov, A., Gromov, A., Korovin, M.: The concept of automation in conventional systems creation applied to the preliminary aircraft design. In: Wiliński, A., et al. (eds.) Soft Computing in Computer and Information Science, Part II. AISC, vol. 342, pp. 155–165. Springer, Switzerland (2015)
3. Smith, S.W.: The Scientist & Engineer's Guide to Digital Signal Processing (1997). ISBN: 978-0966017632

4. Jagadish, H.V., Madar, J., Ng, R.T.: Semantic compression and pattern extraction with fascicles. In: Proceedings of the 1999 International Conference on Very Large Data Bases (VLDB 1999), Edinburgh, UK, pp. 186–197, September 1999
5. Jagadish, H.V., Ng, R.T., Ooi, B.C., Tung, A.K.H.: ItCompress: an iterative semantic compression algorithm. In: Proceedings of the 20th International Conference on Data Engineering, pp. 646–657 (2004)
6. Shustova, D.V.: Approach to the development of the semantic foundations of information systems for the design and manufacture of aviation equipment. Ontol. Des. 5(1), 70–84 (2015). (in Russian)
7. Blessing, L.T.M., Chakrabarti, A.: DRM, a Design Research Methodology, 413 p. Springer, London (2009)
8. Dattoma, V., De Giorgi, M., Giancane, S., Manco, P., Morabito, A.E.: A parametric-associative modeling of aeronautical concepts for structural optimization. Adv. Eng. Softw. 50(1), 97–109 (2012)
9. Kanev, D.: Analysis of automated learning systems architectures and architecture development of intelligent tutoring system based on mobile technology. In: Proceedings of the Interactive Systems: Problems of Human-Computer Interaction, pp. 141–152 (2013)
10. Borgest, N.M., Gromov, A.A., Gromov, A.A., Moreno, R.H., Korovin, M.D., Shustova, D. V.: Robot-designer: fantasy and reality. Ontol. Des. 4(6), 73–94 (2012)
11. Wriggers, P., Siplivaya, M., Joukova, I., Slivin, R.: Intelligent support of engineering analysis using ontology and case-based reasoning. Eng. Appl. Artif. Intell. 20, 709–720 (2008)
12. Olson, J., Olson, G.: The growth of cognitive modeling in human-computer interaction since GOMS. Hum. Comput. Interact. 5, 221–265 (1990)

Real-Time System of Delivering Water-Capsule for Firefighting

Grzegorz Śmigielski[1(✉)], Roman Dygdała[2], Hubert Zarzycki[3],
and Damian Lewandowski[2]

[1] Casimir the Great University in Bydgoszcz, Institute of Mechanics and Applied
Computer Science ul. Kopernika 1, 85-064 Bydgoszcz, Poland
gsmigielski@ukw.edu.pl
[2] Military Institute of Armament Technology ul. Prym. St. Wyszynskiego 7, 05-220
Zielonka, Poland
romdy1@onet.pl, damian_lew2@wp.pl
[3] Department of Computer Science, Wrocław School of Information Technology,
ul. Wejherowska 28, 54-239 Wrocław, Poland
hzarzycki@horyzont.eu

Abstract. The alternative and efficient technique of large-scale fire extinguishment is the method of explosive-produced water aerosol. In this method the water capsule, which is source of aerosol, is delivered by the helicopter near the fire area and then is released and detonated. The efficiency of this technique depends on quality of control system - its determinism, speed of computation of the moment of capsule's release and reliability of components. The article presents design assumptions, selection of the integration step size in numerical method, structure of the real-time system and practical verification of the system.

Keywords: Real-time system · Numerical methods · Fire extinguishment · Water aerosol

1 Introduction

Using water aerosol for extinguishing small-scale fires is an efficient solution known and applied for a few years [13,18]. The explosive method of producing water aerosol consists in the explosion of the explosive charge put inside the water container [11,12]. Removing oxygen form the vicinity of the fire is an additional advantage of this method [14,15]. On the other hand, due to generation of a shock wave during the explosion this method can be applied only for the open area firefighting. A water capsule fastened to the helicopter is fast dispatched near the delivery area. The described system is able to release automatically the water capsule at such a distance from the destination that allows the water capsule to explode at an elevation of a dozen or so meters over the target and produces water spray which covers a circular area of 40 m in diameter.

© Springer International Publishing AG 2017
S. Kobayashi et al. (eds.), *Hard and Soft Computing for Artificial Intelligence, Multimedia and Security*, Advances in Intelligent Systems and Computing 534, DOI 10.1007/978-3-319-48429-7_10

1.1 The Assumptions Concerning System Operation

Determining the moment of the release of the capsule and the time-delay of the signal sent to the explosive charge inside the capsule is based on data concerning the speed and the location of the capsule relative to the centre of the fire. In case of the described system the accuracy of the target hit by the capsule in the horizontal axis was assumed at the level $+/- 10$ m. The optimal height of explosion above the fire is $12 +/- 4$ m. It allows one to extinguish effectively the fire inside a disc of about 20 m in diameter of (which corresponds to the of 314 m^2). The assumed accuracy in the horizontal as well as the vertical axis requires reduction of the uncertainty of position measurements, because the drag coefficients (both the horizontal and the vertical) influence essentially the total error [22]. Finally it was assume, that the time necessary for making the decision to release the capsule should not exceed 20 ms. In such a time interval the water capsule having a capacity about 1200 L, released at height about 200 m covers the distance of 1 m in the horizontal and 2 m in the vertical direction. Since exceeding the above time limit would influence not only the airdrop precision but also the safety it was necessary to build a system based on reliable equipment and software working in the real-time regime [1, 16, 17].

2 Fundamentals of the System Operations

The moment of the capsule release and the time delay of the explosion are determined by the computer using the Runge-Kutta RK(4,4) numerical method [5]. The computations are based on the equations of motion [21, 22], which take into account the mass of the capsule, its initial velocity (at the moment of release) and the drag coefficients. The coordinate system for the capsule released at the height H above the ground mowing with the momentary velocity v(t) and covering the horizontal distance xmax in the air is shown in Fig. 1.

Computations were conducted for a few different values of integration step, the results were compared with the results obtained in Matlab for the step of 10 us. Numerical computations were performed for a 10 s flight of the water capsule, which corresponds to a flight of the capsule from the height of 454 m. Such a large distances were not planned for the real system, and they were used for the sake of the reliability of the numerical tests. The differences in results for the horizontal and vertical axis are shown in Table 1.

As is clear definitely the larger difference for all sizes of integration steps is observed for the coordinates along the vertical axis, which is consisted with the fact that the vertical component of velocity is higher, except for the initial stage of the flight. Based on these tests the maximum integration step dt $= 0.01$ s was assumed.

3 Elements of the System

The core of the system is formed by the controlling computer located on the board of the helicopter. In this role we use the National Instruments

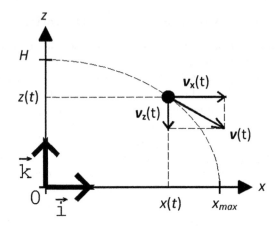

Fig. 1. The coordinate system for the flight of the capsule

Table 1. Differences between the results for the horizontal - OX and vertical - OZ axes, obtained by solving numerically differential equations with the Matlab and the dedicated program written in the C language and using the (RK (4.4)) algorithm depending on the integration step

Integration step	Differences - OX	Differences - OZ
[s]	[m]	[m]
0.00001	$\ll 0.0001$	0.0005
0.0001	$\ll 0.0001$	0.0043
0.001	$\ll 0.0001$	0.0424
0.01	0.0001	0.4233
0.1	0.0004	4.2302
1	0.0032	42.033

CompactRIO specialized controller (Fig. 2) equipped with a 400 MHz 32-bit processor, FGPA programmable system (Fig. 3) and communication modules with serial ports and digital I/O-s. The controller is immune to large amplitude electromagnetic and mechanical perturbations and works well in a broad range of temperature and humidity of the air.

Communication between various components is carried by serial ports or Ethernet (Fig. 4). The helicopter installed GPS receiver provides real-time data on the helicopters position and velocity in the form of the NMEA (National Marine Electronics Association) strings. The communication microcomputer MOXA coupled to a radio-modem allows to receive important information (e.g. targets coordinates) from the commanding center and sending data on the flight parameters [20, 22].

The applied solution disburdens the main computer from the job of controlling radio network. The control appliances are expected only to transmit

Fig. 2. CompactRIO real-time controller and reconfigurable FPGA chassis [3]

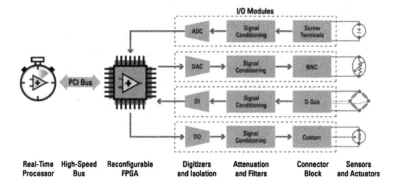

Fig. 3. CompactRIO system architecture [3]

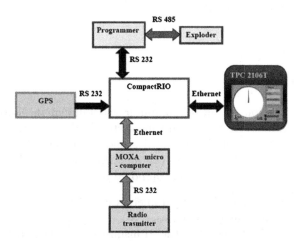

Fig. 4. Components of the water-bag delivery control system

data within the specific protocol. Communication microcomputers perform all tasks associated with data transmission. It allows the controlling computer at the helicopter to use its full computing power for the trajectory calculation. This increases the systems reliability. The controlling computer determines the water-bag trajectory using the Runge-Kutta numerical method. The computed value of the optimum delay-time for the explosion is transmitted to the programmable exploder. All important pieces of information on the current status of the applications are displayed at the pilots control panel (TPC - 2106T). According to the assumptions the maximum time for elaborating the decision on the release of the capsule is 20 ms. Therefore calculations are conducted on the target CompactRIO platform for a few different times of integration which fulfills the earlier conditions $dt \leq 0.01$ s. (Table 2). Eventually an intermediate value dt = 0.005 s was chosen as the integration step, and the computations were performed for the time interval 12 ms. 1@Article, author =, title =, journal =, year =, OPTkey =, OPTvolume =, OPTnumber =, OPTpages =, OPTmonth =, OPTnote =, OPTannote = 2 ms. In Table 2 values of computation times are also given for the industrial PXI computer, which was not considered for construction of the system on account of much higher price and lower reliability.

Table 2. Computation times for three values of integration step

Integration step	Release from 100 m above the ground		Release from 400 m above the ground	
[s]	[m]		[m]	
	CompactRIO	PXI	cRIO	PXI
0.01	2	< 1	3	< 1
0.005	6	< 1	12	1
0.001	30	3	60	6

4   Application

Multithread application of control computer was built in the LabVIEW environment [6,19] and is working in real time system VxWorks. After starting the computer a 1 s pause occurs (frame 0 of sequential external structure). Next frame of the sequential structure (1) contains the endless WHILE loop, inside which the main sequential structure, composed of 5 frame was inserted:

- loading the files containing fixed parameters values used in the application,
- optional loading of configuration files,
- allocation of FIFO queues for the transmission of data between various threads,
- main loop of the program transmitting control to the threads (Fig. 5),
- writing-down information on disc errors.

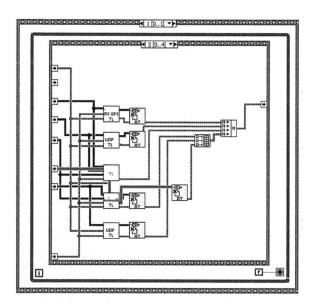

Fig. 5. View of the main application window

The threads concern:

- reception and analysis of data from the satellite receiver (GPS),
- reception of data from the ground computer (the server) playing role of the commanding center,
- transmission of the flight parameters to the commanding center,
- processing the received information according to a suitable algorithm,
- export of data to the ground computer,
- sending to the initiator the information about the optimum time delay of the explosion with respect to the release moment.

The threads are located inside the Timed Loops [6,19]. Such a structure (Fig. 6) enables to adjust Period, Priority, Deadline and Source name for each of threads and in the case of multiprocessor systems one can ascribe separate tasks to particular processors (Processor). For the sake system safety the highest priority was assigned for the threads of communication with the programmable detonator. The threads responsible for the communication with the ground station operate with the lowest priority. From the point of view of the correct functioning of the system those fragments of the program are essential in which the communication with the programmable detonator and the GPS receiver, and computing the time-delay for the detonator take place. For all three above threads the maximum execution times are determined [2,4].

The Fig. 7 shows the fragment the application responsible for receiving the data from GPS. In the RS reading subprogram a contents of the receiving register is checked. A GNGGA string opening the data packet consisting of two NMEA

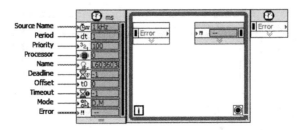

Fig. 6. Timed loop

messages is looked for - GGA and VTG. Finishing the subprogram is possible after completing the entire data packet (twofold occurrence of CR LF symbols) or after exceeding the assumed time limit [7,8]. After the completion of subprogram the checksums of announcements and fields announcing the mode of reception (GGA) and status of received data (VTG) are checked. If the fields contain faulty values or checksums are incorrect, it means the communication between devices is incorrect or the signal of the positioning is absent. All data obtained from such a reading are disregarded. The correct data are written in the FIFO batch. Information about the correctness of the data is also transmitted in the form of the network shared variable [6,19] to the touch panel of the pilot and is displayed on the LED indicator of the CompactRIO controller.

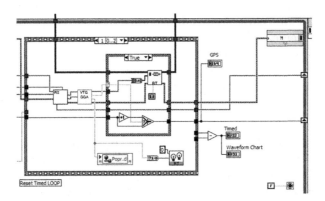

Fig. 7. A part of the application responsible for communication with the GPS receiver inside the timed loop

The airdrop of the capsule is not performed in the following cases:

– the computation time exceeds the set time limit intended for performing the given task,
– the pilot does not allow for the release of the capsule,
– errors in the communication with the programmable detonator appear,

– the data from the GPS are incorrect,
– the angle of approach to the destination exceeds 10°.

The value of the yaw angle from the required direction of flight; the distance to the release point of the capsule, the distance to the target, the velocity and coordinates of the fire are transmitted via Ethernet in the form of shared variables to the touch panel in the pilot's cockpit. The visualization of the yaw angle from the required direction of flight enables the pilot to correct the path of the flight.

5   Verification

The verification was conducted in a few stages. Tests using Execution Tracing Toolkit available under the LabVIEW Real-Time applications allowed to analyze the processor time sharing between various threads and the influence of the assumed priority level and their execution for the functioning of applications. Simulators of GPS receiver and programmable exploder were used for the tests. The built application allows to display the values of execution times for various tasks on the host computer, which allows for the full control of the functioning of the system during the test flights. The application allows to count cases of exceeding execution times too. In this case the output Finished Late was used. During the tests no case of exceeding established times was registered. It gives an evidence that the selection of the value of integration steps was correct. The verification of the correctness of the algorithm consisted in carrying out field tests consisting in release the capsule by the signal given by the control system [9,10]. Coordinates of the mark corresponding to the fire focus and the mass of the water-bag were transmitted via the server to the on-board computer at the beginning of each flight. The trials were recorded with a high-speed camera, and additional data concerning the speed and position of the capsule and the status of the system were registered by the ground control station. The expanded uncertainties (accuracy) of the target hit were $U(x) = 10.5$ m and 5.9 m horizontally and the registered explosion heights above the ground were contained between 3 and 17 m (Fig. 8).

Fig. 8. Explosion of the water-capsule 16 m over the ground - based on the registered film

Acknowledgements. Research were carried on as a part of the research and development project of the Ministry of Science and Higher Education in Poland.

References

1. Angryk, R.A., Czerniak, J.: Heuristic algorithm for interpretation of multi-valued attributes in similarity-based fuzzy relational databases. Int. J. Approximate Reasoning **51**(8), 895–911 (2010)
2. Apiecionek, L., Czerniak, J.M.: Qos solution for network resource protection. In: Informatics 2013: Proceedings of the Twelfth International Conference on Informatics, pp. 73–76 (2013)
3. Apiecionek, Ł., Czerniak, J.M., Dobrosielski, W.T.: Quality of services method as a DDoS protection tool. In: Filev, D., Jabłkowski, J., Kacprzyk, J., Krawczak, M., Popchev, I., Rutkowski, L., Sgurev, V., Sotirova, E., Szynkarczyk, P., Zadrozny, S. (eds.) Intelligent Systems'2014. AISC, vol. 323, pp. 225–234. Springer, Heidelberg (2015). doi:10.1007/978-3-319-11310-4_20
4. Apiecionek, Ł., Czerniak, J.M., Zarzycki, H.: Protection tool for distributed denial of services attack. In: Kozielski, S., Mrozek, D., Kasprowski, P., Małysiak-Mrozek, B., Kostrzewa, D. (eds.) BDAS 2014. CCIS, vol. 424, pp. 405–414. Springer, Heidelberg (2014). doi:10.1007/978-3-319-06932-6_39
5. Bednar, R., Saga, M., Vasko, M.: Effectivity analysis of chosen numerical methods for solution of mechanical systems with uncertain parameters. Komunikacie **13**(4), 40–45 (2011)
6. Bitter, R., Mohiuddin, T., Nawrocki, M.R.: LabView Advanced Programming Techniques, 2nd edn. CRC Press, New York (2007)
7. Czerniak, J., Zarzycki, H.: Application of rough sets in the presumptive diagnosis of urinary system diseases. Artif. Intell. Secur. Comput. Syst. **752**, 41–51 (2003)
8. Czerniak, J.: Evolutionary approach to data discretization for rough sets theory. Fundam. Informaticae **92**(1–2), 43–61 (2009)
9. Czerniak, J.M., Apiecionek, L., Zarzycki, H.: Application of ordered fuzzy numbers in a new OFNAnt algorithm based on ant colony optimization. In: Kozielski, S., Mrozek, D., Kasprowski, P., Małysiak-Mrozek, B., Kostrzewa, D. (eds.) BDAS 2014. CCIS, vol. 424, pp. 259–270. Springer, Heidelberg (2014). doi:10.1007/978-3-319-06932-6_25
10. Czerniak, J.M., Dobrosielski, W., Zarzycki, H., Apiecionek, Ł.: A proposal of the new owlANT method for determining the distance between terms in ontology. In: Filev, D., Jabłkowski, J., Kacprzyk, J., Krawczak, M., Popchev, I., Rutkowski, L., Sgurev, V., Sotirova, E., Szynkarczyk, P., Zadrozny, S. (eds.) Intelligent Systems'2014. AISC, vol. 323, pp. 235–246. Springer, Heidelberg (2015). doi:10.1007/978-3-319-11310-4_21
11. Czerniak, J., Dobrosielski, W., Apiecionek, L.: Representation of a trend in OFN during fuzzy observance of the water level from the crisis control center. In: Proceedings of the Federated Conference on Computer Science and Information Systems. ACSIS, vol. 5, pp. 443–447. IEEE Digital Library (2015)
12. Czerniak, J., Śmigielski, G., Ewald, D., Paprzycki, M.: New proposed implementation of ABC method to optimization of water capsule flight. In: Proceedings of the Federated Conference on Computer Science and Information Systems. ACSIS, vol. 5, pp. 489–493. IEEE Digital Library (2015)

13. Dierdof, D., Hawk, J.: Blast initiated deluge system. An ultra-high-speed fire suppression system. In: Fire Suppresion Detection Research Application Symposium, National Research Council, Orlando (2006)
14. Dygdała, R., Stefański, K., Ingwer-Zabowska, M., Kaczorowski, M., Lewandowski, D.: Aerosol produced by explosive detonation in a water bag as fire extinguishment. In: The 3rd International Conference IPOEX 2006, Ustroń - Jaszowiec, Poland (2006)
15. Dygdała, R., Stefański, K., Śmigielski, G., Lewandowski, D., Kaczorowski, M.: Aerosol produced by explosive detonation. PAK **53**(9), 357–360 (2007)
16. Ewald, D., Czerniak, J.M., Zarzycki, H.: Approach to solve a criteria problem of the ABC algorithm used to the WBDP multicriteria optimization. In: Angelov, P., Atanassov, K.T., Doukovska, L., Hadjiski, M., Jotsov, V., Kacprzyk, J., Kasabov, N., Sotirov, S., Szmidt, E., Zadrożny, S. (eds.) Intelligent Systems'2014. AISC, vol. 322, pp. 129–137. Springer, Heidelberg (2015). doi:10.1007/978-3-319-11313-5_12
17. Kozik, R., Choraś, M., Flizikowski, A., Theocharidou, M., Rosato, V., Rome, E.: Advanced services for critical infrastructures protection. J. Ambient Intell. Humanized Comput. **6**(6), 783–795 (2015). http://dx.doi.org/10.1007/s12652-015-0283-x
18. Liu, Z., Kim, A.K., Carpenter, D.: Extinguishment of large cooking oil pool fires by the use of water mist system. Combustion Institute, Canada Section, Spring Technical Meeting (2004)
19. National Instruments, Austin: CompactRIO Developers Guide (2009)
20. Nawrocki, W.: Measurement Systems and Sensors. Artech House (2005)
21. Śmigielski, G., Dygdala, R., Kunz, M., Lewandowski, D., Stefański, K.: High precision delivery of a water capsule: theoretical model, numerical description, control system and results of field experiment. In: 19th IMEKO World Congress 2009, Lisbon (2009)
22. Śmigielski, G., Toczek, W., Dygdała, R., Stefański, K.: Metrological analysis of precision of the system of delivering water-capsule for explosive production of water aerosol. Metrol. Meas. Syst. **23**(1), 47–58 (2016)

Evaluation of Influence Exerted by a Malicious Group's Various Aims in the External Grid

Kosuke Yamaguchi[✉], Tsutomu Inamoto, Keiichi Endo, Yoshinobu Higami, and Shinya Kobayashi

Graduate School of Science and Engineering, Ehime University, Matsuyama, Japan
yamaguchi@koblab.cs.ehime-u.ac.jp, {endo,higami}@cs.ehime-u.ac.jp,
{inamoto,kob}@ehime-u.ac.jp
http://www.eng.ehime-u.ac.jp/rikougaku/index.cgi?lang=en

Abstract. The external grid is one of grid computing systems. It is composed of numerous computers connected to the Internet. Although the external grid realizes high performance computing, it is necessary to guarantee the robustness against malicious behaviors of the computers. In the previous literature, a technique to protect program codes against such behaviors has been proposed; however, only one type of malicious behavior is considered to evaluate the effectiveness of the technique in the literature. In reality, malicious behaviors vary according to the purpose of malicious groups. The goal of the research in this paper is to guarantee the safety of the external grid in a quantitative way. In order to achieve the goal, we evaluate the effectiveness of concealing processes against several types of malicious behaviors.

Keywords: Grid computing · External grid · Secure processing · Performance evaluation · Safety evaluation · Information disclosure · Actions of malicious computers

1 Introduction

Grid computing is one of distributed computing systems. This uses many computers connected to a network. External grid is one of grid computing systems. There are unlimited computers connected to the Internet. Therefore, it achieves high performance computing quickly and inexpensively. It works for running large scale simulations and executing many tasks.

However, successful and profitable services with the external grid are not known. The reason is that the external grid may contain computers owned by malicious users. As a result, it has not been guaranteed safety.

Moreover, malicious owners may form groups, which increases the risk of the running process being analyzed by them [1]. Thus, we suggested the way to decrease the risk and evaluated performance of it. Particularly, it is concealing processes which a client has required executing for an external grid [2].

© Springer International Publishing AG 2017
S. Kobayashi et al. (eds.), *Hard and Soft Computing for Artificial Intelligence, Multimedia and Security*, Advances in Intelligent Systems and Computing 534, DOI 10.1007/978-3-319-48429-7_11

Although, in the paper [2], we evaluated performance of the way under only one purpose of a malicious group. The performance evaluation was not sufficient because the purposes of malicious group are considered to be various in reality.

The goal of the research in this paper is to guarantee the safety of the external grid in a quantitative way. In order to achieve the goal, we evaluate the effectiveness of concealing processes against several types of malicious behaviors.

The rest of this paper is organized as follows. In Sect. 2, problems of the external grid and solutions for them are mentioned. Section 3 shows the purposes of malicious groups. In addition, actions of malicious owner's computers are explained. Section 4.2 mentions the evaluation of standards and simulation conditions. Finally, in Sect. 5, the results of evaluation are mentioned.

2   External Grid and Previous Research

2.1   External Grid

An external grid achieves a high performance computing quickly and inexpensively. For example, SETI@Home is a famous project using the external grid [5]. However, successful and profitable projects using the external grid are not known. The reason is that it is not a safe system yet.

External grid is composed of many computers connected to the Internet. Almost all these execute a distributed program in an external grid. These computers are called "nodes". Some nodes may be owned by malicious users. These nodes are called "malicious nodes". These give bad effects to an external grid.

2.2   Secure Processing

Secure processing is a technique against cheats by malicious nodes and malicious owners of nodes. There are two types of cheats: one is called "analyze", and the other is called "alter".

analyze
   The cheat is that malicious nodes analyze the intention of the program from a distributed program.
alter
   The cheat is that malicious nodes reply a fake result to the client.
   Already prevailing techniques of network security such as advanced firewalls and collaborative intrusion detection networks cannot protect an external grid from these cheats. The technique against "analyze" is called "program division", and the technique against "alter" is called "multiplex execution".
program division
   In this technique, a program is divided into some codes. The program has been required for a client to execute in an external grid. Each code is distributed to each of the nodes. This technique makes it hard to analyze, because the information which a malicious node can get from a code is less than in the case of getting for them from an original program.

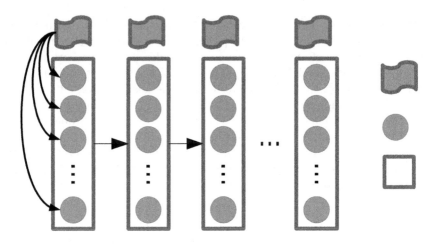

Fig. 1. The sequence of multiplex execution

multiplex execution

This technique achieves reducing the risk of altering. Figure 1 shows the sequence of multiplex execution. A code is replicated and executed on some nodes. The number of times a code is replicated is called "multiplicity". A set of these nodes is called "block".

Moreover, each of the nodes in a block votes by a result which it has replied. Therefore, in a block, the most trusted result is selected. The selected result will be given for the other nodes to execute their code. However, the case is called "split" when a majority result has not been selected in a block. In this case, nodes in a block are disposed, new nodes are selected, and these start to execute code.

2.3   Advance Processing

Advance processing reduces processing time in an external grid. In an external grid with multiplex operations, the processing time with advance processing is shorter than without it. The reason is the earliest replied result in the block is given for other nodes to execute their code.

In previous research, "comprehensive processing" is suggested [3]. It is one of advance processing and reduces processing time in an external grid. Especially processing time with comprehensive processing is shorter than with only advance processing.

This technique is different from only advance processing. In the case with this technique, several executed results in a block are given to other nodes. For detail, these are all mutually different executed result in the block. Therefore, more nodes are needed for an external grid.

2.4 Dependence of Codes and Risk of Conspiring with Malicious Owners

There may be dependence relations between codes [4]. Figure 2 shows a sequence of codes written on a program. Figure 3 shows dependence relations between each of codes from the program. For example, C and D depend on A. In other words, A and C are consecutive codes.

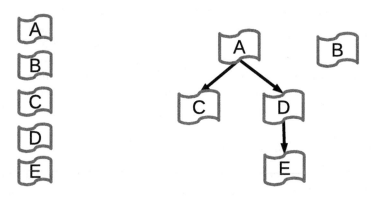

Fig. 2. A program sequence   **Fig. 3.** Dependence relations on program in Fig. 2

Malicious owners analyze a program with code dependence. The more consecutive codes are acquired by them, the easier it is for them to analyze.

In addition, malicious owners may make groups. They can achieve to get more consecutive codes, because they share each of their codes in the group. Making malicious groups means that the risk of analysis increases.

2.5 Concealing Processes

Concealing processes is that each of some codes is given to each of trusted nodes. Trusted nodes are not managed by malicious owners. These codes are called "concealed codes".

This technique achieves reducing the risk of analyzing by malicious owners. As previously mentioned in Sect. 2.4, malicious owners analyze a program with code dependence. This technique forces them to get less consecutive codes, because some of these codes are concealed from them by trusted nodes.

In the paper [2], we evaluated performance of concealing process with a simulation. However, we simulated on only one purpose of a malicious group. The purpose is that they get a lot of information about a program in an external grid. It is not enough evaluation of the technique.

3   Purposes of Malicious Groups and Actions of Malicious Nodes

3.1   Purposes of Malicious Groups

A client probably uses an external grid on his business or research activities. The purposes of the cheats in an external grid by malicious groups are divided into two types.

One purpose is gathering information about activities of a client. For example, malicious owners in a group desire to get information about new products of a rival company. Therefore, they spy on an external grid which it has requested executing a program. They will analyze distributed the program then.

Another purpose is obstructing activities of a client. In particular, they desire to stop or confuse a project. They join the external grid and their nodes will alter results.

3.2   Actions of Malicious Nodes

Actions of malicious nodes are various, because these depend on a purpose of groups which contain these nodes. In this paper, we evaluate performance of the external grid with three intentions of a malicious group.

Spy intention
  The first intention is that malicious nodes gather information in an external grid without being noticed by an owner of it. Then, an action of these is that all malicious nodes return correct results.

Robber intention
  The second intention is that malicious owners get a lot of information about a program in an external grid. Therefore, an action of their nodes is that each of their nodes returns a unique fake. An advantage is that they can gather more information, because more malicious nodes collect in an external grid than without this action. A disadvantage is that the owner of the external grid can notice these easily.

Raider intention
  The third intention is both gathering information about activities and obstructing these. However, obstructing is more important than the other. Then, all malicious nodes in each of blocks return fakes which are each other all the same. This intention achieves a final result of an external grid can lead wrong one if there are enough malicious nodes. Conversely, an owner of it notices these easily when there are not enough these.

4   Evaluation Methods

4.1   Evaluation Standard

In this paper, we set five evaluation standards.

The number of all nodes
This means the number of computers which execute a distributed program during processing of an external grid. The less this is, the more secure the external grid is. The reason is that many malicious nodes are contained in it if there are many nodes.

The maximum number of consecutive codes gained by a malicious group
Malicious owners analyze a program with code dependence. The more consecutive codes are acquired by them, the easier it is for them to analyze. Therefore, the length of consecutive codes obtained by malicious nodes means the difficulty of analyzing.

The rate of acquired codes
This means the amount of information which malicious group has acquired.

Performance of an external grid
We evaluate this by the processing time of an external grid. It is expected to be shorter.

The reliability of an external grid
We evaluate this by the rate of a true result returned finally.

4.2   Conditions of Simulation

We evaluated values mentioned in Sect. 4.1 by a simulator. This supposes the external grid composed only consecutive codes with comprehensive processing.

Parameters characterizing the model of simulation
There are five parameters: the number of codes, the number of nodes in a block, the number of concealed codes, the rate of malicious nodes and the kind of an action of malicious nodes.

The limit on nodes
The number of nodes is unlimited.

Malicious nodes
An owner of an external grid cannot distinguish malicious nodes from others. There are malicious nodes depended on the rate of these in an external grid.

Performance of a node
A performance is the program size which it can execute per a unit time. The performance depends on the gamma distribution. In detail,

The program size
The size is 100.

Actions of malicious nodes
As previously mentioned in Sect. 2.4, actions are classified into three types. The first action is that all malicious nodes return the correct result.

The second action is that each of their nodes returns a unique fake. In the third action, all malicious nodes in each of blocks return fakes which are each other all the same. The each of actions is called as "spy intention", "robber intention" and "raider intention".

The number of trials

This is 1000 on like-for-like bases.

5   Results

In this section, simulation results are shown.

5.1   The Number of All Nodes

Averages of all nodes are shown in Figs. 4 and 5.

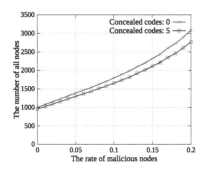

Fig. 4. The relation between the rate of malicious nodes and the number of all nodes under raider intention

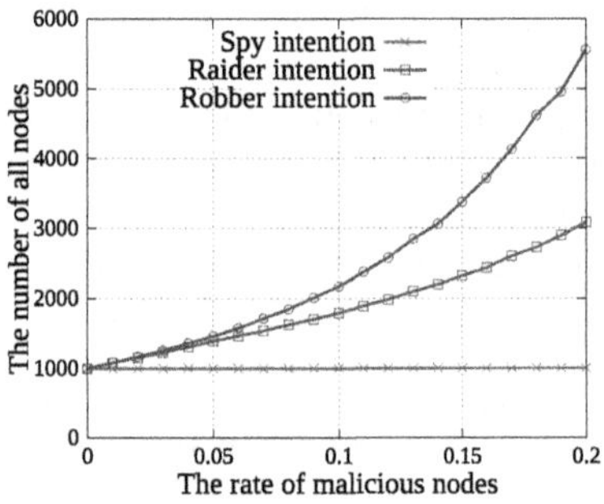

Fig. 5. The relation between the rate of malicious nodes and the number of all nodes. (The number of concealed codes is 0 in this figure.)

Figure 4 shows that changing in the value of averages of all nodes depends on the rate of malicious nodes between the numbers of concealed codes. Figure 5 shows averages of all nodes of differences between three actions of malicious nodes.

According to Fig. 4, in the case of malicious nodes in each of blocks return fakes which are each other all the same, averages of all nodes are less with concealing codes than without it.

In Fig. 5, averages of all nodes in the case where all malicious nodes return fakes are more than in the case where all malicious nodes in each of blocks return fakes which are each other all the same. In addition, these in the case all malicious nodes in each of blocks return fakes which are each other all the same more than each of their nodes return a fake and unique result.

5.2 The Greatest Number of Consecutive Codes Gained by a Malicious Group

Averages of the maximum number of consecutive codes gained by a malicious group are shown in Figs. 6 and 7. Then, the number of codes is 100 and multiplicity is 10. Figure 6 shows that changing in the value of the most consecutive codes gained by a malicious group between the numbers of concealed codes. In Fig. 7, the differences of the most consecutive codes gained by a malicious group between three actions of malicious nodes are shown.

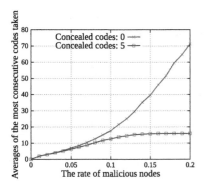

Fig. 6. The maximum number of consecutive codes gained by a malicious group under raider intention.

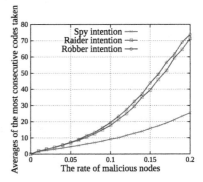

Fig. 7. Averages of the most consecutive codes taken by malicious group. (The number of concealed codes is 0 in this figure.)

According to Fig. 6, the most consecutive codes gained by a malicious group are less with concealing codes than without it in the case of malicious nodes in each of blocks return fakes which are each other all the same. Moreover, the figure shows that concealing processes limit on the most consecutive codes gained by a malicious group.

In Fig. 7, there are difference of the most consecutive codes gained between an action that all malicious nodes return the correct result and the other actions. Otherwise, there are margin difference of these between the two actions of malicious group.

5.3 The Rate of Acquired Codes by a Malicious Group

Figures 8 and 9 show the rate of acquired codes from all of divined program.

Figure 8 shows that changing in the rate of retrieved codes between the numbers of concealed codes. In Fig. 9, the differences of the rate between three actions of malicious nodes are shown.

In Fig. 8, concealing processes make the rate of retrieved codes lower than without the technique.

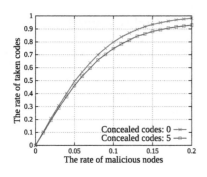

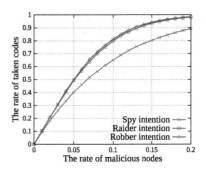

Fig. 8. The rate of acquired codes by a malicious group under raider itention.

Fig. 9. The rate of taken codes by a malicious group. (The number of concealed codes is 0 in this figure.)

According to Fig. 9, there are not significant differences of the rate between in the case of each of their nodes return a unique fake and the case of all malicious nodes in each of blocks return fake which are each other all the same. In an opposite case, in the case of all malicious nodes return the correct result, the rate is lower than in the other cases.

5.4   Performance of the External Grid

Averages of the time from the start to the end of processes in an external grid are shown in Figs. 10 and 11.

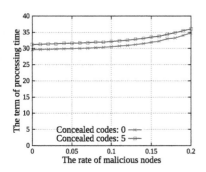

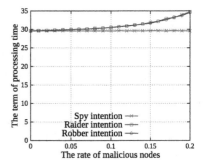

Fig. 10. The term of processing in an external grid under raider intention.

Fig. 11. The term of processing in an external grid. (The number of concealed codes is 0 in this figure.)

In Fig. 10, in the case of all malicious nodes in each of blocks return results which are each other all the same, changing of the term between in the case of the number of concealing nodes is 0 and in the case of the number of concealing nodes is 5 are shown. Figure 11 shows the differences of the terms between three actions of malicious nodes are shown.

In Fig. 10, regardless of the rate of malicious nodes, concealing codes causes the termination of processes in an external grid late.

According to Fig. 11, in the case of all malicious nodes return the correct results, the term of processing in an external grid is almost no change. On the other hand, in the other case, the time increases in accordance with the rate of malicious nodes. Moreover, there is no difference between the other cases.

5.5 The Reliability of an External Grid

We evaluated the reliability of an external grid with the rate of finally correct results of an external grid.

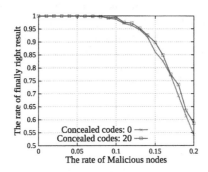

Fig. 12. The rate of finally correct result of an external grid under raider intention.

According to Fig. 12, the rate in the case of the number of concealed codes is 20 is higher than in the case of the number of concealed codes is 0. However, the reliability is light affected if the small number of codes are concealed in an external grid.

6   Conclusion

In this paper, purposes of malicious groups in an external grid have been explained, and three types of malicious actions have been mentioned. In addition, performance of an external grid and effects of concealing processes have been evaluated under the condition that malicious nodes take each of these actions.

The effects of each of actions of malicious nodes are as follows.

In the instance that all malicious nodes return the correct results, the number of all nodes and the term of processing are not affected badly. Furthermore, the number of the greatest length of taken consecutive codes and the rate of taken codes are less than in the other instance.

In the cases that each of malicious nodes return a unique fake and these in each of blocks return fake which are each other all the same, there are not large differences of the amount of information and performances of an external grid. However, these greatly elevated in accordance with the rate of malicious nodes increasing. The reason is that greatly increasing the number of all nodes affects these properties. For all of these reasons, in these cases, a malicious group can take more information about a program in an external grid than in the other case where all malicious nodes return the correct results. Also these actions of malicious nodes affect badly the term of processing in an external grid.

The evaluation results about performance of concealing processes under a case are as follows. The case is that each of malicious nodes return a unique fake.

Concealing processes has an effect which makes increase rate of the number of all nodes low. In addition, the technique achieves limiting the amount of information which malicious groups can take. Moreover, increasing concealed codes attains to raise the reliability of the final result in an external grid. However, the technique affects poorly and marginally the term of processing in an external grid. Otherwise, the difference of the term between these graphs is little.

For all of these reasons, concealing processes achieves improving the safety and reliability of the external grid.

Acknowledgment. This work was supported by MEXT KAKENHI Grant Number 26330105.

References

1. Nakaya, T., Inamoto, T., Higami, Y., Kobayashi, S.: Study on concealmentability of secure processing based on continuity of program flagments (in Japanese). In: Proceedings of DICOMO (Multimedia, Distributed, Cooperative, and Mobile System) 2015, pp. 287–294 (2015)
2. Yamaguchi, K., Inamoto, T., Higami, Y., Kobayashi, S.: A method to reduce the risk of malicious analysis with dependence relations in external grid computing (in Japanese). In: Proceedings of the 78th National Convention of IPSJ, 5R–07 (2016)
3. Hirose, Y., Inamoto, T., Higami, Y., Kobayashi, S.: Evaluation of advance processing in terms of processing time on secure processing (in Japanese). In: Proceedings of 14th Forum on Information Technology, vol. 4, pp. 241–242 (2015)
4. Padua, D.A., Wolfe, M.J.: Advanced computer optimizations for supercomputers. Commun. ACM **29**(12), 1184–1201 (1986)
5. University of California: SETI@Home. http://setiathome.ssl.berkeley.edu/

Subject-Specific Methodology in the Frequency Scanning Phase of SSVEP-Based BCI

Izabela Rejer[(⊠)] and Łukasz Cieszyński

Faculty of Computer Science and Information Technology,
West Pomeranian University of Technology in Szczecin, Szczecin, Poland
irejer@wi.zut.edu.pl, lcieszynski@zut.edu.pl

Abstract. Steady State Visual Evoked Potentials (SSVEPs) often used in Brain Computer Interfaces (BCIs) differ across subjects. That is why a new SSVEP-based BCI user should always start the session from the frequency scanning phase. During this phase the stimulation frequencies evoking the most prominent SSVEPs are determined. In our opinion not only the stimulation frequencies specific for the given user should be chosen in the scanning phase but also the methodology used for SSVEP detection. The paper reports the results of a survey whose aim was to find out whether using subject specific methodology for identifying stimulation frequencies would increase the number of frequencies found. We analyzed three factors: length of time window used for power spectrum calculation, combination of channels, and number of harmonics used for SSVEP detection. According to the outcome of the experiment (performed with 6 subjects) the mean drop in the number of SSVEPs detected with any other but the best combination of factors was very large for all subjects (from 31.52 % for subject S3 to 51.76 % for subject S4).

Keywords: SSVEP · BCI · Brain Computer Interface · EEG

1 Introduction

A Brain Computer Interface (BCI) is a communicating system that links a human brain to a computer or another external device to enable the control of this device or the communication with the environment. One type of BCIs, frequently used in practical applications, is based on Steady State Visual Evoked Potential (SSVEP-based BCI). A visual evoked potential (VEP) is an electrical potential difference, which can be derived from the scalp after a visual stimulus, for example a flash-light. If the stimulation frequency is > 3.5 Hz they are called "steady state" VEPs because the individual responses overlap and result a quasi-sinusoid oscillation with the same frequency as the stimulus [9].

An SSVEP-based BCI is composed of two separate units (Fig. 1). While the task of the first unit is to deliver light stimuli, the second unit records user EEG and processes it to restore the current stimulus frequency. In details, the user interaction with the interface looks as follows. There is a set of light sources flashing at different frequencies $(f_1, f_2 ... f_n)$. Each light source corresponds to a different command, e.g. f_1 - move left, f_2 - move right, f_3 - stop. The user task is to observe the light source corresponding to the

© Springer International Publishing AG 2017

S. Kobayashi et al. (eds.), *Hard and Soft Computing for Artificial Intelligence, Multimedia and Security*, Advances in Intelligent Systems and Computing 534, DOI 10.1007/978-3-319-48429-7_12

command he wants to perform. At the same time EEG signal is recorded and processed in frequency domain in order to find the frequency of the light source observed by the user. When the system recognize one of the stimulus, the command corresponding to this stimulus is executed. For example, if user observes light source flickering at f_1 and the system recognizes this frequency then the command "move left" will be started (Fig. 1.).

The visually evoked potentials are often used to build BCIs since they can be detected in brain waves of most subjects. Moreover, the SSVEP-based BCI does not require special training before use and for most people it is possible to use it with success from the very beginning. This paradigm provides also the high Information Transfer Rate (ITR) and the reasonable stability of results obtained by individuals in various sessions.

According to the neurobiological theory SSVEPs are stable brain responses to a repetitive stimulus which appear at the same frequency as the stimulus [3, 10]. The stability, however, does not mean that stimuli flickering at different frequencies evoke SSVEPs at the same frequencies in different subjects and at different times. Just opposite, many research have proved that different subjects are more sensitive to different stimulation frequencies [2], and moreover the frequencies evoking the most prominent SSVEPs can change even for the same subject [5].

Hence, the stability of SSVEPs means only that they are stable in time [3] providing that nor the subject or the external environment change during the experiment. The slightest distortions in the external conditions or the stimulus delivery can change the perception of the stimuli and hence can disturb the SSVEPs detection. That is why, the scheme that is often followed in SSVEP-based BCI research consists of two steps. First, an off-line analysis is performed with the aim to find out which frequencies of the stimulus evoke the most prominent SSVEP in each subject. Second, the frequencies specific for each subject are used in an on-line BCI sessions [1].

Starting a BCI session with the frequency scanning phase of course improves the BCI performance since the frequencies specific for a given subject in the given external

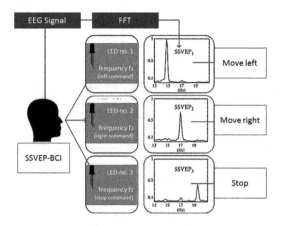

Fig. 1. SSVEP-based BCI.

conditions are detected. Usually the scanning phase is performed by presenting, one by one, a set of stimuli of different frequencies. Next, all EEG data gathered during the stimuli presentation are processed in an off-line session by a chosen algorithm, and the frequencies given the most prominent SSVEP are chosen for on-line session. The question is, however, if this is really the best scheme. Maybe not only the frequencies evoking the best SSVEP are subject-dependent but also the methodology used for their detection? If this is true we could obtain a further improvement of a BCI performance after applying the subject-dependent methodology for identifying the subject-dependent frequencies in the frequency scanning phase.

The idea mentioned above has its source in the previous research carried out in the neuroscience field. For example, one of the possible reasons for intra-subject differences in SSVEPs detection are variations in skull and scalp thickness of different subjects. These variations influence the amplitude of SSVEP - a thicker layer reduces EEG amplitude while a thinner layer increases it [4]. Since these variations can be as well global (regarding the whole skull), as local (regarding only some brain areas), the optimal location of the active electrode in the occipital area can be different for the same frequency for different subjects [6]. Hence, by choosing the channels' combination providing the best results for the given subject in the frequency scanning phase, we can improve the on-line BCI performance.

The aim of this paper is to report the results of the experiment that was performed in order to find out whether using subject specific methodology for identifying subject specific stimulation frequencies would increase the number of frequencies found. The experiment was performed with 6 subjects with respect to three factors: length of time window used for power spectral density calculation (3 levels), combination of channels (8 levels), and number of harmonics used for SSVEP detection (8 levels). For each of 192 combinations of factors ($3 \times 8 \times 8$) we calculated and compared the number of frequencies matching the stimulation frequency separately for each subject. Than we averaged the number of SSVEPs detected with the best combinations of factors over subjects and compared this average with the number of SSVEPs detected per each combination averaged over subjects. The stimuli for evoking SSVEPs were delivered by a single red LED, connected to Arduino board. 85 stimuli of different frequency, fully covering 5.05-31.25 Hz band, were used in the survey.

The rest of the paper is organized as follows. Section 2 describes the setup of the experiment. Section 3 is focused on the methodology used for EEG data processing. Section 4 reports the results of the experiment and Sect. 5 concludes the paper.

2 Experiment Setup

The experiment was performed with six subjects (all man, mean age 28,3, range 25–39 years). Five subjects were right-handed and all had normal or corrected-to-normal vision. None of the subjects had previous experiences with BCI and none reported any previous mental disorders. Written, informed consent was obtained from all subjects.

During the experiment EEG data was recorded from three monopolar channels at a sampling frequency of 256 Hz. Five passive electrodes were used in the experiments. Three of them were attached to the subject's scalp at O1, O2, and Pz positions according to the International 10-20 system [7]. The reference and ground electrodes were located at Fpz and the right mastoid, respectively. The impedance of the electrodes was kept below 5 kΩ. The EEG signal was acquired with Discovery 20 amplifier (BrainMaster) and recorded with OpenVibe Software [8].

The detailed scheme of the experiment with one subject was as follows (Fig. 2). The subject was placed in a comfortable chair and EEG electrodes were applied on his head. In order to limit the number of artifacts, the participant was instructed to stay relaxed and not move. The start of the experiment was announced by a short sound signal and five seconds later EEG recording was started. Every 13 s of the recording the subject was exposed to a new stimulus. The stimuli were delivered by a single red LED (Light Emitted Diode), connected to the Arduino Uno board. LED was flickering with the frequency that was equal to *1000/2/divider*, where *divider* was the parameter, sent to the board from Matlab environment (Matlab R2015a), informing about the length of the period given in milliseconds when the diode had to stayed in one state ("on" or "off").

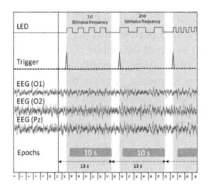

Fig. 2. Setup of the experiment with one subject

Together with the stimulus an acoustic signal was generated to draw subject's attention and to inform him that the frequency would be changed. Two seconds later LED started to flicker with the frequency corresponding to *divider*. After 11 s the stimulation ended and the next divider was sent to the board. The stimuli were picked out in a random order from the whole set of 85 possible stimuli. In order to limit subject fatigue, the experiment was performed in two sessions with a five minutes breaks between them. In the first session 40 stimuli were presented, in the second session - the remaining 45.

3   Methodology

The EEG signal was processed off-line in Matlab environment. First, EEG data recorded from all three channels (O1, O2, Pz) were filtered with a Butterworth band-pass filter of the 4th order in the band 4–35 Hz. Next, the parts of signal corresponding to 10 s of the stimulation periods (from 4th to 13th second) were extracted and stored in three matrixes, one matrix per channel. The size of each matrix was 85×2560 (85 stimulation frequencies $\times$ 10 s).

In order to prepare data for the analysis we added 5 combinations of channels CH4-CH8 to the set of original channels. The definition of succeeding combinations was as follows:

- CH4: O1-O2
- CH5: O1-Pz
- CH6: O2-Pz
- CH7: (O1 + O2)/2
- CH8: (O1 + O2)/2-Pz

To normalize the terminology, we renamed the first three channels: CH1 for O1, CH2 for O2, and CH3 for Pz. Data for each new combination was stored in a matrix of the same structure as in the case of the original channels. Hence, after this step we finished with 8 matrixes storing data from 10 s for each of 85 stimulation frequency. The data for each stimulation frequency in each matrix of channels was processed according to the same scheme. First, three power spectra (PS) were calculated with Fast Fourier Transform (FFT) with different length of time window (2, 5, and 10 s) but the same FFT length, equal to 2560 bins. Next, the two first spectra were averaged in order to obtain one spectrum per each window length. Hence, after calculating the power spectra, three PSD matrixes were obtained for each channel combination. They were indexed using no. of channel combination and window length - for example matrix CH1_W2 contained power spectrum for all 85 stimulation periods for channel O1 calculated for 2-second window.

In order to examine the influence of the third factor used in our analysis (number of harmonics used for SSVEP detection), the amplitudes of all bins from the original power spectrum were enhanced by adding to them the amplitudes of the suitable harmonics/subharmonics. After this transformation 8 final PSD matrixes were obtained from each CH_W matrix:

- CH_W_PS$_F$: PS(F),
- CH_W _PS$_{Fh1}$: PS(F) + PS(h1),
- CH_W _PS$_{Fh2}$: PS(F) + PS(h2),
- CH_W_PS$_{Fsh1}$: PS(F) + PS(sh1),
- CH_W_PS$_{Fh1h2}$: PS(F) + PS(h1) + PS(h2),
- CH_W_PS$_{Fh1sh1}$: PS(F) + PS(h1) + PS(sh1),
- CH_W_PS$_{Fh2sh1}$: PS(F) + PS(h2) + PS(sh1),
- CH_W_PS$_{Fh1h2sh1}$: PS(F) + PS(h1) + PS(h2) + PS(sh1),

where: F stands for the fundamental frequency, $h1$ stands for 1st harmonic, $h2$ stands for 2st harmonic and $sh1$ stands for 1st subharmonic.

At the final stage of data processing the number of stimulation frequencies evoking prominent SSVEP was counted separately per each of 192 CH_W_PS matrixes (8 channel combinations, 3 time windows, 8 number of harmonics) created per each subject.

SSVEP was acknowledged to be found when the power spectrum had its maximum at the frequency equal to the stimulation frequency with a small 0.5 Hz buffer b caused by a frequency spectrum resolution. Because EEG signal power decreases with increasing frequency, the maximum amplitude condition was also relaxed and the band were the maximum was looked for was limited to 10 Hz around the stimulation frequency. Hence, finally, SSVEP was acknowledged to be detected when:

$$PS(fp) = \max(PS(-5\text{Hz} + fs : fs + 5\text{Hz})) \tag{1}$$

$$fs - b \leq fp \leq fs + b, \tag{2}$$

where: fs - stimulation frequency, b - buffer around the stimulation frequency, PS - power spectrum calculated for EEG signal recorded for the given fs, fp - peak frequency in the given part of PS.

4   Results and Discussion

The summary of the results obtained in the experiment is presented in Fig. 3 and in Table 1. The figure shows the stimulation frequencies correctly detected with the best combination of factors for each subject. The compatible results (stimulation frequency = response frequency) are denoted as large black dots. As we can see in the figure the distribution of frequencies for different subjects differs significantly. For one

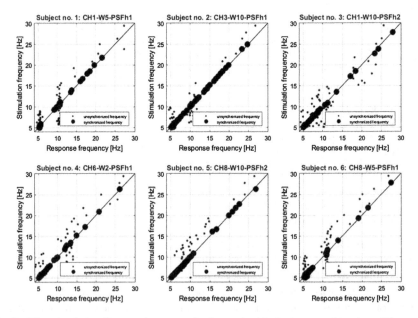

Fig. 3. The stimulation frequencies correctly detected with the best combination of factors for each subject.

subjects it covers equally the wide frequency bands (e.g. a band 5–20 Hz for subject 2 or 5–12 Hz band for subject 3 and 5), for others it is concentrated in single frequency points (e.g. for subject 1 and 6). The only common feature across all subjects are SSVEPs found in a narrow 5–7 Hz band.

Table 1 presents the same information but given as the ratio of the number of SSVEPs detected with the best and the worst combination of factors to the total number of stimulation frequencies. As it can be noticed in the table, for most subjects more than one combination of factors provided the same number of detected SSVEP.

The results presented in Table 1 once again shows that the number of stimulation frequencies allowing for SSVEP detection highly varies across individuals (from 34 % for subject S3 up to 66 % for subject S5). Analyzing the table across the best combinations of factors, it can be noticed that both, the first and the second harmonics are important for a successful SSVEP detection. On the other hand, investigating the worst combinations, a high influence of the first subharmonic can be noticed. It appears in almost all "the worst" combinations. Another conclusion that can be derived from that table, important from of the point of view of the aim of our study, is that the best combinations of factors vary greatly across individuals. In fact only one "best" combination of factors repeats for two subjects - CH6_W2_PSFh1 (subjects S3 and S4) – all others differ in regard to channels, time widows, and PS calculation method.

In order to properly address the question stated in Sect. 1, we calculated the relative drop in the number of detected SSVEPs when any other than the best combination of factors was applied for signal processing. This was done separately per each subject.

Table 1. Summary of the results obtained in the experiment. The numbers in the table present ratio of the number of detected SSVEPs to the total number of stimulation frequencies. The second and third column provides the results for the best combination of analyzed factors (given as the combination index and the ratio of detected SSVEPs), the fourth and fifth column presents the results for the worst combination, and the last column contains the average ratio of detected SSVEPs.

	THE BEST		THE WORST		MEAN
S1	CH1_W5_PSFh1 CH1_W5_PSFh1h2	38 %	CH5_W2_PSFsh1	6 %	19 %
S2	CH3_W10_PSFh1 CH3_W10_PSFh1h2	56 %	CH5_W2_PSFsh1 CH4_W5_PSFsh1 CH1_W5_PSFsh1	11 %	28 %
S3	CH1_W10_PSFh2 CH7_W10_PSF CH6_W2_PSFh1	34 %	CH5_W10_PSFsh1	11 %	23 %
S4	CH6_W2_PSFh1	53 %	CH1_W2_PSFh1h2sh1	6 %	26 %
S5	CH8_W10_PSF CH8_W10_PSFh2	66 %	CH4_W2_PSFsh1	19 %	45 %
S6	CH8_W5_PSFh1	40 %	CH3_W2_PSF CH3_W10_PSFh2 CH3_W2_PSFh2sh1 CH3_W2_PSFh2sh1	9 %	23 %

The results obtained per one subject (Subject 5) are presented in Table 2. Each row of the table represents one combination of channel CH and time window W, each column – one PS variation. The last column and the last row contains the mean values calculated across CH-W combinations and PS variations.

Table 2 illustrates how significant is the negative impact of the selection of the not-optimal methodology when detecting SSVEP. Considering only the rows, we can see that in the case of two worst CH-W combinations (CH4_W2 and CH3_W2) the number of SSVEPs detected for Subject 5 was more than 50 % smaller than in the case of the best combination (CH8_W10). On the other hand, analyzing the variations in power spectra, the larger loss in the number of SSVEPs detected was observed when a first sub-harmonic was included in the calculations (32.7 – 42.9 %).

Table 2. The relative drop in the number of detected SSVEPs when any other than the best combination of factors was applied for signal processing. The table was calculated over the results obtained for Subject 5.

	PS_F [%]	PS_{Fh1} [%]	PS_{Fh2} [%]	PS_{Fsh1} [%]	PS_{Fh1h2} [%]	PS_{Fh1sh1} [%]	PS_{Fh2sh1} [%]	$PS_{Fh1h2sh1}$ [%]	MEAN [%]
CH1_W2	35.71	28.57	42.86	58.93	41.07	51.79	57.14	44.64	**45.09**
CH1_W5	33.93	26.79	48.21	53.57	42.86	44.64	57.14	50.00	**44.64**
CH1_W10	28.57	32.14	37.50	44.64	42.86	42.86	48.21	46.43	**40.40**
CH2_W2	30.36	32.14	30.36	53.57	44.64	46.43	55.36	48.21	**42.63**
CH2_W5	32.14	30.36	30.36	46.43	26.79	33.93	48.21	35.71	**35.49**
CH2_W10	26.79	21.43	26.79	41.07	19.64	30.36	33.93	23.21	**27.90**
CH3_W2	37.50	50.00	44.64	69.64	53.57	55.36	66.07	55.36	**54.02**
CH3_W5	37.50	44.64	35.71	50.00	50.00	48.21	48.21	44.64	**44.86**
CH3_W10	41.07	41.07	39.29	48.21	42.86	39.29	48.21	42.86	**42.86**
CH4_W2	58.93	46.43	58.93	71.43	53.57	60.71	69.64	60.71	**60.04**
CH4_W5	41.07	32.14	46.43	66.07	41.07	57.14	55.36	46.43	**48.21**
CH4_W10	37.50	30.36	41.07	50.00	33.93	46.43	46.43	37.50	**40.40**
CH5_W2	14.29	16.07	12.50	41.07	21.43	33.93	26.79	21.43	**23.44**
CH5_W5	8.93	17.86	8.93	32.14	14.29	25.00	21.43	17.86	**18.31**
CH5_W10	3.57	8.93	1.79	21.43	14.29	19.64	14.29	21.43	**13.17**
CH6_W2	10.71	8.93	10.71	25.00	10.71	19.64	21.43	17.86	**15.62**
CH6_W5	8.93	12.50	8.93	26.79	16.07	25.00	23.21	17.86	**17.41**
CH6_W10	7.14	12.50	5.36	23.21	10.71	17.86	19.64	8.93	**13.17**
CH7_W2	30.36	32.14	42.86	51.79	39.29	44.64	55.36	37.50	**41.74**
CH7_W5	32.14	26.79	30.36	48.21	35.71	39.29	44.64	37.50	**36.83**
CH7_W10	26.79	25.00	26.79	44.64	33.93	33.93	39.29	30.36	**32.59**
CH8_W2	7.14	14.29	5.36	23.21	12.50	19.64	16.07	17.86	**14.51**
CH8_W5	3.57	10.71	3.57	21.43	10.71	21.43	16.07	16.07	**12.95**
CH8_W10	0.00	1.79	0.00	17.86	3.57	10.71	8.93	5.36	**6.03**
MEAN [%]	**24.8**	**25.1**	**26.6**	**42.9**	**29.8**	**36.1**	**39.2**	**32.7**	**32.18**

Table 3. The minimum, maximum, and mean relative drop in the number of detected SSVEP for all subjects. The second and third column presents the combinations of factors providing the minimum drop of the number of detected SSVEP (given as the combination index and the ratio of detected SSVEPs), the fourth and fifth column presents the combination of factors providing the maximum drop, and the last column provides the mean value of detected SSVEPs (averaged over all 192 combinations).

	MIN		MAX		MEAN
S1	CH1_W2_PSFh1h2 CH1_W10_PSFh1 CH7_W5_PSFh1	6.25 %	CH5_W2_PSFsh1	84.38 %	50,44 %
S2	CH3_W5_PSFh1	2.08 %	CH1_W5_PSFsh1 CH4_W5_PSFsh1 CH5_W2_PSFsh1	81.25 %	50,03 %
S3	CH1_W5_PSF CH6_W10_PSFh1	3.45 %	CH5_W10_PSFsh1	68.97 %	31,52 %
S4	CH6_W5_PSFh1 CH8_W10_PSFh1	2.22 %	CH1_W2_PSFh1h2sh1	88.89 %	51,76 %
S5	CH5_W10_PSFh1 CH8_W10_PSFh2	1.79 %	CH4_W2_PSFsh1	71.43 %	32,18 %
S6	CH1_W10_PSFh1h2 CH8_W5_PSFh1h2	2.94 %	CH3_W2_PSFh2sh1 CH3_W10_PSF CH3_W10_PSFh2 CH3_W10_PSFh2sh1	76.47 %	41,59 %

Table 2 was calculated separately for each subject. The aggregated values, given as minimum, maximum and mean relative drop in the number of detected SSVEPs are presented in Table 3. As it can be noticed in the table the mean drop in the number of SSVEPs detected with any other but the best combination of factors was very large for all subjects (from 31.52 % for subject S3 to 51.76 % for subject S4). It may be also noted that for each subject the greatest losses in the number of detected SSVEPs were observed for combinations employing the first subharmonic (68.97–88.89 %). In contrast, the combinations using the first or second harmonic were responsible for the smallest losses in the number of detected SSVEPs (1.79 – 6.25 %).

5   Conclusion

The results of the study presented in the paper clearly demonstrate a large diversity of the number of detected SSVEPs across subjects in regard to the methodology used for their detection. For each of the subjects who took part in the experiment another combination of factors (channel combination, time window, and number of harmonics) delivered the highest number of recognized stimulation frequencies. As Table 1 presents, it was even difficult to find the common features of the "best" methodology used for succeeding subjects, since each of subjects had different CH-W settings.

Regarding the harmonics used, it can be said only that the worst results were obtained for 1st subharmonic. In the case of the best results, the harmonics also differed.

The final question was whether the difference in the number of detected SSVEP is sufficient to justify the extended scanning phase. In our opinion the answer is "yes", since the mean drop in the number of SSVEPs detected with any other but the best combination of factors was very large for all subjects (from 31.52 % for subject S3 to 51.76 % for subject S4). Hence, we believe that our experiment confirms that it is profitable to perform the individual training sessions for each BCI user not only to identify the stimulation frequencies providing the most prominent SSVEP but also to decide on the subject-specific methodology used for SSVEP detection.

References

1. Fernandez-Vargas, J., Pfaff, H.U., Rodriguez, F.B., Varona, P.: Assisted closed-loop optimization of SSVEP-BCI efficiency. Front. Neural Circ. **7**, 1–5 (2013)
2. Allison, B.Z., McFarland, D.J., Schalk, G., Zheng, S.D., Jackson, M.M., Wolpaw, J.R.: Towards an independent brain–computer interface using steady state visual evoked potentials. Clin. Neurophysiol. **119**(2), 399–408 (2008)
3. Vialatte, F.B., Maurice, M., Dauwels, M., Cichocki, A.: Steady-state visually evoked potentials: focus on essential paradigms and future perspectives. Prog. Neurobiol. **90**(4), 418–438 (2010)
4. Cuffin, B.N.: Effects of local variations in skull and scalp thickness on EEG's and MEG's. IEEE Trans. Biomed. Eng. **40**(1), 42–48 (1993)
5. Luo, A., Sullivan, T.J.: A user-friendly SSVEP-based brain–computer interface using a time-domain classifier. J. Neural Eng. **7**(2), 1–10 (2010)
6. Wu, Z., Su, S.: A dynamic selection method for reference electrode in SSVEP-based BCI. PLoS ONE **9**(8), e104248 (2014)
7. Jasper, H.H.: The ten-twenty electrode system of the international federation in electroencephalography and clinical neurophysiology. EEG J. **10**, 371–375 (1958)
8. Renard, Y., Lotte, F., Gibert, G., Congedo, M., Maby, E., Delannoy, V., Bertrand, O., Lécuyer, A.: OpenViBE: an open-source software platform to design, test and use brain-computer interfaces in real and virtual environments. Presence: Teleoperators Virtual Environ. **19**(1), 35–53 (2010)
9. Paulus, W.: Elektroretinographie (ERG) und visuell evozierte Potenziale (VEP). In: Buchner, H., Noth, J. (eds.) Evozierte Potenziale, neurovegetative Diagnostik, Okulographie: Methodik und klinische Anwendungen, pp. 57–65. Thieme, Stuttgart (2005)
10. Regan, D.: Human brain electrophysiology: evoked potentials and evoked magnetic fields in science and medicine. Elsevier, New York (1989)

S-boxes Cryptographic Properties from a Statistical Angle

Anna Grocholewska-Czuryło[✉]

Poznan University of Technology, Poznan, Poland
anna.grocholewska-czurylo@put.poznan.pl

Abstract. A new, strong S-box was required to be designed to be incorporated as the nonlinear element of the PP-2 block cipher designed recently at Poznan University of Technology. This paper presents statistical analysis of the cryptographic criteria characterizing a group of S-boxes generated by inverse mapping with random search elements. Statistical tests used in this research were not pure randomness checks but were related to the most important real cryptographic criteria like nonlinearity, SAC, collision avoidance.

Keywords: Block ciphers · S-box design · Cryptographic criteria · Statistical testing

1 Introduction

S-box design is usually the most important task while designing a new cipher. This is because an S-box is the only nonlinear element of the cipher upon which the whole cryptographic strength of the cipher depends. New methods of attacks are constantly being developed by researchers, and so S-box design should always be one step ahead of those pursuits to ensure cipher's security.

Recently Institute of Control and Information Engineering at Poznań University of Technology proposed three cryptographic ciphers – PP-1, HAF and PP-2 [1–4]. In the process of designing those ciphers a considerable amount of research went into design of the core nonlinear component of the cipher - the S-box. Different requirements of each of those three ciphers have led to an extensive research of S-boxes possessing various properties. The latest S-box designs were the ones for PP-2 cipher and HAF hash function [5, 6].

Cryptographically strong S-box should possess some properties that are universally agreed upon among researchers. Such S-box should be balanced, highly nonlinear, have lowest maximum value in its XOR profile (difference distribution table), have complex algebraic description (esp. it should be of higher degree). They should also have good avalanche characteristics (fulfill Strict Avalanche Criterion). And in case they are used for hash function implementation, they should support collision resistance property of the hash function. Most of the above criteria are dictated by linear and differential cryptanalysis and algebraic attacks.

It is a well-known fact, that S-boxes generated using finite field inversion mapping fulfil these criteria to a very high extent. They are however susceptible to (theoretical)

© Springer International Publishing AG 2017
S. Kobayashi et al. (eds.), *Hard and Soft Computing for Artificial Intelligence, Multimedia and Security*, Advances in Intelligent Systems and Computing 534, DOI 10.1007/978-3-319-48429-7_13

algebraic attacks. To resist algebraic attacks multiplicative inverse mapping used to construct an S-box is composed with an additional invertible affine transformation. This affine transformation does not affect nonlinearity of the S-box, its XOR profile nor its algebraic degree. The best know example of such an S-box is the S-box of AES. It has been publicly known which does not affect its security.

It is generally agreed among researchers that an S-box mapping should ideally be indistinguishable from a perfectly random mapping. In this article we concentrate on and investigate this property using statistics. The use of statistical S-box testing is not a new idea [8] and it has been applied also to AES candidates by the NIST institute [10]. So passing statistical randomness testing is considered one of the most basic properties expected from block ciphers.

In a research for this paper, a package of statistical tests has been implemented as it was applied to the AES finalist in [8]. In author's view such statistical tests based on cryptographic criteria are better suited to cryptographic applications than NIST tests, which are rather pure generic randomness checks. Paper gives detailed test results in form of graphs, compared to a set of random S-boxes as well as the AES S-box.

2   S-boxes Generation

For the purpose of this article, let's call AES-like S-boxes the S-boxes generated using exactly the same inverse mapping procedure that has been used to construct an AES S-box, however with randomly chosen irreducible polynomial defining a Galois Field and random value of c_i in the affine transform. Such S-boxes constitute the base for generating the S-boxes required by PP–1, PP–2 and HAF ciphers.

To generate an inverse mapping in Galois Field (GF(2n)) an irreducible polynomial is needed that defines a Galois Field, and another polynomial that would be a so called generator. In the algorithm presented in this paper, an irreducible polynomial for an S-box being generated is selected in first step at random. Tables of all existing irreducible polynomials for all values of n ranging from 8 to 16 can be precomputed so that finding an irreducible polynomial and a corresponding generator at random is very time efficient. This inverse mapping is different for every irreducible polynomial selected (randomly) for a particular S-box. It doesn't depend on a selected generator - all generators would give the same result for a given irreducible polynomial. Generators are then not taken at random. A simplest generator for a given irreducible polynomial is always selected.

S-boxes based on multiplicative inverse in a finite field have such a peculiar property that all component functions of the S-box are from the same affine equivalence class (all the output functions of the S-box can be mapped onto one another using affine transformations). S-box presented in this paper have been processed to remove this linear redundancy, so that all Boolean functions are from different affine equivalence classes, while still maintaining exceptionally high nonlinearity of inverse mapping.

3 PP-2 Cipher Statistical Testing

Statistical tests of the PP-2 cipher have been based on 4 cryptographic criteria:

1. Strict Avalanche Criterion - SAC, which states that changing a single bit of the cipher input should result in changing every output bit with the probability of 0.5. SAC criterion fulfillment is tested with this procedure.
2. Nonlinearity – a distance of a Boolean function from all affine functions. As already mentioned, that distance should be as large as possible. This statistical test checks for correlation between outputs resulting from highly linearly dependent inputs.
3. Collision avoidance – finding two different inputs that result in the same output should be difficult. Test checks for collisions in subsets of outputs resulting from random inputs.
4. Block ciphers and hash functions should behave as random mapping. This test checks the number of different outputs.

3.1 SAC

As it has been already mentioned, Strict Avalanche Criterion states that for any S-box a change of single input bit should result in change of every output bit with the probability of 0.5. Extending that definition for the entire block cipher the following procedure tests the effect of single input bit change on the cipher output.

In test implementation of the PP-2 cipher, adapted especially for the purpose of these statistical tests, the size of input and output blocks is 64 bits.

Test:

1. Create a 64 × 64 SAC matrix and zero all
2. Encrypt a random 64-bit input value with the cipher and store the cipher output
3. In the random input value from step 2 change the bit i and calculate cipher output again then XOR this output with the output from step
4. For each non-zero output bit j increment SAC table at position (i, j) by one
5. Repeat steps 3 and 4 for all input bits
6. Repeat the whole process (steps 2–5) for 2^{20} times, each time generating a new random input

Table 1. SAC test bins

Bin	Value range	Probability
1.	0–523857	0,200224
2.	523858–524158	0,199937
3.	524159–524417	0,199677
4.	524418–524718	0,199937
5.	524719–1048576	0,200224

Table 2. SAC test expected distribution

Bin	1	2	3	4	5
Probability	0,200224	0,199937	0,199677	0,199937	0,200224
Expected number	820	819	818	819	820

If SAC criterion is fulfilled we should expect that each output bit will change 2^{19} times and the distribution of SAC table values will be binomial with the following probabilities of finding values from the five ranges (bins) (Table 1):

Based on Chi-Squared test one can determine if the actual distribution is the expected one, as given above. We assume that for p values below 0,01 the distribution is not random.

For input and output sizes of 64 bits, the SAC table has 64*64 = 4096 entries. So the expected distribution of values in each bin is the following (Table 2):

Ciphers tested:

- AES cipher
- PP-2 cipher
- PP-2 cipher with a random AES-like S-box
- PP-2 cipher with a random S-box

Tests were performed for reduced number of rounds – 2 to 8 rounds, so that a minimal number of rounds for which the cipher fulfills the SAC criterion can be determined. Results with p values greater than 0.01 have been marked green. Also marked green is the minimal number of rounds for which SAC criterion is fulfilled (Tables 3, 4, 5 and 6).

Table 3. SAC test results – AES cipher

# rounds	RANGE					
	1	2	3	4	5	p
2	1613	160	246	268	1809	0,00000
3	722	725	687	745	1217	0,00000
4	809	792	850	811	834	0,62564
5	817	846	807	803	823	0,84895
6	802	849	789	845	811	0,48609
7	789	862	875	789	781	0,03485
8	852	822	800	830	792	0,59881

Table 4. SAC test results PP-2 cipher

# rounds	RANGE					p
	1	2	3	4	5	
2	3484	131	114	97	270	0,00000
3	2647	467	371	331	280	0,00000
4	1461	728	653	641	613	0,00000
5	873	834	818	815	756	0,06863
6	803	828	813	844	808	0,84060
7	827	856	795	810	808	0,61755
8	812	818	780	833	853	0,49140

Table 5. SAC test results PP-2 cipher with a random AES-like S-box

# rounds	RANGE					p
	1	2	3	4	5	
2	3553	102	114	100	227	0,00000
3	2664	455	372	325	280	0,00000
4	1386	725	686	690	609	0,00000
5	919	828	785	770	794	0,00182
6	817	839	784	803	853	0,46983
7	791	813	820	874	798	0,25248
8	837	772	775	885	827	0,03030

3.2 Linear Span

Nonlinearity is a basic criterion that has to be considered when designing ciphers. The Linear Span test checks the randomness of the block cipher output testing the output as a function of linearly dependent inputs.

Test:

1. Select at random 6 plain text messages and create a dataset of size $m = 2^6 = 64$ comprised of all possible linear combinations of the plain texts.
2. Calculate cipher output for all input data and create a binary matrix of size 64×64.
3. Calculate the rank of the matrix created in step 2 and increment the counter for that rank by one.
4. Repeat steps 1–3 2^{16} times.

Table 6. SAC test results PP-2 cipher with a random S-box

# rounds	RANGE					
	1	2	3	4	5	p
2	3726	122	83	77	88	0,00000
3	2715	470	333	311	267	0,00000
4	1416	715	722	602	641	0,00000
5	953	818	741	828	756	0,00000
6	773	826	830	845	822	0,43945
7	817	779	828	849	823	0,52546
8	836	792	799	846	823	0,63665

Based on the results of Marsaglia and Tsay [10] the probability of a random matrix of size m x m to have a particular rank is known. Their papers state that these probabilities are as follows (Table 7):

Based on Chi-Squared test we can check whether the actual distribution resulting from the test corresponds to the random distribution above. Again, we assume that for values of p below 0.01 indicate that the distribution is not random.

For $2^{16} = 65536$ repetitions we expect the following distribution of values (Table 8):

Performed tests:

- AES cipher
- PP-2 cipher
- PP-2 cipher with AES S-box
- PP-2 cipher with a random S-box

All tested ciphers fulfill Linear Span tests already from round 2. Green color indicates p values greater than 0.01 (Table 9).

Table 7. Matrix rank probability

Rank	m	m − 1	m − 2	≤ m − 3
Probability	0,288788	0,577576	0,12835	0,005286

Table 8. Linear span test expected distribution

Rank:	64	63	62	≤ 61
Probability:	0,288788	0,577576	0,12835	0,005286
Expected value:	18926	37852	8412	346

Table 9. Linear span test results

# rounds	Rank				p	Cipher
	64	63	62	≤ 61		
2	19034	37784	8379	339	0,79895	AES
2	18980	37914	8295	347	0,59644	PP2
2	18659	38187	8323	367	0,02999	PP2 AES
2	18864	37865	8437	370	0,58356	PP2 Random

3.3   Collisions

Resistance to collisions is a very important characteristic when considering hash functions. This criterion states that finding two messages that would result in the same hash should be difficult. Even though this criterion is relevant main in the context of hash functions we can use it for testing block ciphers as well.

The goal of the test is to count the number of collisions using a subset of output bits.

Test:

1. Select a random input text.
2. Using the input generated in step 1 construct a set of 2^{16} input texts by replacing the lowest 16 bits of the input from step 1 by all possible combinations of the 16 bits.
3. Obtain a ciphertext for every plain text from step 2.
4. Repeat step 3 2^{12} times, storing all output texts.
5. Check the number of collisions of outputs generated in steps 3–4 and increment by 1 the corresponding collision counter
6. Steps 1 to 5 are repeated 2^{16} times.

For 2^{12} random values the probability of number of collisions from a particular range is given in a table below (Table 10):

Based on Chi-Squared test we can check whether the actual distribution resulting from the test corresponds to the random distribution above. Again, we assume that for values of p below 0.01 the distribution is not random.

Table 10. Collision test ranges

Bin	Range	Probability
1.	0–116	0,206246
2.	117–122	0,194005
3.	123–128	0,219834
4.	129–134	0,183968
5.	135–4096	0,195947

Table 11. Collision test expected distribution

Bin:	1	2	3	4	5
Probability:	0,206246	0,194005	0,219834	0,183968	0,195947
Expected number:	845	795	899	754	803

For $2^{12} = 4096$ repetitions the expected value distribution is the following (Table 11):

Performed tests:

- AES cipher
- PP-2 cipher
- PP-2 cipher with random S-box with nonlinearity 88

Green color indicates $p > 0,01$ (Tables 12, 13 and 14).

Table 12. Collision test results for AES cipher

	RANGE					
# rounds	1	2	3	4	5	p
2	4096	0	0	0	0	0,00000
3	1353	546	560	469	1168	0,00000
4	855	795	898	728	820	0,84837
5	780	837	907	764	808	0,11376

Table 13. Collision test results for PP-2 cipher

	RANGE					
# rounds	1	2	3	4	5	p
2	0	0	0	0	4096	0,00000
3	0	0	0	0	4096	0,00000
4	211	323	478	607	2477	0,00000
5	800	807	876	789	824	0,25417
6	890	752	866	785	803	0,12529

Table 14. Collision test results for PP-2 + Random cipher

# rounds	Range					p
	1	2	3	4	5	
2	0	0	0	0	4096	0,00000
3	0	0	0	0	4096	0,00000
4	249	388	545	729	2185	0,00000
5	760	855	874	766	841	0,00335
6	897	744	924	758	773	0,08090
7	817	824	847	776	832	0,15364

3.4   Coverage Test

Coverage test checks f function in respect to the size of the output set resulting from the subset of possible inputs. For a random mapping the expected coverage is 63 % ($1-1/e \approx 0,63212$) of the input set. As it was already mentioned, both block ciphers as well as hash functions should behave as random mappings. In case of a block cipher function f is a full permutation so for all possible inputs the coverage will be 100 %. However, if we only consider a subset of input bits then f function should behave like a random mapping with coverage of 63 %.

Coverage test for block ciphers has been proposed by Turan et al. [11], where authors included probabilities and intervals for coverage ranges in this test, shown in the table below (Table 15):

Coverage test is similar to collision test, but instead of counting collisions (output values occurring more than once), a coverage is counted, i.e. all output values that have occurred.

Table 15. Coverage test ranges

No	Range	Probability
1.	1–2572	0,199176
2.	2573–2584	0,204681
3.	2585–2594	0,197862
4.	2595–2606	0,203232
5.	2607–4096	0,195049

Test:

1. Generate one random input text.
2. Based on the input text from step 1 generate 2^{12} input texts by replacing 12 lower bits of input from step 1 by all possible combinations of 12 bits.

3. Calculate output for random input belonging to set generated in step 2.
4. Repeat step 3 2^{12} times, each time storing the calculated output.
5. Count the number of different output values obtained in steps 3–4 and increment by one counter of the respective range.
6. Repeat steps 1 to 5 2^{12} times.

Based on Chi-Squared test we can check whether the actual distribution resulting from the test corresponds to the random distribution. As in previous tests, we assume that values of p below 0.01 indicate that the distribution is not random.

For $2^{12} = 4096$ repetitions the expected value distribution is the following (Table 16):

Table 16. Coverage test expected distribution

Range:	1	2	3	4	5
Probability:	0,199176	0,204681	0,197862	0,203232	0,195049
Expected:	816	839	810	832	799

Ciphers tested:

- AES cipher
- PP-2 cipher
- PP-2 cipher with random S-box with nonlinearity 88

Green color indicates $p > 0,01$ Tables 17, 18 and 19.

Table 17. Coverage test results for AES cipher

	RANGE					
# rounds	1	2	3	4	5	p
2	2058	0	0	0	2038	0,00000
3	949	798	766	787	796	0,00001
4	787	861	772	841	835	0,27623
5	831	807	851	804	803	0,33855

Table 18. Coverage test results for PP-2 cipher

# rounds	RANGE					p
	1	2	3	4	5	
2	4096	0	0	0	0	0,00000
3	4096	0	0	0	0	0,00000
4	2772	651	359	213	101	0,00000
5	902	864	808	778	744	0,00184
6	837	837	772	879	771	0,20184

Table 19. Coverage test results for PP-2 + Random cipher

# rounds	RANGE					p
	1	2	3	4	5	
2	4096	0	0	0	0	0,00000
3	4096	0	0	0	0	0,00000
4	2413	768	454	313	148	0,00000
5	895	832	763	851	755	0,00994
6	825	842	788	856	785	0,80066

4 Summary

Statistical tests described in this chapter show that both PP-2 and AES block ciphers acquire the required property very quickly (for a small number of rounds) and are statistically indistinguishable from a random mapping. Minimum number of rounds required for AES cipher to achieve this random appearance is 4 and for PP-2 cipher it is 6, so in both cases much lower than the standard minimal number of rounds.

Based on the presented results we can state that the PP-2 cipher with its original S-box is comparable to AES cipher. Two round difference between the two ciphers is offset by PP-2's faster processing (starting with 10–11 round).

PP-2 cipher modifications with AES S-box and a random S-box with nonlinearity of 88 have been tested as well. These tests show that the PP-2 cipher S-box has similar properties to the AES cipher S-box as there were no differences in the number of rounds needed to achieve the required random mapping characteristic. Slightly worse results have been observed when PP-2 S-box has been substituted with a random one – in some of the tests an additional round was needed to achieve the required property (Fig. 1).

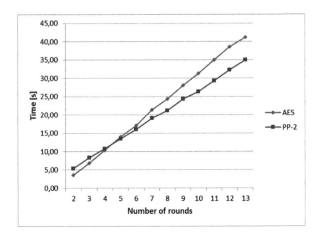

Fig. 1. Execution times for PP-2 and AES ciphers

As part of these tests an encryption time was measured for both AES and PP-2 ciphers as a function of number of rounds. Both implementations were not optimized so results were not definitive. Number of test input texts was 2^19. The graph above summarizes these results.

These measurements show that the PP-2 cipher is faster starting from round 5. Thirteen round PP-2 has performance comparable to 11-round AES.

References

1. Bucholc, K., Chmiel, K., Grocholewska-Czuryło, A., Idzikowska, E., Janicka-Lipska, I., Stokłosa, J.: Scalable PP-1 block cipher. Int. J. Appl. Math. Comput. Sci. **20**(2), 401–411 (2010)
2. Chmiel, K., Grocholewska-Czuryło, A., Stokłosa, J.: Evaluation of PP-1 cipher resistance against differential and linear cryptanalysis in comparison to a DES-like cipher. Fundamenta Informaticae **114**(3–4), 239–269 (2012)
3. Bucholc K., Chmiel K., Grocholewska-Czuryło A., Stokłosa J.: PP-2 block cipher. In: Hof, H.-J., Westphall, C. (eds.) SECURWARE 2013: The Seventh International Conference on Emerging Security Information, Systems and Technologies, Curran Associates, Inc., RedHook, NY, pp.162–168 (2013). http://www.thinkmind.org/download.php?articleid=securware_2013_7_30_30142
4. Bilski, T., Bucholc, K., Grocholewska-Czuryło, A., Stokłosa, J.: Paramiterized hash functions. Ann. UMCS Informatica AI **12**(3), 11–24 (2012). Cryptography and Data Protection
5. Grocholewska-Czuryło, A.: Generating strong S-boxes based on inverse mapping without any linear redundancy. Polish J. Environ. Stud. **16**(5B), 304–330 (2007)
6. Grocholewska-Czuryło, A.: Cryptographic properties of modified AES-like S-boxes. Ann. UMCS Informatica AI **11**(2), 37–48 (2011). Cryptography and data protection
7. Doganaksoy A., Ege B., Kocak O. and Sulak F.: Cryptographic Randomness Testing of Block Ciphers and Hash Functions. http://eprint.iacr.org/2010/564

8. Caelli W., Dawson E., Nielsen N., and Gustafson H.: CRYPT–X statistical package manual, measuring the strength of stream and block ciphers. https://eprint.iacr.org/2010/611
9. Soto, J., Bassham, L.: Randomness testing of the advanced encryption standard finalist candidates. In: NIST IR 6483, National Institute of Standards and Technology (1999)
10. Marsaglia, G., Tsay, L.: Matrices and the structure of random number sequences. Linear Algebra Appl. **67**, 147–156 (2016)
11. Sönmez Turan, M., Çalık, Ç., Saran, N.B., Doğanaksoy, A.: New distinguishers based on random mappings against stream ciphers. In: Golomb, S.W., Parker, M.G., Pott, A., Winterhof, A. (eds.) SETA 2008. LNCS, vol. 5203, pp. 30–41. Springer, Heidelberg (2008)

Wavelet Transform in Detection of the Subject Specific Frequencies for SSVEP-Based BCI

Izabela Rejer[✉]

Faculty of Computer Science and Information Technology, West Pomeranian University of Technology in Szczecin, Szczecin, Poland
irejer@wi.zut.edu.pl

Abstract. One of the paradigms often used to build a brain computer interface (BCI) is a paradigm based on steady state visually evoked potentials (SSVEPs). In SSVEP-based BCI a user is stimulated with a set of light sources flickering with different frequencies. In order to ensure the best performance of the interface built according to this paradigm, the stimulation frequencies should be chosen individually for each user. Usually, during the frequency scanning phase the user-specific stimulation frequencies are chosen according to the power of the corresponding SSVEPs. However, not only the power should be taken into account when choosing the stimulation frequencies. The second very important factor is the time needed to develop the prominent SSVEP. The wavelet transform (WT) seems to be an excellent tool for dealing with this task, since it provides not only the information about the frequency components represented in the signal but also about the time of their occurrence. The aim of this paper is to present a procedure, based on WT, that can be used to determine the user-specific frequencies with respect to the synchronization time and its strength.

Keywords: BCI · SSVEP · Brain computer interface · Wavelet transform · CWT

1 Introduction

A brain computer interface is a system that records user's brain activity and then employs it to control external devices or applications. In order to ensure the stability of the control process, the brain activity patterns that are transformed to commands sent to the outside world should be invariant in time. The problem is, however, that the brain activity is not stable but varies across subjects and also across different conditions for the same subject [1, 2]. That is why the first step to a successful application of a brain computer interface is to adopt it to the brain activity patterns specific for a given subject and to keep readopting it whenever the control accuracy drops below the given level.

One of the paradigm often used in brain computer-interfaces is a paradigm based on evoking steady state visual potentials (SSVEPs) [3–6]. With this paradigm a user is presented with a set of flickering objects. The objects are displayed directly on a computer screen or they are delivered by external light sources e.g. Light Emitting Diodes (LEDs). Each object flickers with a different frequency. As neuroscience experiments have shown, neurons in human visual cortex synchronize their firing rate

© Springer International Publishing AG 2017
S. Kobayashi et al. (eds.), *Hard and Soft Computing for Artificial Intelligence, Multimedia and Security*, Advances in Intelligent Systems and Computing 534, DOI 10.1007/978-3-319-48429-7_14

with the frequency of flickering light, leading to EEG responses which show the same frequency as the flickering stimulus [7, 8]. Hence, in a short time after observing one of the flickering objects, the flickering frequency is reflected in the user's visual cortex and can be observed in EEG signals recorded from the occipital regions. Each flickering frequency is associated with one control signal (one command). This control signal is passed to the application/device controlled by the interface when the corresponding frequency is detected. Summing up, the scheme of the whole process is as follows: the user chooses one of the flickering objects, observes it, the flickering frequency of this object is detected in the EEG signal, and the command assigned to this frequency is executed (Fig. 1).

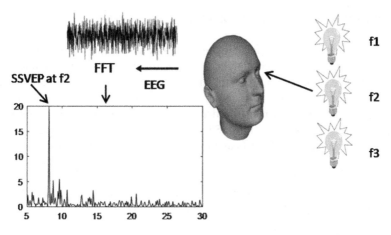

Fig. 1. The scheme of SSVEP-based BCI

The scheme of SSVEP-based BCI seems to be very straightforward but it has one pitfall - not all frequencies are possible to detect in a brain activity of different subjects. Therefore, before the interface is used in everyday routine, first, the flickering frequencies specific for the given subject have to be determined. Usually, the "best" frequencies are determined according to their amplitudes and their dominance over other frequencies. However, not only the amplitude should be taken into account when looking for user-specific frequencies. The other very important parameter is the time required to detect the given frequency in a user brain activity. The short synchronization time is crucial for a successful BCI implementation because it is directly responsible for the speed of the whole interface. If the interface needs 10 s for taking a decision which of the flickering objects is observed by the user, only 6 commands per minute can be executed, but if 2 s are enough for one decision, the interface speed rises to 30 commands per minute.

The aim of this paper is to present an approach that can be used to determine the user-specific frequencies with respect to three factors: the time needed to detect the stimulation frequency, the length of the time interval where the synchronization takes place, and the similarity between the signal and the stimulation frequency. The proposed approach is based on the wavelet analysis that seems to be a suitable tool for dealing with this task, since it allows not only to determine which frequency components are present in the signal in succeeding time intervals, but also to answer the question which frequency components can be detected quicker than the others.

The rest of the paper is organized as follows. Section 2 gives the theoretical background to the rest of the paper and reminds some facts about the wavelet analysis. Section 3 describes the proposed procedure for selecting the user-specific frequencies. Section 4 presents the setup of the experiment that was performed to illustrate the proposed approach and its potential benefits. Section 5 is focused on the results of the experiment and their analysis. Finally, Sect. 6 concludes the paper.

2 Continuous Wavelet Transform

The most popular method to calculate the frequency spectrum of a given time signal is Fourier Transform (FT). After applying FT over time signal, a set of frequency components with the information about the their strength is obtained. The classic form of Fourier Transform has one shortcoming - it does not allow for time localization of detected frequency components. In order to overcome this shortcoming the variation called Short Time Fourier Transform (STFT) can be applied. With this method the time signal is divided into time windows and FT is applied separately per each window. This method provides a way to detect different frequencies in succeeding time windows but the time resolution is rather small and is the same for each frequency.

Another approach to time localization of frequency components provides continuous wavelet transform (CWT) [9], defined as follows:

$$f_\psi(a,b) = \frac{1}{\sqrt{a}} \int\limits_{-\infty}^{\infty} x(t)\psi\left[\frac{t-b}{a}\right] dt \tag{1}$$

where: a is the scale factor, b is the translation factor, $\psi(t)$ is the transformation function, the so-called "Mother Wavelet", and $x(t)$ is the time signal. The family of wavelets is generated from the base wavelet by scaling and translation, it is by changing a and b factors. The transformation factors, a and b, define the time-frequency resolution of CWT. While STFT has a constant resolution at all times and frequencies (Fig. 2(a), CWT has a good time and poor frequency resolution at high frequencies, and good frequency and poor time resolution at low frequencies (Fig. 2(b). As it can be noticed in Fig. 2b each field in the time-frequency plane has the same area but different proportions in time and frequency domains. At lower frequencies the height of the boxes are shorter (which corresponds to a better frequency resolution) but their widths are longer (which corresponds to a worse time resolution). At higher frequencies the boxes become narrower and higher which means that time resolution becomes better while frequency resolution becomes worse [10].

The outcome of the wavelet transform is a set of coefficients, each determines the similarity of a chosen part of a signal and a wavelet of a given a and b value.

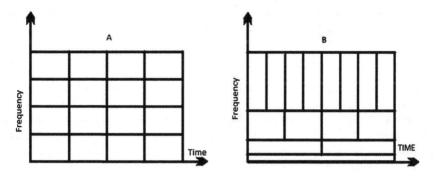

Fig. 2. Resolution in the time frequency for: (A) short time fourier transform; (B) wavelet transform

3 User-Specific Frequency Selection

The selection procedure proposed in this paper is based on the continuous wavelet transform, described in Sect. 2. To be more specific, it is based on three pieces of information that can be derived from the results of wavelet transform - the time of the first synchronization at the stimulation frequency, length of this synchronization, and its strength. The word "synchronization" is used here to point that the stimulation frequency was detected in the signal and was present in it by a given amount of time.

In the first step of the proposed procedure EEG signals corresponding to all stimulation frequencies are processed separately. The signal recorded with each stimulation frequency is processed according to the following scheme. First, the signal is transformed with wavelet transform. After applying CWT over the signal, a two-dimensional matrix of wavelet coefficients is created. The first dimension of this matrix denotes the time changes, the second - the frequency (corresponding to scale) changes. Hence, the size of wavelet coefficient matrix is $N \times M$, where N - no. of time samples, M - no. of frequencies used in wavelet analysis. In order to derived the information about the time, length, and strength of the first synchronization, the wavelet coefficients matrix is transformed to the form of a one dimensional vector *Stim*, informing in which time samples the stimulation frequency (f) was a dominant frequency. In order to build this vector, first the maximum values (*CWT_max*) of wavelet coefficients (*cwt*) in each time moment (i) are calculated ($i = 1...M$).

$$\text{CWT\_max}_i = \{\text{cwt}_{i1}, \dots \text{cwt}_{iN}\} \qquad (2)$$

Next, the vector of frequencies *Freq_max* corresponding to *CWT_max* vector is created. The vector *Freq_max* is of a size *1xM* and contains frequencies corresponding to the maximum wavelet coefficients found in each time sample (f_{peak}). In order to change the vector *Freq_max* to the vector *Stim*, binary values (0 or 1) are assigned to each frequency from *Freq_max*. Value 1 is assigned to those f_{peak} that are recognized as stimulation frequency, value 0 is assigned to the rest f_{peak}. Since the frequencies corresponding to scales in the wavelet transform are usually not exactly the same as stimulation frequencies, in order to recognize f_{peak} as the stimulation frequency, a small buffer around the stimulation frequency has to be created. Hence, f_{peak} is recognized as the stimulation frequency if the following formula is fulfilled:

$$f - b \geq f_{peak} \leq f + b \tag{3}$$

where: f_{peak} – the frequency corresponding to cwt coefficient of a maximum value, f - the analyzed stimulation frequency, b - buffer around the stimulation frequency.

With the binary *Stim* vector and *CWT_max* vector all the information about the time, length and strength of the first synchronization at the processed stimulation frequency can be directly obtained. It should be underlined here once again that in order to avoid accidental synchronizations, the stimulation frequency has to dominate in the signal during a given time to acknowledge that a synchronization took place. Hence, the phrase "first synchronization" does not point the first moment when cwt coefficient was maximal at the stimulation frequency, but the interval where all cwt coefficients were maximal at that frequency.

Two first parameters that are taken into account in the ranking ordering all the stimulation frequencies provided in the survey – time of the first synchronization (T_{FIRST}) and its length (L_{FIRST}) – are straightforward, T_{FIRST} informs when the synchronization has started, and L_{FIRST} informs how long it lasted. The last parameter, the strength of the first synchronization (S_{FIRST}) needs some additional explanation. S_{FIRST} value is based directly on the wavelet coefficients. Wavelet coefficients inform about the similarity of a given part of a signal and a wave of a given frequency. Hence, the higher value of a wavelet coefficient, the higher synchronization of a signal on the frequency of the analyzed wave. Therefore, to calculate the strength of the first synchronization, the absolute values of wavelet coefficients corresponding to the analyzed stimulation frequency, stored in vector *CWT_max,* are averaged. Once again, the vector *Stim* is used to decide which coefficients should be taken into account in the averaging process.

The information about the time, length and strength of the first synchronization are calculated for each stimulation frequency used in the survey. Next, the stimulation frequencies are ordered separately according to the strength criterion (in descending order) and the joined time-length criterion (in ascending order along the time dimension and in descending order along the length dimension).

Finally, the ranks from both rankings are summed up across stimulation frequencies with the same weight for both criteria. Next, the frequencies are ordered once again, this time in an ascending order. The top frequencies are those which should be chosen for the on-line BCI sessions.

4   Experiment Setup

In order to verify the procedure described in previous section, an experiment with a human subject was performed. The subject was a man, aged 29, right-handed with a normal vision and without any previous mental disorders. Before the experiment started, the subject was fully informed about the experiment and signed the informed consent form. The experiment was conducted according to the Helsinki declaration on proper treatments of human subjects.

The subject was placed in a comfortable chair in front of an Arduino board with a red LED attached to it. In order to limit the number of artifacts, the participant was instructed to stay relaxed and not move. The only task of the subject was to observe LED that was flickering with a given frequency. The set of analyzed stimulation frequencies covered 85 frequencies from the band 5–30 Hz obtained from the formula: $f = 1000/2/t$, where f is the flickering frequency, 1000 stand for 1000 ms, 2 is the number of states (on/off) and t is the time of one state.

During the experiment EEG data was recorded from 2 monopolar channels at a sampling frequency of 256 Hz with a Discovery20 amplifier (BrainMaster). The signal (passive) electrode was attached to the subject's scalp at O2 and Pz position according to the International 10–20 system [11]. The reference electrode was placed at the left mastoid and the ground electrode at Fz. The impedance of the electrodes was controlled with BrainMaster Discovery software and was kept below 5 kΩ. The EEG data were recorded with OpenVibe software [12]. The output EEG data set was composed of 85 signals each of 2560 observations (10 s). EEG data from O2 was rereferenced to Pz and filtered with a Butterworth band-pass filter of the 6th order in the band 3–33Hz.

Then, the wavelet transform was performed with Morlet wavelet (N – equal to 2560, M equal to 37). The scales (a parameter) from 6 to 42 were used, and hence the set of frequencies covered the frequencies from the band: 4.95 Hz – 34.67 Hz.

5   Results

In order to find the user-specific frequencies, the procedure described in Sect. 3 was applied to all 85 EEG signals gathered during the experiment. After calculating the values of T_{FIRST}, L_{FIRST}, and S_{FIRST} parameters all frequencies that did not meet the criterion of the sufficient length of the first synchronization interval were discarded from the further analysis. It was assumed that in order to acknowledge that a stimulation frequency was found, it had to be constantly present in the signal by at least 100 ms. From the initial 85 stimulation frequencies provided to the user, only 30 remained after this step.

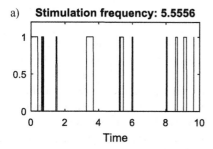
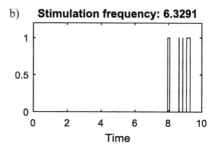

Fig. 3. The best (a) and the worst (b) stimulation frequency according to the time and length of the first synchronization.

The stimulation frequencies that survived the first selection, were ordered according to the two criteria described in the third section (Fig. 3). Table 1 presents the first ranking, where the frequencies were ordered first according to the increasing time of the first appearance of the stimulation frequency and then according to the decreasing length of the synchronization interval. The table contains 10 stimulation frequencies evoking the quickest and longest brain reaction, and also 3 stimulation frequencies that needed the largest amount of time for developing a stable brain response. The first, and also all next, synchronizations obtained for EEG signal recorded during the stimulation at the two extreme stimulation frequencies (of the quickest and the slowest synchronization) are additionally illustrated in Fig. 4 (a – the stimulation frequency of the quickest synchronization, b – the stimulation frequency of the slowest synchronization).

Table 1. The stimulation frequencies evoking the quickest and the longest brain response (10 first places) and the stimulation frequencies evoking the slowest and the shortest brain response (3 last places). The last column S_{FIRST} was added for further analysis.

Rank	Stimulation frequency (f)	The first synchronization start time T_{FIRST} (s)	The first synchronization end time (s)	The first synchronization length (ms)	S_{FIRST}
1	5.56	0.00	0.41	410.16	2675.53
2	5.21	0.00	0.23	230.47	247.51
3	9.80	0.00	0.20	203.13	59.23
4	5.49	0.07	0.27	203.13	66.74
5	5.62	0.10	0.31	214.84	221.62
6	5.75	0.15	0.67	523.44	206.14
7	8.47	0.30	0.54	238.28	213.91
8	7.35	0.39	0.52	128.91	121.94
9	5.05	0.45	0.56	117.19	211.16
10	6.10	0.48	0.87	386.72	91.52
...	...	...	...	...	...
28	8.20	2.56	2.78	218.75	253.19
29	10.20	5.62	5.73	109.38	284.23
30	6.33	8.23	8.36	125.00	20.00

The second ranking, built according to the strength criterion is presented in Table 2. As with the first ranking, the EEG signal recorded during the stimulation at the stimulation frequencies providing the strongest and the weakest response are presented in the figure (Fig. 4), this time via the charts presenting the wavelets coefficients obtained for different a and b parameters (for convenience the y axis was transformed to the frequency domain). Figure 4a presents the wavelet coefficients obtained for the stimulation frequency evoking the strongest brain response, Fig. 4b - the stimulation frequency providing the weakest brain response. For better visualization only this part of the whole chart is presented, where the first synchronization took place (the first two seconds for the stimulation frequency 5.56, and the last two seconds for the stimulation frequency 6.33).

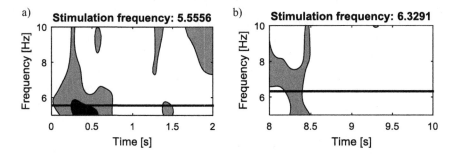

Fig. 4. The wavelet transform of the EEG signal recorded during the light stimulation at the strongest (a) and the worst (b) stimulation frequency (according to the mean strength of the first synchronization).

At the final stage of the survey, the rankings from Tables 1 and 2 were gathered together to build the final ranking of the 30 stimulation frequencies that evoked the synchronization lasting at least 100 ms. In order to get the final rank for each stimulation frequency, both partial ranks were summed up. Next, the frequencies were ordered in an ascending order according to the increasing sum value. The results of the final ranking for 10 best and 3 worst frequencies are presented in Table 3.

Table 2. The stimulation frequencies evoking the strongest brain response (10 first places) and the stimulation frequencies evoking the weakest brain response (3 last places). The last column T_{FIRST} was added for further analysis.

Rank	Stimulation frequency (f)	Mean strength of the first synchronization S_{FIRST}	T_{FIRST} (s)
1	5.56	2675.53	0.00
2	7.81	2227.31	2.07
3	5.88	1924.98	0.74
4	6.94	1107.33	1.29
5	5.81	523.81	1.55
6	7.94	292.81	0.79
7	10.20	284.23	5.62
8	8.20	253.19	2.56
9	9.80	247.51	0.00
10	5.62	221.62	0.10
...	...	...	...
28	10.00	29.48	1.54
29	5.00	26.37	1.28
30	6.33	20.00	8.23

Table 3. The final ranking of 30 chosen stimulation frequencies.

Final rank	Stimulation frequency (f)	$T_{FIRST}+L_{FIRST}$ rank	S_{FIRST} rank	T_{FIRST}	S_{FIRST}
1	5.56	1	1	0.00	2675.53
2	9.80	3	9	0.00	247.51
3	5.62	5	10	0.10	221.62
4	5.88	15	3	0.74	1924.98
5	8.47	7	12	0.30	213.91
6	5.75	6	14	0.15	206.14
7	5.05	9	13	0.45	211.16
8	7.94	16	6	0.79	292.81
9	6.94	21	4	1.29	1107.33
10	5.21	2	25	0.00	59.23
...	...	...	...	...	...
28	6.17	27	22	2.11	112.68
29	10.00	23	28	1.54	29.48
30	6.33	30	30	8.23	20.00

6   Conclusion

Analyzing the outcome of both rankings from Tables 1 and 2, it can be noticed that they are not similar at all. While the first and the last place in both rankings were assigned to the same frequencies, all other ranks are mixed all together across the rankings. For example the frequency that was given the second place in the $T_{FIRST}+L_{FIRST}$ ranking (Table 1), in the S_{FIRST} is not present at all among 10 top frequencies. Just the same, the frequency that was given the second place in the S_{FIRST} ranking (Table 2), is not present in ranking $T_{FIRST}+L_{FIRST}$.

Now, the question arises which ranking should be taken into account if the user that participated in the experiment described in the paper wanted to take part in an online BCI session (for example a session with the interface whose task is to deliver 10 different control commands). If the ranking built according to $T_{FIRST}+L_{FIRST}$ criterion was taken into account, the interface could be quicker (the average T_{FIRST} over 10 top stimulation frequencies from Table 1 – 0.194 s) but more prone to errors (the average S_{FIRST} over 10 top stimulation frequencies from Table 1 – 411.53). On the other hand, if the ranking built according to S_{FIRST} was chosen, the interface could be slower (the average T_{FIRST} from Table 2 – 1.62 s) but more reliable (the average S_{FIRST} from Table 2 – 1059.63).

The final ranking (Table 3), seems to provide a very good compromise between the time and the strength criterion. The top 10 stimulation frequencies from this ranking promise a short synchronization time, only slightly longer than those from T_{FIRST} ranking (average – 0.38 s), and at the same time a sufficiently strong response (average – 716).

References

1. Allison, B.Z., McFarland, D.J., Schalk, G., Zheng, S.D., Jackson, M.M., Wolpaw, J.R.: Towards an independent brain–computer interface using steady state visual evoked potentials. Clin. Neurophysiol. **119**(2), 399–408 (2008)
2. Allison, B., Lüth, T., Valbuena, D., Teymourian, A., Volosyak, I., Gräser, A.: BCI demographics: how many (and what kinds of) people can use an SSVEP BCI? IEEE Trans. Neural Syst. Rehabil. Eng. **18**(2), 107–116 (2010)
3. Cao, T., Wan, F., Mak, P.U., Mak, P.I., Vai, M.I., Hu, Y.: Flashing color on the performance of SSVEP-based brain-computer interfaces. In: Engineering in Medicine and Biology Society (EMBC), Annual International Conference of the IEEE, pp. 1819–1822 (2012)
4. Luo, A., Sullivan, T.J.: A user-friendly SSVEP-based brain–computer interface using a time-domain classifier. J. Neural Eng. **7**(2), 1–10 (2010)
5. Wu, Z., Su, S.: A dynamic selection method for reference electrode in SSVEP-based BCI. PLoS ONE **9**(8), e104248 (2014)
6. RJMG, Tello, Müllerx, S.M.T., Ferreira, A., Bastos, T.F.: Comparison of the influence of stimuli color on steady-state visual evoked potentials. Res. Biomed. Eng. **31**(3), 218–231 (2015)
7. Regan, D.: Human Brain Electrophysiology: Evoked Potentials and Evoked Magnetic Fields in Science and Medicine. Elsevier, New York (1989)
8. Herrmann, C.S.: Human EEG responses to 1–100 Hz flicker: resonance phenomena in visual cortex and their potential correlation to cognitive phenomena. Exp. Brain Res. **137**(3–4), 46–53 (2001)
9. Young, R.K.: Wavelet Theory and Its Applications. Kluwer Academic Publishers, Boston (1993)
10. Polikar, R.: http://users.rowan.edu/~polikar/WAVELETS/WTtutorial.html
11. Jasper, H.H.: The ten-twenty electrode system of the international federation in electroencephalography and clinical neurophysiology. EEG J. **10**, 371–375 (1958)
12. Renard, Y., Lotte, F., Gibert, G., Congedo, M., Maby, E., Delannoy, V., Bertrand, O., Lécuyer, A.: OpenViBE: an open-source software platform to design, test and use brain-computer interfaces in real and virtual environments. Teleoperators Virtual Environ. **19**(1), 35–53 (2010)

Data Scheme Conversion Proposal for Information Security Monitoring Systems

Tomasz Klasa[1(✉)] and Imed El Fray[2]

[1] West Pomeranian Business School, Szczecin, Poland
tklasa@zpsb.pl
[2] West Pomeranian University of Technology, Szczecin, Poland
ielfray@wi.zut.edu.pl

Abstract. Information security monitoring in a highly distributed environment requires gathering and processing data describing state of its components. To allow successful interpretation of that data, they cannot be acquired in any form – numerous meta languages and description schemes are available, but usually only one or few of them is supported by a given data source. A set of those schemes supported by a given device or program is defined by its manufacturer, and because of utilization of proprietary formats, usually it is impossible to apply a single scheme to all data sources. As a consequence, it is necessary to apply data conversion scheme, transforming various incompatible messages to a chosen data scheme, supported by the main repository and the analytic sub-system. Only then it is possible to process data to determine the current state of security of the whole information system.

Keywords: Information security · Security monitoring · Meta-language conversion

1 Introduction

Simple monitoring of a single or few parameters for a single asset can be done locally, with the help of a hardware meter or software tool integrated with that asset. For example, computer network monitoring can be done with the help of functions provided by network's active elements [23], or from a host connected to the network [24]. Also in the case of the integrated systems, despite high level of sophistication, it is possible to apply such a mechanism of control, as is shown e.g. by the SAP R/3. Thanks to the centralized layer of logic, hundreds of events and metrics can be tracked no matter of their real physical location of the user and without the need to send data to external [29].

At the same time, following multiple data sources to conclude about the state of security of the system as a whole, requires data exchange and aggregated processing. In such a case, two scenarios are possible:

- read parameter value and postpone it to the central repository for further evaluation, processing and concluding,

© Springer International Publishing AG 2017
S. Kobayashi et al. (eds.), *Hard and Soft Computing for Artificial Intelligence, Multimedia and Security*, Advances in Intelligent Systems and Computing 534, DOI 10.1007/978-3-319-48429-7_15

- read and evaluate parameter value and then send results to the central repository for further processing and concluding

In both cases above it is required to provide communication between many different data sources and a system component responsible for data processing and concluding about current state of security of the whole system. An information system, which consists of an IT system and organizational components, data sources may differ significantly. Moreover, a distributed structure of the system forces usage of software or hardware agents. What is important, as each of the system components may use its own communication scheme or meta-language, direct communication of such components with reasoning subsystem may be impossible – data sent will be useless, unless a proper way of its interpretation is implemented.

2   Related Work

Information security monitoring is a subject of many research projects - usually focused on a chosen functional area of the information system, e.g.:

- computer networks [5, 16, 21]
- ERP (ERP – Enterprise Resource Planning) class integrated systems [7] or supply chain management systems [8, 19]
- systems in SaaS model (SaaS – Software as a Service) or based on SOA (SOA – Service Oriented Architecture) [2, 25],
- physical and environmental security [27].

As a consequence, numerous solutions are developed, e.g. ConSpec [1], WSDL, OWL-S, BPEL [3, 11], SSDL [15], FAL [30]. Among the others one can point out AVDL, OVAL, WAS, SIML, SDML, VulnXML, CVE, XCCDF [17, 20]. Although they are more adjusted to specific functional areas, it is still necessary to consider dependencies between them [13] and to provide standardized communication mechanisms. Except of the mentioned languages, there are many other meta data schemes. They include formats and languages native for a given family of operating systems, manufacturer of network devices, or specific software, e.g. [18]. A significant challenge arises, when a portfolio of utilized software and hardware becomes diversified. Each vendor or even product family may support a different set of event logs schemes and different access policies. On the basis of a comparative analysis it can be said, that none of the existing solutions is universal enough, to be used in the whole information system (including IT system and organizational areas) [12]. As a consequence, one of the two solutions should be considered:

- definition of a generalized description scheme, applicable to any part of the information system,
- integration of chosen existing solutions to provide support to any part of the information system.

As most of existing and available solutions do not provide support for an external format of exported data, the first of the two solutions above does not solve the problem.

It is necessary to find a way of integration of many existing schemes and languages, supported by applied data sources. A typical approach to this is forcing agents or intelligent sensors not only to acquire data, but also to convert them into a chosen format [6, 10, 26, 28]. What is important, this problem is not specific to monitoring systems. A similar issue is fundamental during any data import to an integrated system, e.g. ERP-class, from other systems [22]. During a one-time migration or implementation of a permanent interface between two systems it is usually a necessity to convert data from one to some other scheme or form.

3   Background

Based on the literature analysis, as a part of the research under reference model of integrated information security monitoring system for virtual organization, a following functional model was created (Fig. 1). Numerous and diversified components of the system stand for sources of data about the state and activity of the system.

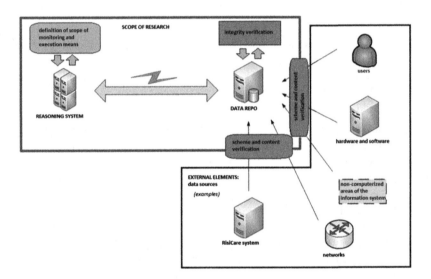

Fig. 1. Functional model of integrated information security monitoring system.

Because of possible significant differences between system components, it is necessary to ensure that transmitted values are exchanged in a form understandable for the integrated monitoring system. This requires conversion mechanism, from data source (agent, metering device, form) native format to data structures used by the repository. At the same time, it is necessary to remember about mechanisms that provide content and scheme verification, to reduce the risk of misinterpretation of received messages.

4   Data Transmission Scheme Conversion Proposal

While in risk assessment it is common that assets of similar functional characteristic or utilization are grouped [9], in the case of security monitoring such grouping becomes more complicated. One of the reasons for that is adoption of non-standardized communication protocols or meta languages by software and hardware vendors. This means that it is often, that grouping protection mechanisms that belong to the same functional family or even group of functional components (according to classification from ISO15408) is impossible. For instance, firewalls – vendors and even product lines offer different set of functions. Moreover, data describing state of system components may be acquired in pure or preprocessed form, which also depends on capabilities of the data source.

There are multiple meta-languages and data schemes. They can be divided into two subsets:

- a set of meta-languages of input data received from data sources (sensors, devices, system components, etc.):

$$\text{MLSet}_{IN} = \{l_1, l_2, l_3, \cdots, l_n\} \tag{1}$$

- a set of meta-languages accepted by the reasoning subsystem, including data repository:

$$\text{MLSet}_{RS} = \{l_1, l_2, l_3, \cdots, l_k\} \tag{2}$$

Input data description languages, that belong to set MLSet_{IN}, are used in data sources – real devices and programs gathering or generating data describing state of some system component. This data can be acquired by dedicated sensors (e.g. temperature, pressure, humidity) or agents, communicating directly with a chosen hardware or software asset. They are then accessible by a dedicated API, gathered in a local database, or even processed by local anomaly detection algorithms.

Due to the number and diversity of languages belonging to set MLSet_{IN}, it is not possible to build a system that processes data regardless of the native language. It is necessary to define a finite set of acceptable languages MLSet_{RS}, with a significantly less members than of a set MLSet_{IN} and convert data from language used by data source to one of the accepted languages. To achieve that, an agent acquiring data to the repository should:

- know, from which real data sources (assets) gather data
- know, which language (description scheme) is used by the data source
- know, which data form (language) is expected by the repository
- verify correctness of data structure before conversion

Top three positions from the list above can be received by the agent in the form of request and order documents, defined in data structure model for the system [14]. In such a case, what remains is to define a way of structure verification, as a conversion prerequisite. Checking obtained data structure can be done by comparison of received sequence of data with a pattern (scheme) of a specified description format. This requires that this scheme contains definition of all data fields and their correct sequence. That validation must be, however, limited to the type of exchanged data and range of their values. The agent should react, if a field meant to contain a number, in fact contains text, or if date has improper format. In more sophisticated form, text fields can be controlled on the basis of content patterns, e.g. to validate document number structures (invoices, orders). Only when received message is confirmed to have a proper structure, conversion itself can be started.

Each field, no matter of the language used, has its meaning – context. For instance, a sequence '12052016' may be a date, ID, amount of money – a correct interpretation is possible only on the basis of the field's meaning. So, conversion from language $l_a \in$ MLSet$_{IN}$ to language $l_b \in$ MLSet$_{RS}$ means that for each field from language $l_b \in$ MLSet$_{RS}$ it is necessary to identify a field from language $l_a \in$ MLSet$_{IN}$ of an analogic meaning. Optional fields may have empty relation between $l_a \in$ MLSet$_{IN}$, and $l_b \in$ MLSet$_{RS}$. In such a case such a field in language $l_b \in$ MLSet$_{RS}$ remains unfilled. This may lead to partial loss of data, only in optional fields.

As a result, each language $l \in$ MLSet$_{IN} \cup$ MLSet$_{RS}$ is a subset of a general set of known fields, which can be shown as:

$$l \in \{f_a, f_b, f_c, \cdots\} \subseteq MLF = \{f_1, f_2, f_3, \cdots, f_z\} \qquad (3)$$

This forms a relation of fields $f_i \in MLF \{f_1, f_2, f_3, \cdots, f_z\}$ to languages $l_i \in \{l_1, l_2, l_3, \cdots\} \subseteq$ MLSet$_{IN} \cup$ MLSet$_{RS}$, which can be shown as a matrix R$_{LMLF}$ (MxN) (where M stands for number of languages, and N number of fields). Relation can take on of the three values: 2 for mandatory fields, 1 for optional fields, 0 for empty relation.

$$R_{LMLF} = \begin{vmatrix} l_1 mlf_1 & l_1 mlf_2 & \cdots & l_1 mlf_n \\ l_2 mlf_1 & l_2 mlf & \cdots & l_2 mlf_n \\ \cdots & \cdots & \cdots & \cdots \\ l_m mlf_1 & l_m mlf_2 & \cdots & l_m mlf_n \end{vmatrix}$$

Each of the fields f_x has assigned exactly one specific meaning mn_y, that belongs to a set MN of known functions of exchanged values (meanings):

$$MN = \{mn_1, mn_2, mn_3, \cdots, mn_g\} \qquad (4)$$

This gives a relation of fields $f_i \in MLF \{f_1, f_2, f_3, \cdots, f_z\}$ to meanings $mn_i \in \{mn_1, mn_2, mn_3, \cdots, mn_g\} = MN$, which can be shown as a matrix R$_{MNF(MxN)}$ (where M stands for a number of meanings and N number of fields).

$$R_{MNF} = \begin{vmatrix} mn_1 f_1 & mn_1 f & \cdots & mn_1 f_n \\ mn_2 f_1 & mn_2 f & \cdots & mn_2 f_n \\ \cdots & \cdots & \cdots & \cdots \\ mn_m f & mn_m f_2 & \cdots & mn_m f_n \end{vmatrix} \qquad (5)$$

Because each field may have only one meaning, each column of matrix R_{MNF} has only one existing relation (equal to 1), while all the others remain empty (marked 0). Conversion is done in the following way:

1. on the basis of the order document[14], determine value $l_a \in MLSet_{IN}$ and value $l_b \in MLSet_{RS}$
2. for language $l_b \in MLSet_{RS}$, on the basis of relation R_{LMLF}, determine a list of mandatory and optional fields
3. on the basis of relation R_{MNF} determine a meaning for each field identified in step 2
4. then, for each field identified in step 2:
   a. on the basis of matrix R_{MNF} determine a list of fields of the same meaning
   b. with the help of matrix R_{LMLF} determine which fields from step 4a are assigned to chosen input language $l_a \in MLSet_{IN}$.
   c. obtain value definition for this field and convert it, if necessary (e.g. change date format)
5. on the basis of definition (scheme) of language $l_b \in MLSet_{RS}$ form a valid sequence of fields, including field separation and structure

Described conversion pattern relies on assumption, that there is some common set of description format fields, and languages are constructed by selection of proper subsets from it. This assumption is fine as long as each field is a unique element of that set. For instance, for a field date that appears in n languages, there should be as many entries in matrix LMLF, as there are possible format variations. For instance, date can be in a form 11042016, 20160411, or 11-04-2016, etc. This condition allows building validation policies that apply to field characteristics in a given language, not a generalized meaning. This is intended to help detect improper formatting of field content, leading to identification of incompatibility with a predefined (expected) description format (scheme).

5   Case Study

Let there be some integrated security monitoring system, which accepts a single scheme of incoming reports and a data model proposed in [14]. Then report structure is as shown on Fig. 2.

Fig. 2. General scheme of agent report about state of security parameters[14].

This means, that:

$$KlasaScheme \in \text{MLSet}_{RS}$$

Let there be two data sources, that use two different description structures. One of them uses some dedicated XML-based structure to save results [4] – it will be referenced as BatRepScheme later on. The other one is a csv file exported periodically from the database of an on-line game system – it will be referenced as GameCSV. Then, a set of input languages look as follows:

$$BatRepScheme, GameCSV \in \text{MLSet}_{IN}$$

The next step is determination of relation of fields to languages for these three languages. As a result, matrix R_{LMLF} takes the following form, shown on Fig. 3. Then, meaning of each of those fields must be determined. Meanings of all identified fields form matrix of relation R_{MNF}, which can be seen on Fig. 4. As can be seen, a number of fields has the same general meaning. Moreover, some of them were classified as "measured value", because, from the perspective of the integrated security monitoring system, they provide data about the state of some component of the system.

	Report ID	Date and time of report	agent ID	order ID	request ID	protection*	value	timestamp	ReportGuid	ReportVersion	ScanTime	LocalScanTime	ReportStartTime	LocalReportStartTime	ReportDuration	UtcOffset	ComputerName	IsNextOnBattery	FullChargeCapacity	Discharge	ChargeCapacity	EntryType	Ac	Duration	LocalTimestamp	Timestamp	sesid	access_time	persistent	uid	data	lockx	locktime
KlasaScheme	2	2	2	2	2	2	2	2	0	0	0	0	0	0	0	0	0	0	0	0	0	0	0	0	0	0	0	0	0	0	0	0	0
BatRepScheme	0	0	0	0	0	0	0	0	2	1	2	1	2	1	2	2	2	2	2	1	1	1	2	1	1	2	0	0	0	0	0	0	0
GameCSV	0	0	0	0	0	0	0	0	0	0	0	0	0	0	0	0	0	0	0	0	0	0	0	0	0	0	2	2	2	2	2	2	2

Fig. 3. Matrix R_{LMLF} for analyzed three languages.

On the basis of those two matrixes, it is now possible to take step 4 of the procedure of language conversion. For each field of the destination language, which is KlasaScheme, its meaning is determined to search for fields of input languages of the same meaning. As can be seen, some of the output fields do not have counterparts with the same meaning in other languages. This is because this language requires that the agent already owns them and does not convert them from values obtained from the data source [14]. Those fields are: agent number, order number, request number. Report ID is also generated by the agent, so it is not converted.

	ReportID	Date and time of report	agedescriptor	orderdescriptor	requestdescriptor	asset	measuredvalue	timestamp	version	ScanTime	ReportScheme	RecordType	SVGscan	LocalScan	RepoState	RepoState	statur	UDDstuf	UcforON	CompotOBEaN	OextchapcarcapT	DescapcT	Enty		Diumreteni	Timmemet	Timmemet	seid	accesstime	persist	acccssis	duraet	lockxis	duacki	lockxtime
report number	1	0	0	0	0	0	0	0	1	0	0	0	0	0	0	0	0	0	0	0	0	0	0	0	0	0	0	0	0	0	0			
date/time	0	1	0	0	0	0	0	0	0	1	1	0	0	0	0	0	0	0	0	0	0	0	0	0	0	0	0	0	0	0	0			
agent number	0	0	1	0	0	0	0	0	0	0	0	0	0	0	0	0	0	0	0	0	0	0	0	0	0	0	0	0	0	0	0			
order number	0	0	0	1	0	0	0	0	0	0	0	0	0	0	0	0	0	0	0	0	0	0	0	0	0	0	0	0	0	0	0			
request number	0	0	0	0	1	0	0	0	0	0	0	0	0	0	0	0	0	0	0	0	0	0	0	0	0	0	0	0	0	0	0			
asset	0	0	0	0	0	1	0	0	0	0	0	0	0	0	0	1	0	0	0	0	0	0	0	0	0	1	0	0	0	0	0			
measured value	0	0	0	0	0	0	1	0	0	0	0	0	0	0	0	1	1	1	1	1	1	0	0	0	1	1	1	0	1	1	1			
timestamp	0	0	0	0	0	0	0	1	0	0	0	0	0	0	0	0	0	0	0	0	0	0	1	0	0	0	0	0	0	0	0			
version	0	0	0	0	0	0	0	0	0	1	0	0	0	0	0	0	0	0	0	0	0	0	0	0	0	0	0	0	0	0	0			
duration	0	0	0	0	0	0	0	0	0	0	0	0	1	0	0	0	0	0	0	0	0	1	0	0	0	0	0	0	0	0	0			

Fig. 4. Matrix R_{MNF} for analyzed three languages.

The rest of the fields is converted on the basis of their common meaning. For example, ScanTime in language BatRepScheme will be converted to field Date and time of report in language KlasaScheme, as well as field date from language GameCSV. When a number of fields from the source language is marked with the same meaning, all of them are transferred to the destination language. In such a case, values are copied together with original field names, to provide their context. Conversion of a sample message in GameCSV scheme is shown on Fig. 5.

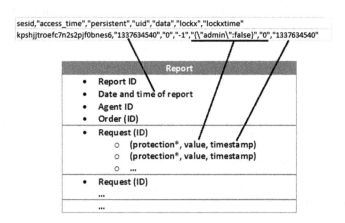

Fig. 5. Sample message in language GameCSV conversion to KlasaScheme.

6 Discussion

Proposed data scheme conversion is supposed to allow gathering data from differentiated data sources. Monitoring security in highly distributed organizations means, that it is necessary to collect and process data from many components. Because there are numerous description languages and schemes and system components support one or few of them only, it is very unlikely to introduce one format, that will be supported by all data sources. This means, that there is a need of data conversion. This can be done once data is received by central repository, or before sending it to the repository. The latter option allows more detailed verification of transmitted content, especially if the system is aware what kind of data is supposed to come from each given source. Moreover, because some of the fields in source data schemes are irrelevant for security monitoring, they can be easily omitted during conversion, which reduces transmission and latter processing costs.

Proposed scheme of data conversion is simple, however it requires significant effort on initial setup. All languages introduced to the system must be examined and each of their fields must be mapped on proper meaning. This process is not automated and is definitely troublesome, but has to be done once only. Afterwards, the whole conversion can be automated.

As an area of further research, there is space for improvement of the model in the case of more sophisticated schemes, e.g. the Conspec language.

7 Conclusion

Proposed data scheme conversion is an attempt to provide compatibility of data, regardless of its source. Data sources differ significantly and usually offer limited functionality in terms of output data formats. This means that it is required to add conversion mechanism so that data gathered from all those sources can be processed easily later on by a central reasoning system. The model uses meaning of the field as a basis for conversion from source to output format.

References

1. Aktung, I., Naliuka, K.: ConSpec – a formal language for policy specification. Electron. Notes Theor. Comput. Sci. **197**, 45–58 (2008)
2. Dhouha, A., Muhammad, A., David, L.-J.: An event processing aproach for threats monitoring of service compositions. In: Internetional Conference on Risks and Security of Internet and Systems (2013)
3. Bai, X., et al.: Model-based monitoring and policy enforcement of services. Simul. Model. Pract. Theory **17**, 1399–1412 (2009)
4. Battery Report. http://schemas.microsoft.com/battery/2012
5. Bodenham, A., Niall, M.A.: Continuous monitoring of a computer network using multivariate adaptive estimation. In: 2013 IEEE 13th International Conference on Data Mining Workshops, pp. 311–318 (2013)

6. Brdys, M.A.: Integrated monitoring, control and security of critical infrastructure systems. Ann. Rev Control **2014**(38), 47–70 (2014)
7. Luca, C., Pierre, G., Achim, B.D.: Business process compliance via security validation as a service. In: 2013 IEEE Sixth International Conference on Software Testing, Verification and Validation (ICST), pp. 455–462 (2013)
8. Du, S., et al.: Towards an analysis of software supply chain risk management. In: World Congress on Engineering and Computer Science, WCECS 2012, Vol. 1. Lecture Notes in Engineering and Computer Science, pp. 162–167 (2013)
9. El Fray, I.: Metoda określająca zaufanie do system informacyjnego w oparciu o process szacowania i postępowania z ryzykiem. Wydział Informatyki, Zachodniopomorski Uniwersytet Technologiczny w Szczecinie, Szczecin (2013)
10. Han, S., et al.: Intrusion detection in cyber-physical systems: techniques and challenges. IEEE Syst. J. **8**, 4 (2014)
11. Hussein, A.A., Ghoneim, A., Dumke, R.R.: An approach for securing and validating business processes based on a defined enterprise security ontology criteria. In: Snasel, V., Platos, J., El-Qawasmeh, E. (eds.) Digital Information Processing and Communications, Part 1. Communications in Computer and Information Science, pp. 54–66. Springer, Heidelberg (2011)
12. Klasa, T.: Information systems security description proposal. In: Swacha, J., Szyjewski, Z. (eds.) Project Management Selected Issues. Uniwersytet Szczeciński, Wydział Nauk Ekonomicznych i Zarządzania, Szczecin (2010)
13. Klasa, T.: Evaluation of influence of identified changes in the state of the information system on information security grade. Found. Comput. Decis. Sci. **36**(3–4), 229–242 (2011)
14. Klasa, T.: Model gromadzenia danych monitorowania bezpieczeństwa informacji w organizacji wirtualnej. Zeszyty Naukowe Studia Informatica **2015**(38), 49–64 (2015)
15. Kwiatkowski, J., Juszczyszyn, K., Kolaczek, G.: An environment for service composition, execution and resource allocation. In: Manninen, P., Öster, P. (eds.) PARA. LNCS, vol. 7782, pp. 77–91. Springer, Heidelberg (2013)
16. Malinowski, T., Arciuch, A.: The procedure for monitoring and maintaining a network of distributed resources. ACSIS **2**, 947–954 (2014)
17. Michalek, P.: Dissecting application security XML schemas, AVDL, WAS, OVAL – state of the XML security standards report. Inf. Secur. Tech. Rep. **9**(3), 66–76 (2004)
18. Microsoft. Configuring Audit Policies. Technet (2015). https://technet.microsoft.com/en-us/library/dd277403.aspx
19. Margherita, P., Irene, S.: Increasing security and efficiency in supply chains: a five-step approach. Int. J. Shipping Transp. Logistics **6**(3), 257–279 (2014)
20. Potter, B.: Security automation. Netw. Secur. **9**(2007), 18–19 (2007)
21. Qin, T., et al.: Robust application identification methods for P2P and VoIP traffic classification in backbone networks. Knowl. Based Syst. **2015**(82), 152–162 (2015)
22. SAP ECC 6 documentation, FICO module (2011)
23. Stallings, W.: Computer Networks Security (2011)
24. Stallings, W.: Operating Systems. Internals and Design. Prentice Hall, Upper Saddle River (2013)
25. Liu, T., Zhao, Y.: A decentralized information flow model for SaaS application security. In: 2013 Third International Conference on Intelligent System Design and Engineering Applications (2013)
26. Van Tan, V., Yi, M.-J.: Design issues and approach to internet-based monitoring and control systems. In: García-Pedrajas, N., Herrera, F., Fyfe, C., Benítez, J.M., Ali, M. (eds.) IEA/AIE 2010, Part I. LNCS, vol. 6096, pp. 478–488. Springer, Heidelberg (2010)
27. Wójcik, A.: System SCS Win. Zabezpieczenia, p. 5 (2009)

28. Wu, M.Z., et al.: Development and Validation on integrated dynamic security monitoring platform. In: 2012 Sixth International Conference on Genetic and Evolutionary Computing (2012)
29. Wun-Young, L., et al.: SAP Security Configuration and Deployment. Syngress, Elsevier, Burlington (2008)
30. Zawoad, S., Mernik, M., Hasan, R.: FAL: a forensics aware language for secure logging. In: Ganzha, M., Maciaszek, L., Paprzycki, M. (eds.) 2013 Federated Conference on Computer Science and Information Systems. IEEE, Kraków, pp. 1567–1574 (2013)

Non-standard Certification Models for Pairing Based Cryptography

Tomasz Hyla$^{(\boxtimes)}$ and Jerzy Pejaś

Faculty of Computer Science and Information Technology,
West Pomeranian University of Technology, Szczecin, Poland
{thyla, jpejas}@zut.edu.pl

Abstract. In the traditional Public Key Infrastructure (PKI), a Certificate Authority (CA) issues a digitally signed explicit certificate binding a user's identity and public key to achieve this goal. The main goal of introducing an identity-based cryptosystem and certificateless cryptosystem was avoiding certificates' management costs. In turn, the goal of introducing an implicit certificate-based cryptosystem was to solve the certificate revocation problem. The certificate and pairing based cryptography is a new technology and at present that technology mainly exists in theory and is being tested in practice. This is in contrast to PKI-based cryptography, which has been an established and is widespread technology. New types of cryptographic schemes require new non-standard certification models supporting different methods of public keys' management, including theirs generation, certification, distribution and revocation. This paper takes a closer look at the most prominent and widely known non-standard certification models, discusses their properties and related issues. Also, we survey and classify the existing non-standard certification models proposed for digital signature schemes that are using bilinear pairings. Then we discuss and compare them with respect to some relevant criteria.

Keywords: Public key cryptography · Certification model · Digital certificate · Digital signature · Key management

1 Introduction

Recently, a pairing based cryptography (PBC) have been a very active research area. The development of this field of cryptography dates back to A. Shamir pioneering work [1] on some proposal of an identity-based cryptosystem, in which the function of the public key plays the user's identity. A. Shamir proposed only requirements for such schemes without specifying how they can be implemented in practice. A satisfactory solutions to Shamir's problem was reported almost simultaneously by three groups of authors: D. Boneh and M.K. Franklin [2], Sakai-Ohgishi-Kasahara [3] and C. Cocks [4]. Interestingly, cryptosystems proposed in the first two works were based on bilinear pairings built on elliptic curves. The creation of the pairing based cryptographic schemes enabled to approach the problem of creating a key management architecture from a new angle.

© Springer International Publishing AG 2017
S. Kobayashi et al. (eds.), *Hard and Soft Computing for Artificial Intelligence, Multimedia and Security*, Advances in Intelligent Systems and Computing 534, DOI 10.1007/978-3-319-48429-7_16

PKI in now widely accepted trust and key management architecture. It supports traditional public key cryptography schemes based on an explicit certificate. PKI standard trust models are very useful for building trust between different entities, but their complexity might cause several problems. The most important problems related to PKI are trust verification of certificate issued by CA and a private key disclosure. When the trust to certificate issued by CA is verified, it is required to build a certification path which lead to the point of a trust. Next, validity of each certificate from the path must be verified. It is not a trivial problem to build and verify a certification path. Because of that, that task is often done imprecisely and its results are the cause of many vulnerabilities, e.g., in web browsers.

The second problem results from poor resistance of PKI trust models to a private key disclosure (belonging to CA or a user). In such case, there is no simple way to solve this problem or at least to limit the effects of the key compromise. Can that problems be solved using other trust models? The answer to this question is not obvious. Big expectations were and are still associated with non-standard PKI trust models, but the first attempts to practical application of these models has shown not only their advantages, but unfortunately also disadvantages.

Non-standard (alternative) trust models in public-key cryptography allow managing asymmetric keys, which are related to the users' identities. The two basic classes of such models are trust models in public-key cryptography based on the identity and based on the implicit certificate.

Trust to the public key depends on the level of trust to services related to the management of trust tokens (e.g., certificates). Certificates bind a public key with an identity of a trusting entity. According to M. Girault [5], three general levels of trust to services which manages trust tokens can be specified:

(a) trust level 1: the service provider knows or can easily compute the private key of the entity and therefore can impersonate the owner of the key and act on its behalf;

(b) trust level 2: the service provider does not know or cannot easily compute the entity' private key; however, the provider can generate a fake pair of asymmetric keys and falsify trust tokens related to them;

(c) trust level 3: the service provider does not know or cannot easily compute an entity' private key; the provider can still impersonate other relying parties, by generating a fake pair of asymmetric keys and falsifying related trust tokens of trust; such fraud can be detected.

Services provided at trust levels 1 and 2 are unacceptable in many applications, algorithms and cryptographic protocols. A classic example are digital signature schemes that for these levels do not provide non-repudiation of a signature and thus its legal significance. In practice, the aim of the service provider is to achieve the third level of trust. This trust level can be ensured by providers operating within a traditional PKI.

1.1 Contribution

Four techniques of linking identity with public keys are used in Public Key Cryptography (PKC) schemes. As a result, PKC schemes can be divided into four categories [6–8]:

(a) category I – public key cryptography schemes based on explicit certificates, e.g., *Standard Explicit Certificate-based Public Key Cryptography* (SEC-PKC); in this category are traditional *Public Key Infrastructure* (PKI) systems;
(b) category II – *Identity-based Public Key Cryptography* (ID-PKC) schemes; in that category, the linking mechanism is an identity mapping; a public key equals to the identity of an entity;
(c) category III - public key cryptography schemes based on implicit certificates, e.g., *Certificateless Public Key Cryptography* (CL-PKC) or *Implicit Certificate-based Public Key Cryptography* schemes (IC-PKC);
(d) category IV - public key cryptography schemes based on an implicit and explicit certificates, e.g., *Implicit and Explicit Certificate-based Public Key Cryptography* schemes (IEC-PKC).

In the case of certification models supporting traditional PKC schemes, CA authority issues mainly explicit certificates for public keys. The keys can be related to any supported PKC scheme. The tasks of a trusted authority (TA) that supports PKC schemes belonging to categories II, III and IV (these schemes will be further called non-standard PKC schemes, while certification models will be called non-standard certification models, see Sect. 3) are diametrically opposed. TA main task is key generation (single component keys or parts of two component keys) using its system parameters. Optionally, TA can issue explicit certificates in schemes from categories II and III, but this is done beyond PKC schemes, i.e. within these schemes does not exist algorithm that can generate an explicit certificate. In opposite to this, in scheme from IV category, explicit certificates are a part of the schemes and are not optional.

In the paper, certification models related to categories I-IV of PKC schemes are presented. The starting point of the analysis is a standard PKI certification model. Several problems are related to that model, among others, a certificate revocation problem. Hence, it would seem that if the use of certificates in the PKI infrastructure cause obvious practical problems, then the certificates should be eliminated. In the paper, we show that it is not so obvious, because the problems related to certificates are replaced by new ones, e.g., public key distribution problems. A compromise and a practical solution is a certification model in which are issued both explicit (as in I category schemes) and implicit certificates (as in III category schemes).

1.2 Paper Structure

The remainder of this paper is organized as follows. Section 2 contains description of different models of digital signature schemes. In Sect. 3 non-standard certification models are described and analysed. Next, in Sect. 4 we compare non-standard certification models with a standard PKI certification model. The paper ends with conclusion.

2 Models of Digital Signature Schemes

In this section, we describe four different models of digital signature schemes that are using bilinear pairings:

(a) ID-Based Signature scheme - Pairing-Based Cryptography (IBS-PBC);
(b) Certificateless Signature scheme - Pairing-Based Cryptography (CLS-PBC);
(c) Implicit Certificate-Based Signature scheme - Pairing-Based Cryptography (ICS–PBC);
(d) Implicit and Explicit Certificate-Based Signature scheme - Pairing-Based Cryptography (IECS-PBC).

A digital signature based on identity (IBS-PBC) is a specific type of signature introduced in 1984 by A. Shamir [1]. A characteristic feature of the IBS-PBC scheme is a signature validity verification procedure. The procedure is implemented based on TA's public key and on user's identification data that are considered as his public key. Sometimes TA is called Private Key Generator (PKG). Its main function is generation of private (signing) keys based on identities.

The problem of having to entrust the user's key to TA (i.e., key escrow problem) has been solved by Al-Riyami and Paterson in [9]. They proposed certificateless signature scheme (CLS-PBC) in which a singing entity has a private key consisting of two components: the partial private key ppk_{ID} generated by the TA and the secret key generated by the signer psk_{ID}. The ppk_{ID} key depends on, among others, the identity of the signer, but is not associated with the signer's public key. Because of that, the scheme is open to public key replacement attacks, e.g., [10].

Implicit certificate-based signature scheme (ICS-PBC) is very similar to the model of the CLS-PBC scheme. The main difference is a method used to generate a partial private key ppk_{ID}. In the ICS-PBC scheme, it depends on the identity of the signer $ID \in \{0, 1\}^*$ and his public key pk_{ID}. A user who has the private key and the corresponding certificate can sign the message. If the result of signature verification is positive, it indirectly means that the public key is implicitly certified and without explicit verification (actually, without such possibility) the certificate should be assumed to be valid. According to W. Wu et al. [11], to generate a traditional valid signature based on a certificate, it is required to have two components in ICS-PBC schemes: a signing entity certificate and a secret key. It is assumed that both components are secret [12–14]. Generally, the analysis of the method used to generate a certificate binding the public key with the signer's identity shows that there is no need to keep the implicit certificate $iCert_{ID}$ secret. However, in this case the ICS-PBC scheme should be classified as the scheme with an explicit certificate $eCert_{ID}$.

In 2014, Hyla et al. [7] have introduced a new paradigm called Implicit and Explicit Certificates-Based Public Key Cryptography (IEC-PKC) and proposed a concrete implicit and explicit certificates-based encryption scheme (IE-CBE). Next, in 2016 the IE-CBE scheme was generalised for implicit and explicit certificate-based signature scheme IECS-PBC (Hyla et al. [8]). Both schemes preserve all advantages of Implicit Certificate-based Public Key Cryptography (IC-PKC), i.e., every user is given by the TA an implicit certificate as a part of a private key and generates his own secret key and corresponding public key. In addition, in the IC-PKC scheme the TA has to generate an explicit certificate for a user with some identity and a public key. The implicit and explicit certificates should be related with each other, is such a way that no one, even the entity of those certificates and their issuer (TA authority), should not be able to recreate an implicit certificate using the explicit certificate.

3   Non-standard Certification Models

Certification models of public keys corresponds to different ways of public keys' management (Fig. 1), including their generation, certification, distribution and revocation. Their architecture consists of set of components, their functionality and relations between them, including trust relations.

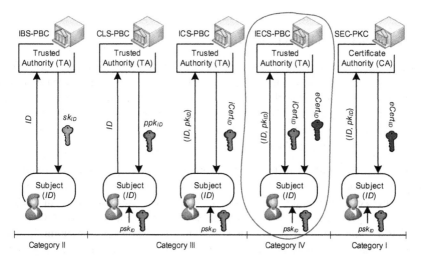

Fig. 1. Category IV versus other categories of public key cryptography schemes (prepared on the basis [15]) (Legend: ID – user's identity, sk_{ID} – single-component private key, psk_{ID} – secret key, ppk_{ID} – partial secret key, pk_{ID} –public key, $eCert_{ID}$ – explicit certificate, $iCert_{ID}$ – implicit certificate)

The main purpose of introducing non-standard certification models was to eliminate explicit certificates and problems related to them. Category II and III schemes are based only on identity, which is used to generate one- or two-component private keys. However, only in the case of schemes based on the implicit certificates ICS-PBS, the key (more precisely the implicit certificate) depends on the subscriber' public key. This situation requires careful identity management, which is just as troublesome as in the case of certificates' management.

Without going into details, it can be stated that in the case of category II and III schemes, it is tactically assumed that implicit certificates bind or include only the identifier specifying the subscriber's identity and its public key. This assumption was quickly verified in practice (see Fig. 1). In realistic schemes, certification is more than just linking between the identity and the key - it contains the metadata that describes the identity and public key system parameters scheme, as well as descriptions of the same metadata. An example of such metadata is ASN.1 [16] used to describe certificates compliant with X.509 format.

Without metadata, it is difficult in practice to verify whether a digital signature meets the requirements of a specific technical specification or a legal act and is, for example, a qualified signature. Moreover, the assumption that in the event of a positive verification, the implicit certificate has been issued and is valid, in many cases is difficult to accept. For example, how the verifier will know that the signer used a proper private key? Especially, when the same TA issues keys for different applications (email signing, web authentication, etc.). In such cases, it is said that the certificate plays a technical role. Therefore, a certificate is not drawback of both traditional as well as non-standard signature schemes.

3.1 Base PKI Certification Model

Models of digital signature schemes described in Sect. 3 correspond to different non-standard certification models. These models can be derived from a base six-party PKI certification model. PKI that allows to implement basic trust services (apart from the base services, additional services like timestamping or authentication and authorization services are available [17]), called – base public key infrastructure, consists of six types of components (roles) [18]:

(a) *Certification Authority* (CA) - the component is responsible for issuing and revoking certificates; the certificate binds subscriber identification data with his public key. Optionally, the component provides information about certificate status (e.g., using a service like Online Certificate Status Protocol (OCSP) [19]);

(b) *Registration Authority* (RA) - the component is responsible for submitting to the CA reliable user IDs; it guaranties that relation between user identity and the public key is authentic;

(c) *Repository* – the component is responsible for storing and sharing public key certificates and certificate revocation lists (CRLs);

(d) *Key Creation Authority* (KCA) – the component is responsible for generation of asymmetric key pairs;

(e) *Subscriber* – the owner of a public key certificate; the certificate was issued to him and can be used to create a digital signature, to verify it or to decryption; the subscriber cannot provide trust services;

(f) *Relaying party* – a user who use certificate during signature verification or during encryption of confidential information; in both cases, it is required to build a certification path which leads to a trust point from a subscriber certificate; then, the trust level must be verified.

Standard PKI certification model consist of those six components. Also, the model is called as six-party public key certification model. Because the model require mutual trust relationships between components, is also called as six-party public key certification trust model. Components (roles) can be combined and implemented by one party. Roles of CA and RA are very often combined which result in four-party certification model. In this model keys are generated by the subscriber or by CA (which is CA third role).

3.2   Base PKI Certification Model vs. Non-standard Models

Base components necessary to generate, distribute and revoke keys in the base PKI model in relation to non-standard models are compared in Table 1.

Table 1. Comparison of the base PKI and non-standard certification models' components

No.. Components	Type of non-standard PKI certification model (plus PKI)				
	PKI	IBS-PBC	CLS-PBC	ICS-PBC	IECS-PBC
1. Private key creation trusted authority (TA-KCA)	$\perp$[1]	+	+[2]	+[2]	+[2]
2. Certificate management trusted authority (TA-CMA)	+	$\perp$[3]	$\perp$[3]	$\perp$[3]	+[4]
3. Registration authority (RA)	+	+	+	+	+
4. Repository of public keys (rPK)	-	-	+	+	$\perp$[5]
5. Repository of certificates (rCert)	+	-	-	-	+
6. Repository of identity (rID)	-	+	+	+	$\perp$[5]
7. Repository of revoked public keys (rCRL)	+	$\perp$[6]	$\perp$[6]	$\perp$[6]	$\perp$[6]

Legend:

„+" – is present, „-" – is not present, „$\perp$" – is optional (see details below)

[1] Also, keys can be generated by a subscriber or RA.

[2] In case of some signature schemes, TA-KCA takes part in private keys or its partial generation.

[3] TA-CMA can inform about a status of registered user identity (IBS-PBC and CLS-PBC) or about a status of an implicit certificate (ICS-PBC and IECS-PBC).

[4] TA-CMA generates the explicit certificates and can additionally inform about theirs status.

[5] Repository does not occur when both an implicit certificate and an explicit certificate are generated.

[6] Repository is not present in the certification model, if in a signature scheme certificate revocation is not available; the revocation should be applied in systems with high trust level.

In the Table 1 a subscriber and a relaying party are omitted, because they must be present in every certification model. A repository has be divided into four parts:

(a) a repository of public keys (rPK) – contains subscribers public keys; for these keys certificates are nor issued;

(b) a repository of certificates (rCert) – contains explicit public key certificates;

(c) a repository of identity (rID) – contains subscribers identification data, which can be included in implicit certificates or are used during key generation;

(d) a repository of revoked public keys (rCRL) – contains lists of revoked public keys to which certificates were issued (explicit or implicit ones); if a scheme does not contain revocation mechanism, the repository is not present in a certification model.

Additionally, the more general names were assigned to certificate and key creation authorities: certificate management trusted authority (TA-CMA) that plays the role similar to a certification authority in PKI, private key creation trusted authority (TA-KCA). If TA-CMA plays at the same time the role of TA-KCA, it is assumed that they are a part of a trusted authority and denoted as TA.

In Table 1, optional components in given certification model are indicated in grey colour. It can be noticed, that the IECS-PBC schemes have characteristics of both the ICS-PBC certification model and features similar to PKI certification model. This observation confirms unique features of IECS-PBC schemes, i.e., the possibility to work in modes of ICS-PBC schemes and PKI based schemes.

3.3   Certification Models in ID-Based Signature Schemes

Public Key Cryptography schemes based on the identity eliminate the need to issue the certificates. This is due to the fact that the public key is an identity (ID) of the user (subscriber) and the secret keys are extracted from the ID by the key generation authority (TA-KCA). Essentially, the correctness of the schemes based on the identity result from the fact, that the TA-KCA in IBS-PBC scheme binds user's identity with a public key of local or a global trust authority TA. TA has the function of a trust point.

Theoretically, the IBS-PBC scheme to work requires a certification model which consists of: a subscriber, a relaying party, a trust authority TA = TA-CMA + TA-KCA, while a set of function of TA-CMA can be empty. In practice, this is not sufficient. There are several reasons which cause that situation:

(a) the entity which verify the signature must have access to an authentic set of public parameters *params* and to authentic identity of the signer ID_{Reg} registered in the RA; information should be publicly available repository of identity (rID);
(b) in the IBS-PBC scheme it is assumed that unambiguous relation between a subscriber and unique information (that defines subscriber digital identity) exists; this type of relation prior identity confirmation by a registration authority, then registering it, assuring that ID_{reg} is authentic and storing it in a repository; these actions are very similar to RA functions in the standard PKI certification model;
(c) in every signature scheme, a public key revocation mechanism should be available (also for registered identity ID_{reg}); because of that, periodic distribution of authentic registered information is necessary (e.g., using a protocol similar to OCSP [19]); these function can be assigned to TA-CMA; it is possible to resign from publishing information about status of a public key, when it is registered for short period (e.g., [12, 13]), such revocation method is not always adequate in IBS-PBC schemes when high credibility of the signature is required; moreover, the method can be applied to the standard PKI certification model and theoretical advantage of IBS-PBC schemes is eliminated).

Problems related to certificates management are indicated as fundamental flaw of the standard PKI certification model. Does the problem was solved in the IBS-PKC model? Rather not, the certification management problem was replaced with an identity (public key) management problem.

3.4 Certification Model for Certificateless Digital Signature Schemes

The main drawback of the IBS-PBC signature scheme (and at the same time the certification model) is entrusting key to the TA-KCA authority. This contradicts with the idea of signature non-repudiation (i.e., in M. Girault scale IBS-PBC has a trust level 1). In the PKI certification model, asymmetric keys can also be generated by TA-KCA and transmitted to subscribers. However, in this case, it is required that the keys are randomly generated within a secure device (e.g., a smartcard with a crypto-processor), in such a way that a private key never leaves the card in overt form. Theoretically, the same solution can be used for IBS-PBC model. But in this case, the media protects the key, but TA-KCA still has the possibility to recreate the private key based on the preserved key material (i.e., using TA-KCA private key and the subscriber's identity).

The key escrow problem in IBS-PBC scheme can be solved in different ways (e.g., [20]). One of the solution is a certificateless signature scheme (CLS-PBC). In CLS-PBC model, TA-KCA generates a partial secret key ppk_{ID}, which depends on: TA-KCA master secret key, a subscriber identity, and optionally subscriber' public key pk_{ID}. It is easy to notice, that if a partial secret key ppk_{ID} is not linked with the public key pk_{ID}, than a subscriber can have only one private key issued in that certification model.

Registered identities and public keys related to them are stored in a repository rID and rPK. Similarly to IBS-PBC certification model, authenticity must be ensured for both parameters. However, even in that case, the model will not achieve a third level of trust which is available in the standard PKI certification model. In fact, TA = TA-TA-CMA + KCA authority may cheat by exposing fake key pair or by replacing the existing public keys. This attack is harder than in CLS-PBC certification model, because TA must execute active attack on the keys, while in IBS-PBC model the attack is passive. CLC-PBC scheme ensures trust level 2 or between 2 and 3 in case, when a partial secret key is bonded with a subscriber's public key.

3.5 Certification Model for Signature Schemes Based on Implicit Certificates

ICS-PBC certification model (Fig. 2) using signature schemes based on implicit certificates is very similar to CLS-PBC model. Also, it has many common features with the standard PKI certification model, e.g., a subscriber creates his own random pair of keys. A public key from this pair and an identity is registered by the RA (after proof of the private key' ownership is presented). On that basis, TA-KCA authority with cooperation of TA-CMA generates a short or a long-term implicit certificate $iCert_{ID}$ that strongly depends on a subscriber identity and his public key. The short-term certificate cannot be revoked, but after its expiration period, it can be automatically

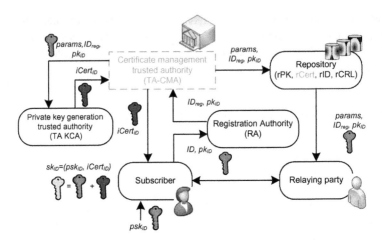

Fig. 2. Certification model for signature schemes based on implicit certificates (ICS-PBC) (Legend: *ID* and *ID_{reg}* – not registered and registered subscriber identity, *psk_{ID}* – subscriber secret key, *pk_{ID}* – subscriber public key, *sk_{ID}* – a subscriber private key, *$Cert_{ID}$* – implicit certificate, *params* – system parameters; grey colour: component is not present or has limited functionality)

renewed for the next period. The long-term certificate requires revocation in situations when it is necessary.

The certificate *$iCert_{ID}$* generation process is very similar to the process of a partial secret key generation in CLS-PBC model. If we additionally assume that the certificate *$iCert_{ID}$* is secret, this similarity is significant. The secrecy of the certificate means that if a verifier assumes by default that certificate related to signature exists, then mathematical correctness of the certificate allows to recognize the authenticity of the certificate and signer identity, but not certificate validity. Such a solution does not allow transferring the certificate to verifiers of signature, but implies that verifier need access to information about the validity of the certificate. From this point of view, ICS-PBC scheme is not different from solutions used in the standard PKI certification models.

If information about the status of the certificate is available, then ICS-PBC model achieves trust level 3 in M. Girault taxonomy. Such trust level ensures non-repudiation of signatures. However, it is required that public keys are obligatory published in rCRL repository and their status is published.

3.6 Certification Model for Implicit and Explicit Certificate-Based Signature Schemes

IECS-PBC certification model (Fig. 3) combines the features of traditional PKI and ICS-PBC certification models. This certification model is based on two certificates: an implicit *$iCert_{ID}$* and an explicit *$eCert_{ID}$* certificates (see Hyla *et al.* [7, 8]).

The implicit certificate *$iCert_{ID}$* generation process is very similar to the process of an implicit certificate generation in ICS-PBC certification model. In this process an

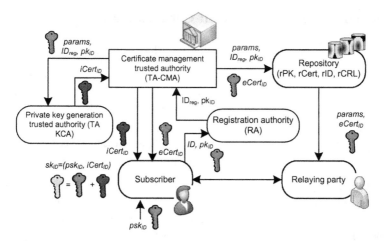

Fig. 3. Certification model for signature schemes based on implicit and explicit certificates (IECS-PBC) (Legend: *ID* and ID_{reg} – not registered and registered subscriber identity, psk_{ID} – subscriber secret key, pk_{ID} – subscriber public key, sk_{ID} – a subscriber private key, $iCert_{ID}$ – implicit certificate, $eCert_{ID}$ – explicit certificate, *params* – system parameters)

algorithm takes as input TA-KCA master private key, its master public key, system parameter *params* and subscriber's parameter (i.e. registered subscriber identity ID_{reg} and his public key pk_{ID}). A TA-KCA runs the algorithm once for each subscriber and the corresponding implicit certificate is distributed to that user through a secret channel.

The explicit certificate $eCert_{ID}$ generation process depends on the same parameters. However, this time the explicit certificate $eCert_{ID}$ is generated by TA-CMA and is strongly related to the implicit certificate $iCert_{ID}$. Both certificates are issued to a signing entity (a subscriber), but only an explicit certificate is made public. An explicit certificate is used first to directly verify the authenticity of public key assignments to a signing entity and next to direct digital signature verification. The last operation also results in verifying the private key assignment to a signing entity. Hence, an explicit certificate has two advantages:

(a) The explicit certificate allows indicating proper, unambiguous and direct bindings with public keys, which are necessary during signature verification; hence, an explicit certificate has all the functionality of a conventional PKI certificate and can be used explicitly as proof of identity certification and a public key binding.
(b) The verifier has *a key element* available, which, when used directly in a verification process, allows receiving the signature status (a valid signature), whose binding to the authenticity of an explicit certificate is obvious for a signer (the certificate is or is not bound to a signer); *a key element* also allows indirectly verifying implicit certificate authenticity.

From the perspective of recently perceived needs, the features presented above are important to build into IECS-PBC scheme mechanisms, allowing revoking signers' keys (e.g., if their private keys get compromised) and thus certificate validity verification.

4 Comparison of Non-standard Certification Models

Different models of certification (including the PKI standard certification model) are compared in Table 2.

Table 2. Comparison of the certification models

No.	Property	Type of a certification model				
		PKI	IBS-PBC	CLS-PBC	ICS-PBC	IECS-PBC
1.	explicit certificates	yes	no	no	no	yes
2.	implicit certificate	no	no	yes	yes	yes
3.	requires repository	yes	yes	yes	yes	yes
4.	public key registration	yes	yes [1]	yes	yes	yes
5.	requires CRL	yes	$\perp$ [2]	$\perp$ [2]	$\perp$ [2]	$\perp$ [2]
6.	difficulty of keys update	low	high	low	low	low
7.	difficulty of public keys distribution	high	high	medium	medium	medium
8.	trust level	3	1	2	3 [3]	3
9.	source of private key might be a subscriber	yes	no	$\perp$ [4]	$\perp$ [4]	$\perp$ [4]
10.	key escrow	no	yes	no	no	no
11.	non-repudiation	yes	no	yes [2,5]	yes [2,5]	yes [2]
12.	TA secret key protection	high	high	medium	medium	medium
13.	without authenticated channel	yes	no	no	yes	yes
14.	without secure channel	yes [6]	no	no	no	no
15.	resistance to DoSV [7]	yes	no	no	no	yes

Legend:
„$\perp$" – means that property is present depending on the context, described in note below
[1] identity is registered and it simultaneously functions as a public key
[2] non-repudiation of the signature requires delivering information about a status of registered identity (IBS-PBC) or explicit/implicit certificates
[3] depends on the specific scheme
[4] a subscriber generates a part of his private key
[5] when TA cannot replace a public key
[6] when TA is not key source
[7] Denial of Signature Validation (see Hyla et al. [7, 8])

Fifteenth different properties were compared, where four of them should be emphasised: difficulty in public keys distribution, presence of an authenticated channel or a secure channel, and signature non-repudiation. Public key distribution concerns the method of transferring keys to an entity which encrypts a message or verifies a signature. This action is relatively difficult in cases when a public key is built into a certificate. First, everyone must find and download a proper certificate, verify the issuer' signature, verify its validity period and status.

An authenticated channel is required in the certification model when exchanged information are not authenticated and authentication data must be added on the fly. Because of that, information from an authenticated channel is from known source and was not modified during transmission. In turn, a secure channel additionally ensures confidentiality. This property is essential during key transfer from TA to a subscriber. Non-repudiation of a signature makes it difficult to deny earlier signatures (in long term, validity of a signature must be maintained, e.g., [21–25]). Due to this property, it is possible to settle any disputed that may arise between the signatories, and the verifier. It is easy to notice, that non-repudiation of a signature is difficult to achieve in a simple key distribution mechanism.

The comparison in Table 2 is not so clearly disadvantageous to the PKI model, and therefore also for the explicit certificate. Despite many theoretical advantages of non-standard certification models, in practical deployment the advantages diminish. This observation is similar to opinion of Y. Desmedt and M. Burmester [26]: *"We argued that to securely deploy identity-based cryptography, one needs a structure as complex as Public Key Infrastructures"*.

5   Conclusion

Trust models can be built for each described certification model. In every certification model, which is a part of a trust model, it is necessary to indicate a certification model (in case of non-standard models denoted as TA trust authority) which will enter into trust relationships with TA belonging to another certification model.

It is possible to build different PKI trust models using different trust mechanisms. Is it possible to build a corresponding non-standard trust model for each of these models? Leaving aside the question whether it is always reasonable, the answer is: no. Because of the fact that in standard trust models the trust relations between TAs belonging to different certification models are built using classic certificates (e.g., in X.509 format [18]), that authenticates only public keys. In non-standard certification models, even if certificates are present (explicit or implicit ones), there are no classic certificates, but numerical values used. They are used in cryptographic operations, e.g., in signature verification algorithm, under the appropriate algebraic group.

The algebraic groups are defined by system parameters specific to a particular signature scheme. The trust models for selected certification model can be built only under the same system parameters (in particular determining the bilinear pairings groups for the groups of the same order). Any attempt to go beyond this limitation leads to standard PKI trust models and/or beyond signature schemes based on bilinear pairings (e.g., [27]).

The results presented in this paper authorize us to declare that when designing a signature scheme of PBC, we should bear in mind the certification model, which it supports, and the trust model that we plan to build on it. Due to this approach, the cryptographic schemes belonging to each of four categories PKC schemes can be easily mapped into corresponding certification models. They can be also mapped into one of the proposed trust models. The components that are obtained allow us to build trust and key management architectures with features similar to traditional PKI infrastructures.

References

1. Shamir, A.: Identity-based cryptosystems and signature schemes. In: Blakely, G.R., Chaum, D. (eds.) CRYPTO 1984. LNCS, vol. 196, pp. 47–53. Springer, Heidelberg (1985)
2. Boneh, D., Franklin, M.: Identity-based encryption from the weil pairing. In: Kilian, J. (ed.) CRYPTO 2001. LNCS, vol. 2139, pp. 213–229. Springer, Heidelberg (2001)
3. Sakai, R., Ohgishi, K., Kasahara, M.: Cryptosystems based on pairing (in Japanese). In: Symposium on Cryptography and Information Security – SCIS, Okinawa, (2000)
4. Cocks, C.: An identity based encryption scheme based on quadratic residues. In: Honary, B. (ed.) Cryptography and Coding 2001. LNCS, vol. 2260, pp. 360–363. Springer, Heidelberg (2001)
5. Girault, M.: Self-certified public keys. In: Davies, D.W. (ed.) EUROCRYPT 1991. LNCS, vol. 547, pp. 490–497. Springer, Heidelberg (1991)
6. Pejaś, J.: Implicit and explicit certificates-based digital signature schemes in infrastructure with multiple trust authorities (in Polish). Wyd. Stowarzyszenie Przyjaciół Wydziału Informatyki w Szczecinie. Seria: Monografie Informatyczne, Tom II, Szczecin (2013)
7. Hyla, T., Maćków, W., Pejaś, J.: Implicit and explicit certificates-based encryption scheme. In: Saeed, K., Snášel, V. (eds.) CISIM 2014. LNCS, vol. 8838, pp. 651–666. Springer, Heidelberg (2014)
8. Hyla, T., Pejaś, J.: A hess-like signature scheme based on implicit and explicit certificates. Comput. J. (2016). doi:10.1093/comjnl/bxw052, http://comjnl.oxfordjournals.org/cgi/reprint/bxw052
9. Al-Riyami, S.S., Paterson, K.G.: Certificateless public key cryptography. In: Laih, C.-S. (ed.) ASIACRYPT 2003. LNCS, vol. 2894, pp. 452–473. Springer, Heidelberg (2003)
10. Huang, X., Susilo, W., Mu, Y., Zhang, F.T.: On the security of certificateless signature schemes from Asiacrypt 2003. In: Desmedt, Y.G., Wang, H., Mu, Y., Li, Y. (eds.) CANS 2005. LNCS, vol. 3810, pp. 13–25. Springer, Heidelberg (2005)
11. Wu, W., Mu, Y., Susilo, W., Huang, X.: Certificate-based signatures revisited. J. Univ. Comput. Sci. 15(8), 1659–1684 (2009)
12. Gentry, C.: Certificate-based encryption and the certificate revocation problem. In: Biham, E. (ed.) EUROCRYPT 2003. LNCS, vol. 2656, pp. 272–293. Springer, Heidelberg (2003)
13. Kang, B.G., Park, J.H., Hahn, S.G.: A certificate-based signature scheme. In: Okamoto, T. (ed.) CT-RSA 2004. LNCS, vol. 2964, pp. 99–111. Springer, Heidelberg (2004)
14. Li, J., Huang, X., Mu, Y., Susilo, W., Wu, Q.: Certificate-based signature: security model and efficient construction. In: López, J., Samarati, P., Ferrer, J.L. (eds.) EuroPKI 2007. LNCS, vol. 4582, pp. 110–125. Springer, Heidelberg (2007)
15. Al-Riyami, S.S.: Cryptographic Schemes based on elliptic curve pairings. Ph.D. thesis. Information Security Group, Department of Mathematics, Royal Holloway, University of London (2004)
16. Dubuisson, O.: ASN.1 - Communication Between Heterogeneous Systems. Academic Press, San Diego (2001)
17. Fray El, I., Hyla, T., Maćków, W., Pejaś J.: Authentication and authorization in multilevel security systems for public administration. Pomiary Automatyka Kontrola, vol. 56, no. 8, pp. 983–987 (2010)
18. Cooper, D. et al.: RFC 5280 - Internet X.509 Public Key Infrastructure Certificate and Certificate Revocation List (CRL) Profile (2008)
19. Santesson, S., et al.: RFC 6960 - X.509 Internet Public Key Infrastructure Online Certificate Status Protocol - OCSP (2013)

20. Libert, B., Quisquater, J.-J.: What is possible with identity based cryptography for PKIs and what still must be improved? In: Katsikas, S.K., Gritzalis, S., López, J. (eds.) EuroPKI 2004. LNCS, vol. 3093, pp. 57–70. Springer, Heidelberg (2004)
21. Hyla, T., Bielecki, W., Pejaś, J.: Non-repudiation of electronic health records in distributed healthcare systems. Pomiary, Automatyka, Kontrola, vol. 56, no. 10, pp. 1170–1173 (2010)
22. Pejaś, J.: Signed electronic document and its probative value in certificate and certificateless public key cryptosystem infrastructures. Elektronika 11, 30–34 (2009)
23. Hyla, T., El Fray, I., Maćków, W., Pejaś, J.: Long-term preservation of digital signatures for multiple groups of related documents. IET Inf. Sec. 6(3), 219–227 (2012)
24. Hyla, T., Pejaś, J.: A practical certificate and identity based encryption scheme and related security architecture. In: Saeed, K., Chaki, R., Cortesi, A., Wierzchoń, S. (eds.) CISIM 2013. LNCS, vol. 8104, pp. 190–205. Springer, Heidelberg (2013)
25. Hyla, T., Pejaś, J.: Certificate-based encryption scheme with general access structure. In: Cortesi, A., Chaki, N., Saeed, K., Wierzchoń, S. (eds.) CISIM 2012. LNCS, vol. 7564, pp. 41–55. Springer, Heidelberg (2012)
26. Desmedt, Y., Burmester, M.: Identity-based key Infrastructures (IKI). In: Deswarte, Y., Cuppens, F., Jajodia, S., Wang, L. (eds.) Security and Protection in Information Processing Systems, IFIP International Federation for Information Processing, vol. 147, pp. 167–176. Springer, Dordrecht (2004)
27. Kiltz, E., Neven, G.: Identity-based signatures. In: Joye, M., Neven, G. (eds.) Identity-Based Cryptography, pp. 31–44. IOS Press, Amsterdam (2009)

Multimedia Systems

An Algorithm for the Vandermonde Matrix-Vector Multiplication with Reduced Multiplicative Complexity

Aleksandr Cariow$^{(\boxtimes)}$ and Galina Cariowa

West Pomeranian University of Technology, Żołnierska 52,
71-210 Szczecin, Poland
{acariow, gcariowa}@wi.zut.edu.pl

Abstract. In this chapter an algorithm for computing the Vandermonde matrix-vector product is presented. The main idea of constructing this algorithm is based on the using of Winograd's formula for inner product computation. Multiplicative complexity of the proposed algorithm is less than multiplicative complexity of the schoolbook (naïve) method of calculation. If the schoolbook method requires MN multiplications and $M(N-1)$ additions, the proposed algorithm takes only $M + N(M + 1)/2$ multiplications at the cost of extra additions compared to the naïve method. From point of view its hardware realization on VLSI where the implementation cost of multiplier is significantly greater than implementation cost of adder, the new algorithm is generally more efficient than a naïve algorithm. When the order of the Vandermonde matrix is relatively small, this algorithm will have smaller multiplicative complexity than the well-known "fast" algorithm for the same task.

Keywords: Vandermonde matrix · Matrix-vector multiplication · Fast algorithms

1 Introduction

Vandermonde matrices play an important role in many theoretical and practical applications, such as automatic control theory, polynomial interpolation, digital signal processing, coding theory, and wireless communications [1–8]. The most time-consuming operation involving the use of the Vandermonde matrix is its multiplication by an arbitrary vector. Therefore, finding ways to reduce the number of multipliers in the implementation of vector-matrix multiplier is an extremely urgent task. It is known that a main kernel during the calculation of vector-matrix product is the inner product operation. In calculating a vector-matrix product the inner product of vectors is computed as many times as rows the matrix contains. To reduce the number of multiplications required to implement this kernel, we use the Winograd's formula for calculating inner product of two vectors.

© Springer International Publishing AG 2017
S. Kobayashi et al. (eds.), *Hard and Soft Computing for Artificial Intelligence, Multimedia and Security*, Advances in Intelligent Systems and Computing 534, DOI 10.1007/978-3-319-48429-7_17

2 Preliminary Remarks

A Vandermonde $M \times N$ matrix is completely determined by M arbitrary numbers $v_1, v_2, \ldots, v_M$ in term of which its elements are integer powers v_m^n, $m = 0, 1, \ldots, M - 1$, $n = 0, 1, \ldots, N - 1$:

$$
\begin{bmatrix} y_0 \\ y_1 \\ \vdots \\ y_{M-1} \end{bmatrix} = \begin{bmatrix} v_0^0 & v_0^1 & \cdots & v_0^{N-1} \\ v_1^0 & v_1^1 & \cdots & v_1^{N-1} \\ \vdots & \vdots & \ddots & \vdots \\ v_{M-1}^0 & v_{M-1}^1 & \cdots & v_{M-1}^{N-1} \end{bmatrix} \begin{bmatrix} x_0 \\ x_1 \\ \vdots \\ x_{M-1} \end{bmatrix} \tag{1}
$$

The challenge is to multiply the Vandermonde matrix $\mathbf{V}_{M \times N}$ by an arbitrary vector

$$
\mathbf{Y}_{M \times 1} = \mathbf{V}_{M \times N} \mathbf{X}_{N \times 1} \tag{2}
$$

where $\mathbf{X}_{N \times 1} = [x_0, x_1, \ldots, x_{N-1}]^{\mathrm{T}}$ - is an input data vector, and $\mathbf{Y}_{M \times 1} = [y_0, y_1, \ldots, y_{M-1}]^{\mathrm{T}}$ - is an output data vector.

The schoolbook (naïve) way of implementing the calculations in accordance with (2) requires MN multiplications and $M(N - 1)$ additions. The actual problem is the task of finding ways to reduce the computational complexity of the implementation of this operation.

We shall present the algorithm, which reduce multiplicative complexity to $M + N(M + 1)/2$ multiplications at the cost of many more additions.

It should be noted that the scientific literature contains mention of some "fast" algorithm for $M \times N$ Vandermonde matrix-vector multiplication that provides the computational complexity of order $O((M + N) \log_2^2(M + N))$ [9]. Nevertheless, we have not found any practically useful paper, which would clearly and accessible presented the peculiar properties and nuances for constructing such an algorithm. We however will assume by default that such an algorithm exists. But even with this assumption, our algorithm has lower multiplicative complexity for small order Vandermonde matrix-vector multiplication. At the same time, it must be emphasized that namely the small order Vandermonde matrices are used in the many practical applications. In these cases, applying of our algorithm may be more preferable.

It should also be noted that during the construction of algorithms the extensive use of the principles of parallelization and vectorization of data processing will be made. Therefore, it is assumed that the synthesized procedures will be implemented on the hardware platforms with parallel processing.

In the general case a fully parallel hardware implementation of arbitrary $M \times N$ matrix-vector multiplication requires MN multipliers. Minimizing the number of multiplications is especially important in the design of specialized VLSI mathematical or DSP processors because reducing the number of multipliers also reduces the power dissipation and lowers the cost implementation of the entire system being implemented. Moreover, a hardware multiplier is more complicated unit than an adder and occupies much more chip area than the adder [10]. Even if the chip already contains embedded multipliers, their number is always limited. This means that if the implemented algorithm has a large number of multiplications, the projected processor may not always fit into the chip and the problem of minimizing the number of multiplications remains relevant It is known that a main kernel during the calculation of arbitrary matrix-vector product is the inner product operation. In calculating a matrix-vector product the inner product of vectors is computed as many times as rows the matrix contains. To reduce the number of two-input multipliers required to implement Vandermonde matrix-vector multiplier, we use the Winograd's formula for calculating inner product of two vectors. Below, we briefly review the essence of computing the vector-matrix product using the Winograd's inner product identity.

3   Brief Background

Let $\mathbf{X}_{N \times 1} = [x_0, x_1, \ldots, x_{N-1}]^T$ and $\mathbf{Y}_{M \times 1} = [y_0, y_1, \ldots, y_{M-1}]^T$ - are N-point and M-point one dimensional data vectors respectively, and $m = 0, 1, \ldots, M - 1, n = 0, 1, \ldots, N - 1$.

The problem is to calculate a product

$$\mathbf{Y}_{M \times 1} = \mathbf{A}_{M \times N} \mathbf{X}_{N \times 1} \tag{3}$$

where

$$\mathbf{A}_{M \times N} = \begin{bmatrix} a_{0,0} & a_{0,1} & \cdots & a_{0,N-1} \\ a_{1,0} & a_{1,1} & \cdots & a_{1,N-1} \\ \vdots & \vdots & \ddots & \vdots \\ a_{M-1,0} & a_{M-1,1} & \cdots & a_{M-1,N-1} \end{bmatrix},$$

and $a_{m,n}$- are l-bit fixed-point variables.

According to Winograd's formula for inner product calculation each element of vector $\mathbf{Y}_{M \times 1}$ can be calculated as follows [11]:

$$y_m = \sum_{k=0}^{\frac{N}{2}-1} [(a_{m,2k} + x_{2k+1})(a_{m,2k+1} + x_{2k})] - c_m - \xi(N) \tag{4}$$

where

$$c_m = \sum_{k=0}^{\frac{N}{2}-1} a_{m,2k} \cdot a_{m,2k+1}, \xi(N) = \sum_{k=0}^{\frac{N}{2}-1} x_{2k} \cdot x_{2k+1}$$

and N is even.

Let us now consider an example of calculating the 8×8 Vandermonde matrix-vector product:

$$
\begin{bmatrix} y_0 \\ y_1 \\ y_2 \\ y_3 \\ y_4 \\ y_5 \\ y_6 \\ y_7 \end{bmatrix} = \begin{bmatrix} v_0^0 & v_0^1 & v_0^2 & v_0^3 & v_0^4 & v_0^5 & v_0^6 & v_0^7 \\ v_1^0 & v_1^1 & v_1^2 & v_1^3 & v_1^4 & v_1^5 & v_1^6 & v_1^7 \\ v_2^0 & v_2^1 & v_2^2 & v_2^3 & v_2^4 & v_2^5 & v_2^6 & v_2^7 \\ v_3^0 & v_3^1 & v_3^2 & v_3^3 & v_3^4 & v_3^5 & v_3^6 & v_3^7 \\ v_4^0 & v_4^1 & v_4^2 & v_4^3 & v_4^4 & v_4^5 & v_4^6 & v_4^7 \\ v_5^0 & v_5^1 & v_5^2 & v_5^3 & v_5^4 & v_5^5 & v_5^6 & v_5^7 \\ v_6^0 & v_6^1 & v_6^2 & v_6^3 & v_6^4 & v_6^5 & v_6^6 & v_6^7 \\ v_7^0 & v_7^1 & v_7^2 & v_7^3 & v_7^4 & v_7^5 & v_7^6 & v_7^7 \end{bmatrix} \times \begin{bmatrix} x_0 \\ x_1 \\ x_2 \\ x_3 \\ x_4 \\ x_5 \\ x_6 \\ x_7 \end{bmatrix} \tag{5}
$$

Using Winograd's inner product formula this product can be rewritten as follows:

$$
y_0 = \sum_{i=0}^{7} v_0^i x_i = (x_0 + v_0^1)(x_1 + v_0^0) + (x_2 + v_0^3)(x_3 + v_0^2) + (x_4 + v_0^5)(x_5 + v_0^4) + (x_6 + v_0^7)(x_7 + v_0^6) - c_0 - \xi(8),
$$

$$
y_1 = \sum_{i=0}^{7} v_1^i x_i = (x_0 + v_1^1)(x_1 + v_1^0) + (x_2 + v_1^3)(x_3 + v_1^2) + (x_4 + v_1^5)(x_5 + v_1^4) + (x_6 + v_1^7)(x_7 + v_1^6) - c_1 - \xi(8),
$$

$$
y_2 = \sum_{i=0}^{7} v_2^i x_i = (x_0 + v_2^1)(x_1 + v_2^0) + (x_2 + v_2^3)(x_3 + v_2^2) + (x_4 + v_2^5)(x_5 + v_2^4) + (x_6 + v_2^7)(x_7 + v_2^6) - c_2 - \xi(8),
$$

$$
y_3 = \sum_{i=0}^{7} v_3^i x_i = (x_0 + v_3^1)(x_1 + v_3^0) + (x_2 + v_3^3)(x_3 + v_3^2) + (x_4 + v_3^5)(x_5 + v_3^4) + (x_6 + v_3^7)(x_7 + v_3^6) - c_3 - \xi(8),
$$

$$
y_4 = \sum_{i=0}^{7} v_4^i x_i = (x_0 + v_4^1)(x_1 + v_4^0) + (x_2 + v_4^3)(x_3 + v_4^2) + (x_4 + v_4^5)(x_5 + v_4^4) + (x_6 + v_4^7)(x_7 + v_4^6) - c_4 - \xi(8),
$$

$$
y_5 = \sum_{i=0}^{7} v_5^i x_i = (x_0 + v_5^1)(x_1 + v_5^0) + (x_2 + v_5^3)(x_3 + v_5^2) + (x_4 + v_5^5)(x_5 + v_5^4) + (x_6 + v_5^7)(x_7 + v_5^6) - c_5 - \xi(8),
$$

$$
y_6 = \sum_{i=0}^{7} v_6^i x_i = (x_0 + v_6^1)(x_1 + v_6^0) + (x_2 + v_6^3)(x_3 + v_6^2) + (x_4 + v_6^5)(x_5 + v_6^4) + (x_6 + v_6^7)(x_7 + v_6^6) - c_6 - \xi(8),
$$

$$
y_7 = \sum_{i=0}^{7} v_7^i x_i = (x_0 + v_7^1)(x_1 + v_7^0) + (x_2 + v_7^3)(x_3 + v_7^2) + (x_4 + v_7^5)(x_5 + v_7^4) + (x_6 + v_7^7)(x_7 + v_7^6) - c_7 - \xi(8),
$$

where

$$
c_0 = v_0^0 v_0^1 + v_0^2 v_0^3 + v_0^4 v_0^5 + v_0^6 v_0^7 = v_0^1 + v_0^5 + v_0^7 (v_0^2 + v_0^6),
$$

$$
c_1 = v_1^0 v_1^1 + v_1^2 v_1^3 + v_1^4 v_1^5 + v_1^6 v_1^7 = v_1^1 + v_1^5 + v_1^7 (v_1^2 + v_1^6),
$$

$$
c_2 = v_2^0 v_2^1 + v_2^2 v_2^3 + v_2^4 v_2^5 + v_2^6 v_2^7 = v_2^1 + v_2^5 + v_2^7 (v_2^2 + v_2^6),
$$

$$
c_3 = v_3^0 v_3^1 + v_3^2 v_3^3 + v_3^4 v_3^5 + v_3^6 v_3^7 = v_3^1 + v_3^5 + v_3^7 (v_3^2 + v_3^6),
$$

$$
c_4 = v_4^0 v_4^1 + v_4^2 v_4^3 + v_4^4 v_4^5 + v_4^6 v_4^7 = v_4^1 + v_4^5 + v_4^7 (v_4^2 + v_4^6),
$$

$$
c_5 = v_5^0 v_5^1 + v_5^2 v_5^3 + v_5^4 v_5^5 + v_5^6 v_5^7 = v_5^1 + v_5^5 + v_5^7 (v_5^2 + v_5^6),
$$

$$
c_6 = v_6^0 v_6^1 + v_6^2 v_6^3 + v_6^4 v_6^5 + v_6^6 v_6^7 = v_6^1 + v_6^5 + v_6^7 (v_6^2 + v_6^6),
$$

$$
c_7 = v_7^0 v_7^1 + v_7^2 v_7^3 + v_7^4 v_7^5 + v_7^6 v_7^7 = v_7^1 + v_7^5 + v_7^7 (v_7^2 + v_7^6).
$$

$$
\xi(8) = x_0 x_1 + x_2 x_3 + x_4 x_5 + x_6 x_7.
$$

Assume now for simplicity (and in fact without loss of generality) that $N = 2p$ for some positive integer p.

Equation (4) can for Vandermonde matrix can rewrite as follows:

$$y_m = \sum_{i=0}^{N-1} v_m^i x_i = \sum_{k=0}^{\frac{N}{2}-1} (x_{2k} + v_m^{2k+1})(x_{2k+1} + v_m^{2k}) - c_m - \xi(N) \tag{6}$$

where

$$\xi(N) = x_0 x_1 + x_2 x_3 + \cdots + x_{N-2} x_{N-1}.$$

.

If we look carefully at the equations for $\{c_m\}$ and remember that the $v_m^r v_m^s = v_m^{r+s}$ we can rewrite this equations in two following unified form:

$$c_m = \sum_{l=0}^{\delta-1} v_m^{4l+1} + v_m^{N-1} \sum_{l=0}^{\delta} v_m^{4l}, \text{ if } p \text{ is odd} \tag{7}$$

or

$$c_m = \sum_{l=0}^{\delta-1} v_m^{4l+1} + v_m^{N-1} \sum_{l=0}^{\delta-1} v_m^{4l+2}, \text{ if } p \text{ is even} \tag{8}$$

$\delta = \lfloor N/4 \rfloor$ - is the largest integer not greater than $N/4$.

It is easy to see that the computation of c_m requires only one multiplication. This fact allows us to construct an efficient algorithm for computing the Vandermonde matrix-vector product. Let us consider the synthesis of the proposed algorithm in more detail.

4   The Algorithm

First, we splits vector $\mathbf{X}_{N \times 1}$ into two vectors $\mathbf{X}_{\frac{N}{2} \times 1}^{(1)}$ and $\mathbf{X}_{\frac{N}{2} \times 1}^{(2)}$ containing only even-numbered and only odd-numbered elements respectively:

$$\mathbf{X}_{\frac{N}{2} \times 1}^{(1)} = [x_0, x_2, \ldots, x_{N-2}]^{\mathrm{T}}, \mathbf{X}_{\frac{N}{2} \times 1}^{(2)} = [x_1, x_3, \ldots, x_{N-1}]^{\mathrm{T}}.$$

Then from the elements of the matrix $\mathbf{V}_{M \times N}$ we form two super-vectors of data:

$$\mathbf{V}_{\frac{MN}{2} \times 1}^{(1)} = [\widehat{\mathbf{V}}_{M \times 1}^{(0)}, \widehat{\mathbf{V}}_{M \times 1}^{(1)}, \ldots, \widehat{\mathbf{V}}_{M \times 1}^{(\frac{N}{2}-1)}]^{\mathrm{T}} \tag{9}$$

$$\mathbf{V}_{\frac{MN}{2} \times 1}^{(2)} = [\widecheck{\mathbf{V}}_{M \times 1}^{(0)}, \widecheck{\mathbf{V}}_{M \times 1}^{(1)}, \ldots, \widecheck{\mathbf{V}}_{M \times 1}^{(\frac{N}{2}-1)}]^{\mathrm{T}} \tag{10}$$

where

$$\hat{\mathbf{V}}_{M\times1}^{(k)} = [v_0^{2k+1}, v_1^{2k+1}, \ldots, v_{M-1}^{2k+1}]^{\mathrm{T}} \tag{11}$$

$$\breve{\mathbf{V}}_{M\times1}^{(k)} = [v_0^{2k}, v_1^{2k}, \ldots, v_{M-1}^{2k}]^{\mathrm{T}}, k = 0, 1, \ldots, \frac{N}{2}-1, \tag{12}$$

We also define the vectors

$$\mathbf{C}_{M\times1} = [-c_0, -c_1, \ldots, -c_{M-1}]^{\mathrm{T}} \tag{13}$$

$$\mathbf{E}_{M\times1} = [-\underbrace{\xi(N), -\xi(N), \ldots, -\xi(N)}_{M \text{ times}}] \tag{14}$$

Next, we introduce some auxiliary matrices:

$$\mathbf{P}_{\frac{MN}{2}\times\frac{N}{2}} = (\mathbf{I}_{\frac{N}{2}} \otimes \mathbf{1}_{M\times1}), \Sigma_{M\times\frac{MN}{2}} = (\mathbf{1}_{1\times\frac{N}{2}} \otimes \mathbf{I}_M) \tag{15}$$

where $\mathbf{1}_{M\times N}$ - is an $M \times N$ matrix of ones (a matrix where every element is equal to one), $\mathbf{I}_{N}$- is an identity $N \times N$ matrix and sign "$\otimes$" denotes tensor product of two matrices [12].

Using the above matrices the rationalized computational procedure for calculating the constant matrix-vector product can be written as follows:

$$\mathbf{Y}_{M\times1} = \mathbf{E}_{M\times1} + \{\mathbf{C}_{M\times1} + [\Sigma_{M\times\frac{MN}{2}}\mathbf{D}_{\frac{MN}{2}}(\mathbf{A}_{\frac{MN}{2}\times1}^{(1)} + \mathbf{P}_{\frac{MN}{2}\times\frac{N}{2}}\mathbf{X}_{\frac{N}{2}\times1}^{(1)})]\} \tag{16}$$

where

$$\mathbf{D}_{\frac{MN}{2}} = diag\,(\mathbf{D}_M^{(0)}, \mathbf{D}_M^{(1)}, \ldots, \mathbf{D}_M^{(\frac{N}{2}-1)}) \tag{17}$$

and

$$\mathbf{D}_M^{(k)} = diag(s_0^{(k)}, s_1^{(k)}, \ldots, s_{M-1}^{(k)})\, k = 0, 1, \ldots, \frac{N}{2}-1 \tag{18}$$

If the elements of $\mathbf{D}_{\frac{MN}{2}}$ placed vertically without disturbing the order and written in the form of the vector $\mathbf{S}_{\frac{MN}{2}\times1} = [s_0^{(k)}, s_1^{(k)}, \ldots, s_{M-1}^{(k)}]^{\mathrm{T}}$, then they can be calculated using the following vector-matrix procedure:

$$\mathbf{S}_{\frac{MN}{2}\times1} = \mathbf{V}_{\frac{MN}{2}\times1}^{(2)} + \mathbf{P}_{\frac{MN}{2}\times\frac{M}{2}}\mathbf{X}_{\frac{N}{2}\times1}^{(2)} \tag{19}$$

For example, suppose $M = N = 8$. Then procedures (17)–(18) take following form:

$$\mathbf{Y}_{8\times1} = \mathbf{E}_{8\times1} + \{\mathbf{C}_{8\times1} + [\mathbf{\Sigma}_{8\times32}\mathbf{D}_{32}(\mathbf{V}_{32\times1}^{(1)} + \mathbf{P}_{32\times4}\mathbf{X}_{4\times1}^{(1)})]\} \tag{20}$$

$$\mathbf{S}_{32\times1} = \mathbf{V}_{32\times1}^{(2)} + \mathbf{P}_{32\times4}\mathbf{X}_{4\times1}^{(2)} \tag{21}$$

where

$$\mathbf{X}_{4\times1}^{(1)} = [x_0, x_2, x_4, x_6]^{\mathrm{T}}, \mathbf{X}_{4\times1}^{(2)} = [x_1, x_3, x_5, x_7]^{\mathrm{T}},$$

$$\mathbf{V}_{32\times1}^{(1)} = [\widehat{\mathbf{V}}_{8\times1}^{(0)}, \widehat{\mathbf{V}}_{8\times1}^{(1)}, \widehat{\mathbf{V}}_{8\times1}^{(2)}, \widehat{\mathbf{V}}_{8\times1}^{(3)}]^{\mathrm{T}}, \mathbf{V}_{32\times1}^{(2)} = [\breve{\mathbf{V}}_{8\times1}^{(0)}, \breve{\mathbf{V}}_{8\times1}^{(1)}, \breve{\mathbf{V}}_{8\times1}^{(2)}, \breve{\mathbf{V}}_{8\times1}^{(3)}]^{\mathrm{T}},$$

$$\widehat{\mathbf{V}}_{8\times1}^{(k)} = [v_0^{2k+1}, v_1^{2k+1}, \ldots, v_7^{2k+1}]^{\mathrm{T}}, \breve{\mathbf{V}}_{8\times1}^{(k)} = [v_0^{2k}, v_1^{2k}, \ldots, v_7^{2k}]^{\mathrm{T}},$$

$$\mathbf{C}_{8\times1} = [-c_0, -c_1, \ldots, -c_7]^{\mathrm{T}}, c_m = \sum_{l=0}^{1} v_m^{4l+1} + v^7 \sum_{l=0}^{1} v_m^{4l+2},$$

$$\mathbf{E}_{8\times1} = [\underbrace{-\xi(8), -\xi(8), \ldots, -\xi(8)}_{8\ times}],$$

$$\mathbf{P}_{32\times4} = (\mathbf{I}_4 \otimes \mathbf{1}_{8\times1}) = \begin{bmatrix} \begin{matrix}1\\1\\ \vdots \\1\end{matrix} & \mathbf{0}_{8\times1} & \mathbf{0}_{8\times1} & \mathbf{0}_{8\times1} \\ \mathbf{0}_{8\times1} & \begin{matrix}1\\1\\ \vdots \\1\end{matrix} & \mathbf{0}_{8\times1} & \mathbf{0}_{8\times1} \\ \vdots & \vdots & \ddots & \begin{matrix}\vdots\\1\\1\end{matrix} \\ \mathbf{0}_{8\times1} & \mathbf{0}_{8\times1} & \mathbf{0}_{8\times1} & \begin{matrix}\vdots\\1\end{matrix} \end{bmatrix}$$

$$\mathbf{D}_{32} = diag\,(\mathbf{D}_8^{(0)}, \mathbf{D}_8^{(1)}, \mathbf{D}_8^{(2)}, \mathbf{D}_8^{(3)}), \mathbf{D}_8^{(k)} = diag(s_0^{(k)}, s_1^{(k)}, \ldots, s_7^{(k)}), k = 0, 1, \ldots, 3.$$

$$\mathbf{\Sigma}_{8\times32} = (\mathbf{1}_{1\times4} \otimes \mathbf{I}_8) = \begin{bmatrix} 1 & & & 1 & & & & 1 & & \\ & 1 & & & 1 & & \cdots & & 1 & \\ & & \ddots & & & \ddots & & & & \ddots \\ & & 1 & & & 1 & & & & 1 \end{bmatrix}$$

Figure 1 shows a data flow diagram of the proposed algorithm for computation of the Vandermonde matrix-vector product for case $M = N = 8$ in accordance to procedure (20). Figure 2 shows a data flow diagram of the process for calculating the

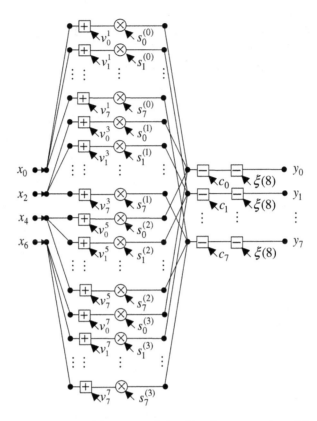

Fig. 1. The data flow diagram of the proposed algorithm for computation of the Vandermonde matrix-vector product for $M = N = 8$ in accordance to procedure (20)

matrix $\mathbf{D}_{32}$ (vector $\mathbf{S}_{32 \times 1}$) elements. Straight lines in the figures denote the operations of data transfer. The squares and circles in these figures indicate operations of additions (substractions) and multiplications, respectively. The appropriate summands and the multiplicands are attributed near with incoming arrows. Points where lines converge denote summations.

5 Computation Complexity

Now let us consider the computational complexity needed by the proposed algorithm in terms of the number of arithmetical operations. We calculate how many multiplications and additions are required by our algorithm, and compare this with the number required for a direct implementation of matrix–vector product in Eq. (1). The number of multiplications required for realization of the proposed algorithm is $M + N(M + 1)/2$. Thus using the proposed algorithm the number of multiplications to implement the Vandermonde matrix-vector product is significantly reduced. The number of additions required by our algorithm is $[N(4M + 1)/2] - 1$.

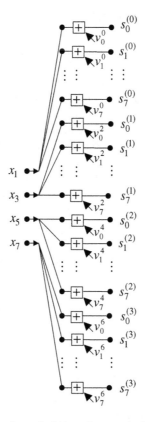

Fig. 2. The data flow diagram for calculating the matrix $\mathbf{D}_{32}$ elements in accordance to procedure (21) for the same example

The number of multiplications required for implementation of "schoolbook" method for matrix-vector multiplication is MN. This implementation also requires $M(N-1)$ additions. Thus, our proposed algorithm saves almost 50 % multiplications (through the using of Winograd's identity for the inner product of two vectors) but significantly increases number of additions compared with direct method of matrix-vector product computation. This allows concluding that the suggested solution may be useful in a number of cases and have practical application allowing to minimize the Vandermonde matrix-vector multiplier's hardware implementation cost.

6 Conclusion

In this paper, we have presented an original hardware-oriented algorithm allowing to multiply the Vandermonde matrix by arbitrary vector with reduced multiplicative complexity. Unfortunately, the proposed algorithm generally needs more additions compared to the direct way of computing the Vandermonde matrix-vector product. Today, the multiplications can be performed almost as fast as additions, but the hardware required to perform multiplication at the maximum possible speed is much more expensive than the hardware required to perform an addition. Then, reducing the number of multiplications is especially important in the VLSI design because the minimization of the number of multipliers also reduces the power dissipation and lowers the power consumption of the entire implemented system.

It should also be noted that the proposed algorithm can be implemented with different degrees of parallelism. It is not necessarily a complete parallelization of computations. Is possible to use parallel-serial implementation computing mode when M multiplier-accumulators are working in parallel, recurrently. Another way of implementing the calculation may use $N/2$ multipliers and $N/2$-input adder. There are also other ways of implementation of the proposed algorithm. Each such way has both advantages and disadvantages. Either way, the proposed algorithm offers developers some additional possibilities in the design of the Vandermonde matrix-vector multiplier.

References

1. Joseph, J.: Rushanan: on the vandermonde matrix. Am. Math. Mon. **96**(10), 921–924 (1989)
2. Kailath, T.: Linear Systems. Prentice Hall, Inc., Englewood Cliffs (1980)
3. Waterfall, J.J., Joshua, J., Casey, F.P., Gutenkunst, R.N., Brown, K.S., Myers, C.R., Brouwer, P.W., Elser, V., Sethna, J.P.: Sloppy model universality class and the Vandermonde matrix. Phys. Rev. Lett. **97**(15), 150601 (2006)
4. Sampaio, L., Kobayashi, M., Ryan, Ø., Debbah, M.: Vandermonde frequency division multiplexing. In: 9th IEEE Workshop on Signal Processing Advances for Wireless Applications, Recife, Brazil (2008)
5. Cardoso, L.S., Kobayashi, M., Debbah, M., Ryan, Ø.: Vandermonde frequency division multiplexing for cognitive radio. arXiv:0803.0875v1 [Cs. IT], 6 March 2008
6. Wang, Z., Scaglione, A., Giannakis, G., Barbarossa, S.: Vandermonde-Lagrange mutually orthogonal flexible transceivers for blind CDMA in unknown multipath. In: Proceedings of IEEE-SP Workshop on Signal Processing Advances in Wireless Communications, May 1999, pp. 42–45 (1999)
8. Lacan, J., Fimes, J.: Systematic MDS erasure codes based on vandermonde matrices. IEEE Commun. Lett. **8**(9), 570–572 (2004)
9. Pan, V.: Structured Matrices and Polynomials: Unified Superfast Algorithms. Birkhäuser, Boston (2001)
10. Cariow, A., Cariowa, G.: A parallel hardware-oriented algorithm for constant matrix-vector multiplication with reduced multiplicative complexity. Measure. Autom. Monit. **60**(7), 510–512 (2014)
11. Winograd, S.: A new algorithm for inner product. IEEE Trans. Comput. **17**(7), 693–694 (1968)
12. Granata, J., Conner, M., Tolimieri, R.: The tensor product: a mathematical programming language for FFTs and other fast DSP operations. IEEE Sig. Process. Mag. **9**(1), 40–48 (1992)

The Use of the Objective Digital Image Quality Assessment Criterion Indication to Create Panoramic Photographs

Jakub Peksinski[1(✉)], Grzegorz Mikolajczak[1], and Janusz Kowalski [2]

[1] Faculty of Electrical Engineering, West Pomeranian University of Technology,
Sikorskiego St. 37, 70-313 Szczecin, Poland
{jpeksinski,grzegorz.mikolajczak}@zut.edu.pl
[2] Faculty of Medicine, Pomeranian Medical University,
Rybacka St. 1, 70-204 Szczecin, Poland
janus@sci.pum.edu.pl

Abstract. In the article, the authors presented a method of creating panoramic photographs developed by themselves. For its operation, the method uses a matching algorithm developed by the authors, which is based on the analysis of the indication of a popular digital image assessment quality measure - Universal Quality Index. The result of applying the suggested algorithm is an effective match of the sequence of partial digital photographs which make the panoramic photograph.

Keywords: Digital image · Quality measure · Matching · Synchronisation · Rotation axis

1 Introduction

Development and popularity of digital photography, which occurred in the last decade, facilitated the design and introduction of still newer, more effective and sophisticated methods and algorithms used for analysis and digital image processing [9, 10]. The developed tools and algorithms are implemented both in photographic equipment and in commercial computer software like: Corel or Adobe Photoshop as well as in free software created by enthusiasts. All these algorithms and methods are included in a very large group of the so called digital image processing algorithms.

The concept of digital image processing is very wide and includes many issues beginning from simple point operations on images and ending with very complex operations including in their scope many aspects of the digital image.

In digital image processing, there is a group of algorithms used for assessment of digital images [11]. This group includes many classic measures, which have been used for many years, such as popular measures like: Mean Square Error (MSE), Normalized Mean Square Error (NMSE) etc. [1, 2], as well as new, recently introduced measures.

The group of new measures includes a very popular and more frequently used, yet relatively new measure, which is Universal Quality Image Index (Q) [3, 4] described with formula 1.

© Springer International Publishing AG 2017
S. Kobayashi et al. (eds.), *Hard and Soft Computing for Artificial Intelligence, Multimedia and Security*, Advances in Intelligent Systems and Computing 534, DOI 10.1007/978-3-319-48429-7_18

Analysing the [Q] measure described with formula 1, it is possible to notice that its indications depend on three aspects: correlation, luminance and contrast.

This is extremely significant since the measure indication is, therefore, strongly correlated with human image perception.

$$Q = \frac{4\sigma_{xy}\bar{x}\bar{y}}{(\sigma_x^2 + \sigma_y^2)(\bar{x}^2 + \bar{y}^2)} \tag{1}$$

where:

$$\bar{x} = \frac{1}{N}\sum_{i=1}^{N} x_i; \ \bar{y} = \frac{1}{N}\sum_{i=1}^{N} y_i; \ \sigma_x^2 = \frac{1}{N-1}\sum_{i=1}^{N} (x_i - \bar{x})^2$$

$$\sigma_y^2 = \frac{1}{N-1}\sum_{i=1}^{N} (y_i - \bar{y})^2 \ ; \ \sigma_{xy} = \frac{1}{N-1}\sum_{i=1}^{N} (x_i - \bar{x})(y_i - \bar{y})$$

In the article, the authors presented the possibilities of using the [Q] measure indication not for the purposes it was intended for - i.e. assessment of digital image quality with the use of a reference image, but for creating of panoramic photographs from several digital photographs presenting a given scene.

Creation of panoramic images has fascinated photographers since the very beginning of cameras. They got interested in methods that enable extending of the visual field of a common camera, reflecting what a human eye can see and overcoming of the initial technical limitations (which enabled them to obtain a photograph showing a view that is larger than the standard one). This resulted in development of several methods of creating digital photography [5]. A popular and widely used method is the one used while creating the principles governing circular panoramas. It involves a series of pictures taken by rotating a camera around its axis after each exposition by a certain angle determined in advance. In this manner, by making a full turn, the edges of the captured image converge together. Such editing resembled a panorama consisting of flat images.

In the article, the authors used the above mentioned idea to create panoramic photographs, i.e. subsequent photographs for the panorama were taken from one location and the camera was rotated by a certain constant angle.

2   The Method of Creating a Panoramic Photograph with the Use of the Q Measure

To make the algorithm for creating panoramic photographs proposed in the article effective, it is necessary to provide proper conditions while taking partial photographs making the panoramic photograph, which are as follows:

- photographs must be taken with the same camera;
- photographs must be taken from the same location, the same point;
- photographs must differ only with rotation around one's own axis, i.e. without leaning or changing the location where the first photograph was taken;

a)   b)

Fig. 1. A series of two photographs presenting a given scene with marked overlapping parts of the photos (Color figure online)

- photographs must overlap each other;
- the exposure and zoom of subsequent photographs must be the same. The camera must be set to manual settings.

If the above conditions are met, it is possible to begin the creation of panoramic photographs. The principle and sequence of creating a panoramic photograph with the use of the algorithm proposed by the authors is as follows.

Firstly, a series of photographs is taken of the scene of which we want to obtain a panoramic photograph. The subsequent photographs must be taken in such a way that the following photograph must partly overlap the preceding one. This is presented in Fig. 1 showing a series of two photographs taken from the same location where the overlapping fragments in both photographs are marked in red [12].

Having taken a series of photographs, the next step is joining them together. To this end, the neighbouring photographs must be properly synchronised. To achieve mutual matching of the photographs, the authors used the algorithm developed by themselves based on analysis of the [Q] measure indication. The whole process of matching photographs was as follows.

Out of the fragments of neighbouring photographs which overlap each other, a fragment of the image is selected with dimensions $N \times N$ pixels that is located in the same place on the X and Y plane both in photograph a and b. This is presented in Fig. 2.

In this way, we obtain a pair of fragments of image 1a–b located in the same places on the X and Y plane both in image "a" and in image "b".

In order to create a proper panoramic image out of photographs "a" and "b" we have to match (synchronise) them. They must be connected in the places where the photographs overlap each other. If we do not perform this process in a precise manner, many defects will appear on the panoramic photograph like e.g. the so called parallax phenomenon. This means that the photographs must be matched in the following manner:

- taking into consideration the translation in the X and Y plane;
- taking into consideration the rotation axis between the photographs.

The process of synchronising images begins with checking the indication of the Q measure between the selected pairs of fragments of images a and b, which will enable determination whether the photographs are correlated against each other and whether

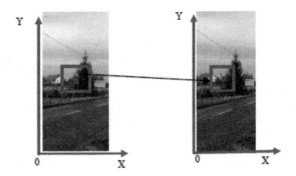

Fig. 2. The overlapping areas with three marked fragments

the photographs require synchronisation. For fragment 1a-1b of the image presented in Fig. 2 the value of the Q measure indication is presented in Table 1.

As it can be seen, the measure indication deviates from value 1 (exactly the same images), which means that the photographs are not correlated with each other. It is true because, analysing formula 1, it can be seen that one of the elements of the [Q] measure is the correlation coefficient. This means that indication of the above mentioned measure depends very strongly on translation between two images. The problem has been presented and described in publications [6, 7].

Thanks to specifying the initial indication value, we know that the selected fragments are not properly synchronised and the matching process is necessary.

The synchronisation process of the studied fragments begins with finding the translation of the studied fragments in the X and Y plane. To this end, the selected fragments 1a-1b are divided into a range of segments with dimensions $M \times M$ each - this is presented in Fig. 3.

In each of such segments, the value of the [Q] measure indication is calculated. For the image fragment in Fig. 2, the [Q] measure indication in individual segments 1-1″, ..., 8-8″ is presented in Table 2.

The average [Q] measure indication value from individual segments was *0.63*, which is a value close to the indication value obtained for the overall indication, which was *0.65*. This also confirms that the images are not properly correlated.

Having determined the [Q] measure indication in individual squares, the square with the highest [Q] measure indication is selected (in this case, it is 5-5′) and the translation in the X and Y plane is found within it with the use of the algorithm described in article no. [12].

Thanks to the adopted assumptions, it is possible to assume that the translation found in the X and Y plane in the segment where the [Q] measure indication is the highest is also the rotation axis sought.

Table 1. Q measure indication for photograps in Fig. 2

Pair of images	Q measure indication
Pair 1a-1b	0.65

1	2	3
4	5	6
7	8	9

1'	2'	3'
4'	5'	6'
7'	8'	9'

Fig. 3. Division of an image fragment into segments

Table 2. Q measure indication for photograps in Fig. 2

Pair of images	Q measure indication
Segment 1-1'	0.63
Segment 2-2'	0.68
Segment 3-3'	0.59
Segment 4-4'	0.62
Segment 5-5'	0.70
Segment 6-6'	0.65
Segment 7-7'	0.61
Segment 8-8'	0.59

Having specified the rotation axis, the next step is finding the rotation angle between the studied images.

To this end, the edge of the studied segment is taken which is the most distant from the rotation axis found (O). In this way, a section of length (a) is obtained which begins in the rotation axis and ends in point X. Next, a mask with dimensions N × N pixels is selected. The mask is shifted along the horizontal edge by one pixel at a time beginning with point X. The [Q] measure indication is calculated in each location of the mask. Point Y is marked where the [Q] indication obtains the maximum value. In this way, a section of length (b) is obtained beginning in point X and ending in point Y. The whole process is illustrated in Fig. 4.

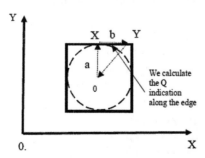

Fig. 4. The principle of finding rotation

In this way, we receive a right-angled triangle with the edge lengths equal to the length of sections a and b found. Using the trigonometric dependency, the α angle is calculated according to formula 2.

$$\alpha = \arctan\left(\frac{b}{a}\right) \tag{2}$$

The effectiveness of finding the rotation angle is confirmed by the experimental test results presented in Table 3. For the experimental purposes, the rotation angle was artificially predefined.

Analysing the results presented in Table 3, it is possible to conclude explicitly that the rotation axis is found with very high precision, while the rotation angle is found with one hundred per cent precision. Having found the angle and axis of rotation between the studied fragments, it is only enough to combine the photographs into one whole, which will create a panoramic photograph, by using typical point operations on digital images such as rotation and translation.

Figure 5 presents an example panorama obtained by joining two photographs presented in Figs. 1a and b.

For comparison, Fig. 6 presents a panorama created in a popular commercial programme *Adobe Photoshop* with the use of a function implemented in it for creating of panoramic photographs - *photomerge*.

3   Experimental Test Results

In order to confirm the effectiveness of the algorithm, the authors carried out a range of experimental tests where they compared panoramic photographs created with the use of the algorithm proposed by the authors of the article with panoramic photographs created with the use of the commercial programme *Adobe Photoshop* (*photomerge* option) as well as panoramic photographs created with the use of the algorithm developed by the authors and presented in publication no. [12].

For evaluation of the photographs, the authors used subjective and objective assessment with the use of the [Q] measure indication.

The subjective assessment involved showing partial photographs to 10 people before creating a panorama as well as the panoramic photograph made of them.

Table 3. Example results of experimental tests of finding the rotation axis and angle

Pairs of images	Predefined axis		Axis found		Predefined rotation angle	Rotation angle found
	X	Y	X	Y		
Pair 1a-1b	1421	1519	1419	1522	3	3
Pair 2a-2b	1224	1201	1223	1201	2	2
Pair 3a-3b	2221	2457	2221	2456	4	4
Pair 4a-4b	1339	1567	1337	1567	3	3
Pair 5a-5b	1400	1500	1400	1500	4	4

Fig. 5. An example panorama obtained with the use of the algorithm proposed in the article

Fig. 6. An example panorama obtained with the use of Adobe Photoshop

The panoramic image was created with the use of the algorithm proposed in the article as well as with the use of a ready function for creating panoramas offered by the commercial programme *Adobe Photoshop* (*photomerge* option).

Each of the assessors awarded each panoramic photograph with a mark from 1 to 10 (where 10 meant a very pretty photograph without any defects). Next, the sum of the marks was divided by 10 according to formula 3.

$$subiect = \frac{\sum_i S_i}{10} \tag{3}$$

Where: S_i – a partial mark within the scope from 1 to 10.

In order to make the evaluation more credible, so that none of the assessors was prompted in any way (which could influence their assessment), the assessors did not know which panoramic photograph was created with the algorithm proposed by the authors and which with the use of *Photoshop*.

Table 4 presents the results of subjective assessment made by a group of 10 people for panoramic images created with the use of the proposed algorithm. Table 5 presents the results of subjective assessment for panoramic photographs created with *Photoshop*.

As it can be seen, subjective assessments of panoramic photographs with the use of the algorithm proposed by the authors of the article are the most satisfactory. Subjective assessments for the method proposed by the authors are, in many cases, slightly better than subjective assessments of panoramic photographs created with *Photoshop*.

Table 4. Subjective assessment of panoramas created with the use of algorithm proposed in the article

Pair of images	Subjective assessment
Panorama 1	8
Panorama 2	9
Panorama 3	8
Panorama 4	7
Panorama 5	7

Table 5. Subjective assessment of panoramas created with the use Photoshop

Pair of images	Subjective assessment
Panorama 1	8
Panorama 2	9
Panorama 3	8
Panorama 4	7
Panorama 5	7

The results presented in Tables 5 and 6 confirm that panoramic photographs created with the use of the algorithm proposed by the authors and photographs created with the use of *Photoshop* differ from each other, but the difference is not significant. This is compliant with the expectations as both methods are based on different matching algorithms. Simultaneously, it must be pointed out that the algorithm proposed in the article is much better than the one proposed by the authors previously in the article described in publication [11].

For comparison of how much the photographs created with the use of the algorithm proposed in the article differ from the photographs created with Photoshop, the authors used the [Q] measure indications. The results of this comparison are presented in Table 7. The table also presents the [Q] measure indications as regards the comparison of the photographs obtained with the use of Photoshop with panoramic photographs obtained with the use of the algorithm presented by the authors in publication [11].

The [Q] measure indications correlate also with the subjective assessment, i.e., for example, panoramas no. 2 and 3, according to the [Q] measure indication, are the most similar, which is confirmed by the subjective assessment being the closest to mark 10.

Table 6. Subjective assessment of panoramas created with the use of algorithm proposed in the article [11]

Pair of images	Subjective assessment
Panorama 1	8
Panorama 2	9
Panorama 3	8
Panorama 4	7
Panorama 5	7

Table 7. Q measure indications for panoramas created with the algorithm proposed in the article and the algorithm from publication [11] compared to panoramas created with Photoshop

Panoramas	Q measure indications for the algorithm described in the article	Q measure indication for the algorithm from publication 12
Panorama 1	0.88	0.80
Panorama 2	0.97	0.89
Panorama 3	0.95	0.85
Panorama 4	0.87	0.79
Panorama 5	0.86	0.82

The quality of the obtained panoramic photographs depends to a large extent from the scene presented in the photograph. Photographs with significant amount of details are encumbered with more errors than photographs where there are fewer details.

4 Conclusion

The method of finding the axis and angle of rotation between images and creating panoramic photographs proposed by the authors is effective and enables creation of panoramic photographs of good quality.

The proposed method of synchronisation presented in the article may successfully be used for other issues concerning digital image processing where the process of matching two images is necessary, e.g. assessment of the progress of lesions in retina [13], preparation of the input data in the neural network learning process generating the output FIR filter [4]. The proposed method will be improved continuously and, during their further work on the issue, the authors will try to use other measures for digital image quality assessment.

References

1. Wang, Z., Bovik, A.C.: Mean squared error: Love it or leave it? A new look at signal fidelity measures. IEEE Sig. Process. Mag. **26**(1), 98–117 (2009)
2. Peksiński, J., Mikołajczak, G.: Generation of FIR filters by using neural networks to improve digital images. In: IEEE Conference: 34th International Conference on Telecommunications and Signal Processing (TSP), Budapest, Hungary, 18–20 August 2011, pp. 527–529 (2011)
3. Wang, Z., Bovik, A.C.: A universal image quality index. IEEE Sig. Process. Lett. **9**, 81–84 (2002)
4. Pęksiński, J., Mikołajczak, G.: Generation of a FIR filter by means of a neural network for improvement of the digital images obtained using the acquisition equipment based on the low quality CCD structure. In: Nguyen, N.T., Le, M.T., Świątek, J. (eds.) ACIIDS 2010. LNCS, vol. 5990, pp. 190–199. Springer, Heidelberg (2010)
5. Gora-Klauzinska, A., Benicewicz-Miazga, A., Klauzinski, E.: The history of panoramic photography. Kwartalnik Internetowy CKfoto.pl ISSN 2080-6353 No. CK 11/2012 (I-III 2012) (publication in Polish)

6. Peksinski, J., Mikolajczak, G.: The preparation of input data in digital image processing, Przegląd Elektrotechniczny R. **89**(3b), 282–284 (2013)
7. Peksinski, J., Mikolajczak, G.: Using a single-layer neural network to generate a FIR filter that improves digital images using a convolution operation. In: Zhu, Min (ed.) Business, Economics, and Financial Sci., Manag. AISC, vol. 143, pp. 675–682. Springer, Heidelberg (2012)
8. Gonzalez, R.C., Woods, R.E.: Digital Image Processing, 3rd edn. Hardcover –Edition, 3rd August 31 2007
9. Chris, S., Breckon, T.P.: Fundamentals of Digital Image Processing. John Wiley and Sons Ltd., December 2012
10. Wu, H.R., Rao, K.R.: Digital Video Image Quality and Perceptual Coding, Series: Signal Processing and Communications. CRC Press (2005). Published: November 18
11. Kowalski, J., Peksinski, J., Mikolajczak, G.: Using the Q measure to create panoramic photographs. In: 38th International Conference on Telecommunications and Signal Processing TSP2015, pp. 560–563 (2015)
12. Kowalski, J., Peskinski, J., Mikolajczak, G.: Controlling the progression of age-related macular degeneration using the image quality index and the reference image. Elektronika i Elektrotechnika **21**(6), 70–74 (2015)

Human Face Detection in Thermal Images Using an Ensemble of Cascading Classifiers

Paweł Forczmański[✉]

Faculty of Computer Science and Information Technology, West Pomeranian
University of Technology, Szczecin, Żołnierska Str. 52, 71–210 Szczecin, Poland
pforczmanski@wi.zut.edu.pl

Abstract. The paper addresses the subject of thermal imagery in the context of face detection. Its aim is to create and investigate a set of cascading classifiers learned on thermal facial portraits. In order to achieve this, an own database was employed, consisting of images from IR thermal camera. Employed classifiers are based on AdaBoost learning method with three types of low-level descriptors, namely Haar–like features, Histogram of oriented Gradients, and Local Binary Patterns. Several schemes of joining classification results were investigated. Performed experiments, on images taken in controlled and uncontrolled conditions, support the conclusions drawn.

Keywords: Thermovision · Biometrics · Face detection · Haar–like features · Histogram of Oriented Gradients · Local Binary Patterns · AdaBoost

1 Introduction

Nowadays, biometric identification systems became easily accessible to average citizens and they are no longer considered only as a sophisticated way of guarding military secrets. Even low-end laptops and some middle-end smartphones are equipped with simple cameras and bundled software capable of face recognition. It is because, human face is still one of the most obvious biometric features, easy to capture, describe and recognize [7]. In many situations biometric information replaces identifiers and passwords as means of securing our both real and virtual assets. However, there are still some situations when typical biometric approaches are not enough, e.g. when environmental conditions are not fully under our control [10] or a higher level of security is required. One of the possible solutions, in such case, is thermal imaging [2]. Data captured from infrared or thermal sensors can be used to perform face recognition without the necessity to properly illuminate the subject [9] with visible light. Some of related problems are presented in Fig. 1 (courtesy of Piotr Jasiński). Additionally, thermal imaging can also provide high level of protection against spoofing attempts (e.g. using a photo or video stream instead of a real face [21]), because heat signatures of artificial objects are completely different from real facial ones [16].

© Springer International Publishing AG 2017
S. Kobayashi et al. (eds.), *Hard and Soft Computing for Artificial Intelligence, Multimedia and Security*, Advances in Intelligent Systems and Computing 534, DOI 10.1007/978-3-319-48429-7_19

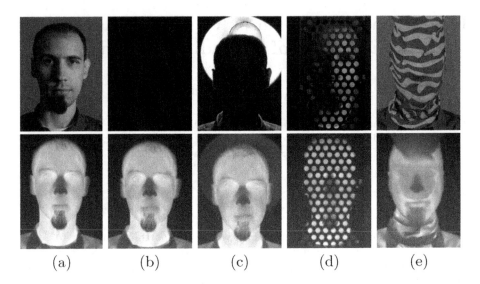

(a)   (b)   (c)   (d)   (e)

Fig. 1. Problems with imaging in the visible light (upper row) are removed using imaging in the thermal spectrum (lower row): (a) directional illumination, (b) no illumination, (c) strong backlight, (d) partial occlusion, (e) face cover with a cloth

1.1   Problem Definition

Generally speaking, extracting human faces from a static scene involves the determination of the image part containing the searched face (or faces). The main issue in such case is a proper selection of distinguishable features used to build a face model. The other important aspects are the mechanism for feature matching and the method for scanning the source image. It should be noted that the process of detecting faces in static images assumes that certain useful information, namely the probable face number, their sizes and locations are not available. Additional difficulties result from changes in the scene illumination and face appearance modifications resulted from a perspective changes. Many methods employ so called sliding window approach where the detection is performed by the scanning of the image and matching the selected image parts with the templates collected in the training set. If there is no information about probable face position and size, the detection requires to perform search process in all possible locations, taking into consideration all probable window (or image) scales, which increases overall computational overhead. In the above mentioned approach, in each image location the features of a particular part are calculated and later used during the classification process in order to indicate the most probable face location.

In the paper we focus on static scene feature extractors that enable proper facial portrait representation for the purpose of human detection. The algorithms were selected taking into consideration the computational complexity, the simplicity of implementation and the accuracy. Hence, we have selected three low-level characteristics that use Haar-like features, Histogram of Oriented Gradients and Local Binary Patterns. The classification is performed using Viola-Jones detector based on AdaBoost [4,22]. Selected algorithms are highly popular and constantly applied to different tasks of object detection.

1.2 Related Works

Although the problems of human face detection and recognition in visible light have been investigated many times, the problems of detecting and recognizing faces in thermal spectrum are less represented. However, there are some works related to this problem [1,12,18]. The lack of established methods is caused by the characteristics of thermal images which change the appearance of human faces depending on the surrounding temperature. In this context, the main drawback specific to the thermal sub-band images (or thermograms, as they are often referred to), the most often used sub-band of the IR spectrum, stems from the fact that the heat pattern emitted by the face is affected by a number of confounding variables, such as ambient temperature, air flow conditions, exercise, postprandial metabolism, illness and drugs [20]. Moreover, the thermal imaging equipment is quite expensive and not very wide spread. Hence, recently, we proposed a unified capture procedure based on a Matlab application and a photographic stand of our own design. Through its use it is possible to create a database of fused visual and thermal facial images, ready to serve any testing or research purposes [15].

The methods of human face detection in thermal imagery can be classified as appearance-based and model-based. While both groups of methods are used for face detection in visible light, the second one is more popular, e.g. [5,6].

Histogram of Oriented Gradients [3] was proposed as an approach for human silhouette detection. It extracts local image characteristic features and stores them in a compact form as a feature vector. Since it is based on gradient information it requires larger size of the detection window, in comparison to other methods. Such window provides a significant amount of context information that helps in detection [3]. The HOG representation analyses the direction of brightness changes in local areas and is derived in five main steps. Firstly, gamma and colour normalization are performed, and secondly the directional filters are applied to calculate oriented gradients (actually, vertical and horizontal edges are detected). Then, the length and orientation of each gradient is estimated. In the third step, a whole image is decomposed into cells and frequencies of oriented gradients in each cell are calculated. In the fourth step, cells are grouped into larger overlapping spatial blocks and normalized separately. Single cell is normalized several times, each time with respect to a different block. Such procedure makes the features resistant to local variations in illumination and contrast. Furthermore, it significantly improves the performance [3]. The final HOG representation is obtained in the fifth step by concatenating oriented histograms of all blocks. Thanks to the histogram representation, the features are invariant to translation within an image plane – if an object changes its position the directional gradients and histogram remain the same.

Local Binary Patterns has been originally proposed as a feature used for texture classification [19], and it is a particular case of the Texture Spectrum model proposed in [13]. The goal of LBP is to label pixels by thresholding their neighbourhood. In our case we analyse pixels that have eight neighbours. In order to create the LBP representation, several steps have to be followed. In the first step, the analysed window is divided into equal cells, and each pixel in a cell is compared with its eight neighbours in a clockwise or counter-clockwise direction (the direction and the starting point are common for all cells). In the second step, the pixel's value is examined and all neighbourhood values are calculated as follows. If the central pixel's value is equal or greater than the neighbouring pixels, then a number '1' is stored. Otherwise, '0' is saved. Such procedure produces a binary sequence, usually converted to decimal. As a third step, the histogram is computed over the cell, based on the occurrence frequency of each number. Finally, the histograms of all cells are concatenated, forming a feature vector for the whole window. It should be remembered, that the original LBP representation is not invariant to object rotation within an image plane. However, it is robust to monotonic grey-scale changes resulted from variations in illumination. In case of human face detection, the rotations in the image plane are not taken into consideration.

Original Haar-like features were proposed and applied together with a variant of AdaBoost in [22] for face detection in images taken in visible light. They were also successfully applied in the problem of human silhouettes detection [8]. While individual Haar-like features are weak classifiers, their combination in the form of the cascade leads to a very efficient solution. During training, using AdaBoost technique, the most important image features are selected and only these are used for detection. Features are selected in a way enabling to reject negative regions (not containing interesting object) at early stage of recognition, what reduces the number of calculations. During the classification subsequent features of an unknown object are calculated only if the result from previous feature is equal to the learned value. Otherwise, the examined region is rejected. A simple rectangular Haar-like feature is defined as the difference of the sum of pixels of areas inside the rectangle, which can be characterized by any position and scale in an input image. For instance, using two-rectangle feature, the borders can be indicated based on the pixel intensities under the white and black rectangles of the Haar-like feature. Such features could be calculated fast using the integral image approach.

The derived feature vectors (HOG, LBP or Haar-like features) are further used in the process of classification using e.g. Support Vector Machine or, in our case, AdaBoost. The original AdaBoost (Adaptive Boosting) was proposed in [11]. It is a machine learning meta-algorithm–a method of training boosted classifier that during the training process selects only those features that would be able to improve the prediction. For each weak classifier from a cascade AdaBoost determines an acceptance threshold, which minimizes the number of false classifications. However, a single weak classifier is not able to classify objects with low error rate. The boosting algorithm used for learning the classifier uses positive and negative samples as an input. The initial weights are equal to 1. In each iteration the weights are normalized and the best weak classifier is selected based on the weighted error value. In the next step weight values are updated and it is determined if an example was classified correctly or not. After all iterations a set of weak classifiers characterized by a specified error rate is selected and a resulting strong learned classifier is obtained. The classification is performed iteratively and its effectiveness depends on the number of learning examples. During classification, an image is analysed using a sliding window approach. Features are calculated in all possible window locations. The window is slid with a varying step, which depends on the required accuracy and speed.

In the algorithm presented in this paper we propose to use a combination of above three low-level descriptors, since each of them posses different properties. While Haar-like features effectively detect frontal faces, LBP and HOG allow for slight face angle variations. By combining the results from three independent detectors, yet learnt on the same dataset, we can achieve higher face detection rate, while decreasing false detections.

2 Experiments

2.1 Cascades Training

In order to train three individual cascades, our proprietary database was used [15]. It consists of 505 images of 101 subjects (each subject was photographed in 5 head orientations). However, only frontal faces (101 images) were used for training. At that stage, each face has been cropped using an automatic algorithm (employing face proportions and registered eyes positions). No other preprocessing has been employed. All the faces used for training are presented in Fig. 2.

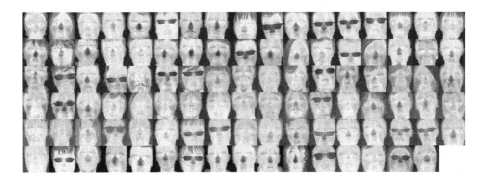

Fig. 2. Images used for cascade learning

During learning standard boosted cascade algorithm was applied, namely AdaBoost (Gentle AdaBoost variant), and varying values of the following parameters: number of positive and negative samples, the number of learning stages. They are presented in Table 1.

Table 1. Cascades training parameters

Features	Haar	HOG	LBP
Window size	32 × 32		
Pos. samples	800	800	800
Neg. samples	2000	2000	3000
No. stages	18	18	16

The experiments were performed using Matlab 64-bit frame-work and OpenCV 2.4.10 implementation of Viola-Jones detector.

2.2   Benchmark Datasets

To show that the algorithm is unbiased for the training dataset, we select a training set, which is completely different than the test set. We evaluated the algorithm's performance on four different datasets:

1. WIZUT database (grouped in 5 subsets: 'a'–'e') containing all face's orientations (where 'b' is a frontal face subset, 'a' and 'c' are faces with $\pm 20°$ and 'd' and 'e' with $\pm 45°$ rotations, respectively);
2. Caltech (Courtesy NASA/JPL-Caltech) [14];
3. OTCBVS Terravic Facial IR Database [17];
4. Uncontrolled images taken from the Internet;

Faces collected in the above datasets include full frontal (except WIZUT 'a','c','d','e' and UC), some faces include glasses, head cover, or both. The exemplary images taken from all testing databases are presented in Fig. 3. Each two rows of the figure contain images from above presented datasets. The details of datasets are given in Table 2.

It should be noted that in case of OTCBVS Terravic Facial IR Database, from a total number of 23262 images, 12255 images containing faces rotated not more than $20°$ were selected, from which 2000 were randomly chosen for the evaluation. The images from WIZUT and OTCBVS are normalized in terms of temperature range, while the rest are images with variable temperature ranges. It leads to the very different representation of temperatures in the pixel's intensities. The images forming UC set were taken using various equipment, in variable conditions, indoor/outdoor, hence they are the most problematic data for the detection.

2.3   Results

Several experiments have been carried out in order to verify the effectiveness of the selected detectors in the task of face detection in static scenes. Selected detectors are based on Haar-like features, Histogram of Oriented Gradients and Local Binary Patterns. The detection involved trained detectors based on three

Table 2. Benchmark datasets characteristics

DB	WIZUT	Caltech	OTCBVS	UC (Uncontrolled)
No. images	505	64	2000	63
No. faces	505	64	2000	107
No. subjects	101	n/a	20	n/a
Image width	320	285	320	$167 - 1920$
Image height	240	210	240	$129 - 1215$
Rotation angle	$\{-45°, -20°, 0°, +20°, +45°\}$	$\pm5°$	$\pm20°$	n/a

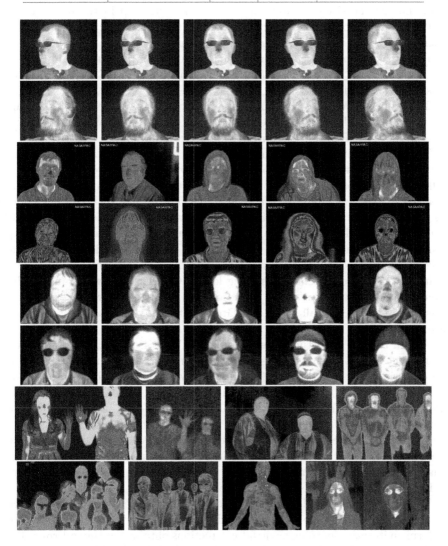

Fig. 3. Example images taken during experimental run

Table 3. Detector parameters

Nominal window size	32×32
Min size	16×16
Max size	360×360
Scaling factor	1.02
No. neighbors	$\{1, 4\}$

learned image features. The parameters of all detectors are presented in Table 3. They include the size of search window, the scaling step in the pyramid and candidates rejection based on the number of adjacent detections (number of neighbouring detection indicating true detection).

During experiments we estimated the face detection performance in terms of its rough location. If a detector marked most of the expected face area, it was considered successful.

The time of detection using heterogeneous environment (Matlab and OpenCV binary libraries) equalled less than a second per frame (Core i7, 2nd generation, 8 MB RAM, SSD Hard drive, 64-bit Windows).

The experiments were aimed at selecting the best classifier (in terms of image features) taking into consideration face detection rate (true acceptance rate) as well as stability of the results. We performed a detection using single feature type (Haar-like, HOG and LBP) and combinations of them (in terms of combining individual results, using AND operator), namely Haar+HOG, Haar+LBP, HOG+LBP and all of them. The above mentioned rule is based on the assumption, that all of the classifiers in the set should give positive detection in order to consider it successful. As for the comparison, we provide the results of combining the results of all classifiers using logical OR criterion, which means that if only one classifier gives a true detection, the result of the whole ensemble is considered positive. In all cases, the later solution gives the highest detection rate.

As can be seen from the presented tables (Tables 4 and 5), face detection in static scenes gives varying results, depending on the benchmark dataset. While the detection rate for standardized images is very high (in some cases almost perfectly 100 %), for images taken "in the wild" drops. When safety is the main priority, the AND criterion for three classifiers should be used. On the other hand, when we want a detector that detect most of the faces, OR criterion is the best choice. When considering the computational complexity, LBP features should be applied, thanks to their high-speed learning and fast classification (thanks to integer number calculations). In all cases, the four neighbors aggregation criterion lowers the detection rate, while reducing false detections.

The presence of a small number of false detections as well as some missed faces cause that the solution should be modified. One of the possible modifications is to perform an iterative learning using false detections obtained at the test runs. This would result in higher detection rate and decrease in the false detections number.

Table 4. Face detection rate [%] in case of 4 aggregated neighbours

DB	Haar	HOG	LBP	Haar +HOG	Haar +LBP	HOG +LBP	Joint (AND)	Joint (OR)
WIZUT (a)	99	98	98	98	98	96	95	100
WIZUT (b)	100	99	100	99	100	100	99	100
WIZUT (c)	94	97	100	92	99	93	92	100
WIZUT (d)	98	68	100	67	98	68	67	100
WIZUT (e)	79	30	96	24	82	23	23	99
Caltech	47	52	63	34	45	39	33	77
OTCBVS	95	97	99	93	94	97	92	100
UC	42	51	50	30	36	36	28	68

Table 5. Face detection rate [%] in case of no neighbor aggregation

DB	Haar	HOG	LBP	Haar +HOG	Haar +LBP	HOG +LBP	Joint (AND)	Joint (OR)
WIZUT (a)	100	99	100	99	100	99	99	100
WIZUT (b)	100	99	100	99	100	99	99	100
WIZUT (c)	99	97	100	97	99	97	97	100
WIZUT (d)	98	77	97	77	95	75	75	100
WIZUT (e)	86	46	97	43	83	46	43	100
Caltech	63	70	78	48	58	61	47	91
OTCBVS	96	98	99	94	95	98	94	100
UC	61	62	67	43	50	51	42	82

The other way is to employ an architecture of multi-tier cascade of classifiers employing different features and a coarse-to-fine strategy. Proposed classifier, in order to be applicable, should focus on single face orientation and first detect as many potential faces as possible, even if there are false detections. This task could be performed using Haar-like features, since it is the most tolerant approach. Then the LBP features to resulting objects should be applied. The final stage should involve HOG features in order to fine-tune the results.

2.4 Comparison with State of the Art

The comparison with other approaches is not straightforward, since most of the research involve different benchmark datasets. However we try to provide a comparison with the algorithms that use dataset of similar characteristics. The authors of [23] presented a method employing head curve geometry to extract the face from the images. Unfortunately, there are several limitations of the presented method, namely there should be only one person facing the camera and no other external heat-emitting objects are captured by the camera. Achieved accuracy is equal to 91.4 %.

The authors of [5] presented an idea of detecting faces using Hough transform and skin region segmentation. Reported accuracy for frontal faces (474 test images of 18 subjects) is equal to 96.35 % in case of the absence of glasses and 72.25 % for subjects wearing glasses. In case of rotated faces (±30°) is is significantly lower - 66.88 % and 67.40 %, respectively.

3   Summary

In the paper a dedicated, simple solution for detecting faces in the thermograms was presented. Its architecture consisting of three independent classifiers may be employed in biometric identification system. Thanks to the use of thermal signatures, such system may be resistant to the attacks involving substitution with images or videos of faces taken in the visible light. The results of the experiments show that combining all the presented classifiers leads to the very high detection rate.

References

1. Bebis, G., Gyaourova, A., Singh, S., Pavlidis, I.: Face recognition by fusing thermal infrared and visible imagery. Image Vis. Comput. **24**, 727–742 (2006)
2. Chang, H., Koschan, A., Abidi, M., Kong, S.G., Won, C.-H.: Multispectral visible and infrared imaging for face recognition. In: 2008 IEEE Computer Society Conference on Computer Vision and Pattern Recognition Workshops, pp. 1–6 (2008)
3. Dalal, N., Triggs, B.: Histograms of oriented gradients for human detection. In: IEEE Computer Society Conference on Computer Vision and Pattern Recognition, vol. 1, pp. 886–893 (2005)
4. Davis, J.W., Keck, M.A.: A two-stage template approach to person detectionin thermal imagery. In: Proceedings of the Seventh IEEE Workshop on Applicationsof Computer Vision (WACV/MOTION 2005) (2005)
5. Dowdall, J., Pavlidis, I., Bebis, G.: A face detection method based on multi-band feature extraction in the near-IR spectrum. In: IEEE Workshop on Computer Vision Beyond the Visible Spectrum (2001)
6. Dowdall, J., Pavlidis, I., Bebis, G.: Face detection in the near-IR spectrum. Image Vis. Comput. **21**(7), 565–578 (2001)
7. Forczmański, P., Kukharev, G.: Comparative analysis of simple facial features extractors. J. Real-Time Image Process. **1**(4), 239–255 (2007)
8. Forczmański, P., Seweryn, M.: Surveillance video stream analysis using adaptive background model and object recognition. In: Bolc, L., Tadeusiewicz, R., Chmielewski, L.J., Wojciechowski, K. (eds.) ICCVG 2010. LNCS, vol. 6374, pp. 114–121. Springer, Heidelberg (2010). doi:10.1007/978-3-642-15910-7_13
9. Forczmański, P., Kukharev, G., Kamenskaya, E.: Application of cascading two-dimensional canonical correlation analysis to image matching. Control Cybern. **40**(3), 833–848 (2011)

10. Forczmański, P., Kukharev, G., Shchegoleva, N.: Simple and robust facial portraits recognition under variable lighting conditions based on two-dimensional orthogonal transformations. In: Petrosino, A. (ed.) ICIAP 2013. LNCS, vol. 8156, pp. 602–611. Springer, Heidelberg (2013). doi:10.1007/978-3-642-41181-6_61
11. Freund, Y., Schapire, R.E.: A decision-theoretic generalization of on-line learning and an application to boosting. In: Proceedings of the 2nd European Conference on Computational Learning Theory, pp. 23–37 (1995)
12. Ghiass, R.S., Arandjelovic, O., Bendada, H., Maldague, X.: Infrared face recognition: a literature review. In: International Joint Conference on Neural Networks, 2013 Subjects: Computer Vision and Pattern Recognition (cs.CV) arXiv:1306.1603 [cs.CV] (2013)
13. He, D.C., Wang, L.: Texture unit, texture spectrum, and texture analysis. IEEE Trans. Geosci. Remote 28, 509–512 (1990)
14. Hermans-Killam, L.: Cool Cosmos/IPAC website, Infrared Processing and Analysis Center. http://coolcosmos.ipac.caltech.edu/image_galleries/ir_portraits.html. Accessed 10 May 2016
15. Jasiński, P., Forczmański, P.: Combined imaging system for taking facial portraits in visible and thermal spectra. In: Choraś, R.S. (ed.) Image Processing and Communications Challenges 7. AISC, vol. 389, pp. 63–71. Springer, Heidelberg (2016). doi:10.1007/978-3-319-23814-2_8
16. Kukharev, G., Tujaka, A., Forczmański, P.: Face recognition using two-dimensional CCA and PLS. Int. J. Biometrics 3(4), 300–321 (2011)
17. Miezianko, R.: IEEE OTCBVS WS Series Bench - Terravic Research Infrared Database. http://vcipl-okstate.org/pbvs/bench/. Accessed 20 May 2016
18. Mostafa, E., Hammoud, R., Ali, A., Farag, A.: Face recognition in low resolution thermal images. Computer Vis. Image Underst. 117, 1689–1694 (2013)
19. Ojala, T., Pietikinen, M., Harwood, D.: Performance evaluation of texture measures with classification based on Kullback discrimination of distributions. In: Proceedings of the 12th International Conference on Pattern Recognition, vol. 1, pp. 582–585 (1994)
20. Prokoski, F.J., Riedel, R.: Infrared identification of faces and body parts. In: BIOMETRICS: Personal Identification in Networked Society. Kluwer (1998)
21. Smiatacz, M.: Liveness measurements using optical flow for biometric person authentication. Metrol. Measur. Syst. 19(2), 257–268 (2012)
22. Viola, P., Jones, M.J.: Robust real-time face detection. Int. J. Comput. Vis. 57(2), 137–154 (2004)
23. Wong, W.K., Hui, J.H., Lama, J.A.K., Desa, B.M., J., Izzati, N., Ishak, N.B., Bin Sulaiman, A., Nor, Y.B.M.: Face detection in thermal imaging using head curve geometry. In: 5th International Congress on Image and Signal Processing (CISP 2012), pp. 1038–1041 (2012)

Accuracy of High-End and Self-build Eye-Tracking Systems

Radosław Mantiuk$^{(\boxtimes)}$

West Pomeranian University of Technology, Szczecin,
Żołnierska Street 52, 71-210 Szczecin, Poland
rmantiuk@wi.zut.edu.pl

Abstract. Eye tracking is a promising technology for human-computer interactions, which is however rarely used in practical applications. We argue that the main drawback of the contemporary eye trackers is their limited accuracy. There is no standard way of specifying this accuracy what leads to underestimating the accuracy error by eye tracker manufacturers. In this work we perform a subjective perceptual experiment measuring the accuracy of two typical eye trackers: a commercial corneal reflection-based device mounted under a display and a head-mounted do-it-yourself device of our construction. During the experiment, various conditions are taken into consideration including viewing angle, human traits, visual fatigue, etc. The results indicate that eye tracker accuracy is observer-referred and measured gaze directions exhibit a large variance. Interestingly, the perceptually measured accuracy of the low-cost do-it-yourself device is close to the accuracy of the professional device.

Keywords: Eye tracker accuracy · Eye tracking · Gaze tracking · Human computer interactions · User interfaces

1 Introduction

Eye tracking is a technique of capturing the *gaze direction* of human eyes. Interestingly, although there is a choice of eye-tracking devices, they are rarely used in practical applications. The main drawback of the contemporary eye trackers is their limited accuracy. A average precision below 0.5° of the viewing angle (roughly 20 pixels on a 22" display observed from 65 cm distance) is possible to obtain only using very expensive and/or intrusive eye trackers together with the chin-rest or bite bar. Moreover, accuracy of eye trackers is observer-referred and difficult to reproduce even for the same observer in the subsequent eye tracking sessions.

To the best of our knowledge, there are no any formal standard, which specifies how to measure the accuracy of eye tracker [1]. This leads to underestimating the accuracy error by eye trackers manufacturers. In this work we propose a technique which measures the accuracy in a straightforward manner and can be applied for any type of eye tracker. This technique does not interfere with eye trackers' software and hardware components, in particular, we use the native calibration procedure delivered by the manufacturers and measure the accuracy based on the raw

© Springer International Publishing AG 2017
S. Kobayashi et al. (eds.), *Hard and Soft Computing for Artificial Intelligence, Multimedia and Security*, Advances in Intelligent Systems and Computing 534, DOI 10.1007/978-3-319-48429-7_20

and unfiltered gaze data. As a proof of concept we measure the accuracy of two typical eye trackers: commercial RED250 from SensoMotoric Instruments [2] and low-cost *Do-It-Yourself* (DIY) eye tracker of our construction [3].

We perform a perceptual experiment, in which people are asked to look at the markers of the known locations on the screen. Their gaze direction is captured by eye tracker and the accuracy of the device is analysed in comparison to the reference direction. Additionally, we evaluate the accuracy for different viewing angle and declared level of the visual fatigue of the observer's eyes.

In Sect. 2 we present a basic information related to the eye tracking technology. A proposed accuracy measurement procedure is introduced in Sect. 3. We provide details on the perceptual experiment in Sect. 4 and analyse its results in Sect. 5.

2   Background

2.1   Tracking of Human Gaze Direction

The human viewing angle spans more than 180° horizontally and 130° vertically, although, the binocular field of vision is limited to about 120° horizontally. The eye can rotate about 45° in all directions, although it is natural to change head position for angles higher than 20° rather than move the eyes [4, Sect. 4.2.1], [5, Sect. 24]. Therefore, it is a reasonable that an eye tracker should cover only 40° (±20°) of the human viewing angle (it is equivalent of a 21" monitor observed from about 74 cm distance). Tracking the full range of the viewing angle for larger displays should be supported by head tracking devices.

Eye trackers capture two types of eye movement: *smooth pursuit* and *saccades*. Smooth pursuit is active when eyes track moving target and are capable of matching its velocity. Saccades represent a rapid eye movement used to reposition the fovea to a new location [6]. Other types of the eye movements like vestibular, miniature eye movements (*drift, nystagmus, microsaccades*), or vergence have a smaller effect on the determination of the gaze direction [7–9].

The main goal of the gaze tracking is to capture a single location an observer intents to look at. This process is known as a *visual fixation* [10]. A point of fixation can be estimated as a location where saccades remain stable in space and time. Different computational techniques are used to estimate location of this point based on analysis of the smooth pursuit and saccadic movements (see [10–13] for survey). The fixation is controlled by cognitive processes in human brain and identification of its location strongly depends on the top-down visual mechanism (a task given to observer before eye tracking, [14]). Therefore, it is difficult to propose a general fixation computation model suitable for every application. Interestingly, parameter settings in fixation algorithms have crucial meaning for their operation [13]. Therefore, in our studies, we decided to base the accuracy measurement rather on the raw gaze data than the data processed by the fixation identification algorithms.

2.2 Eye Tracking Technology

Devices called *video eye tracker* [11, Chap. 5.4] usually consists of an infrared camera and an infrared light source, which are directed at the eye. The camera captures the image of the eye with the dark circle of the pupil and the bright corneal glint, which is a reflection of the infrared light from the outer surface of the cornea. The pupil follows the gaze direction during eye movement while the corneal reflection stays in the same position. The relative position between the reflection and the center of the pupil is used to estimate the gaze direction. This type of eye trackers is called the pupil-corneal reflection (P-CR). Another option is to use a chin rest to stabilized the observer's head. Assuming that the head is not moving relative to the display, only pupil center location is estimated to compute the gaze direction.

To find a *gaze point* (called also point-of-regard, POR), at which an observer looks on the screen, the mapping between the camera image and the screen surface must be determined. This mapping is difficult to compute implicitly because of unknown initial position of the head and complex geometry of the eye movement. Therefore, eye trackers employ the calibration based on the non-linear approximation technique [9,15,16]. In this technique an observer is asked to look at a set of target points displayed one by one in different positions on the screen. The relation between the calculated position of the pupil centre and known position of the target points is used to approximate coefficients a_{0-5} and b_{0-5} of a pair of second-order polynomials:

$$\begin{cases} screen_x = a_0 + a_1x + a_2y + a_3xy + a_4x^2 + a_5y^2 \\ screen_y = b_0 + b_1x + b_2y + b_3xy + b_4x^2 + b_5y^2, \end{cases} \tag{1}$$

where $(screen_x, screen_y)$ denotes the gaze point coordinates on the screen and (x,y) are the coordinates of the centre of the pupil in image coordinates. At least, six target points are needed to compute twelve unknown coefficients. Additional target points are used to balance diversity between centre and corners of a screen (larger inaccuracies are induced by wide viewing angles to the target points located in monitor corners).

Accuracy of the mapping presented in Eq. 1 depends on exact estimation of pupil positions during observation of the target points. An observer is asked to look at a point for a few seconds to ensure that she/he gazes directly at this point and some of this data is filtered out.

3 Accuracy Measurement

Eye tracker accuracy is defined as average angular offset (distance in degrees of viewing angle) between the gaze direction measured by an eye tracker and the corresponding direction to a known reference target. The main task of *accuracy measurement* is to compare the gaze direction captured using an eye tracker with the *reference gaze direction*. The reference gaze direction is a direction from observer's eyes to the known point (called *target point*).

3.1   Procedure

The accuracy measurement procedure consists of three steps: calibration, validation, and data analysis (see Fig. 1).

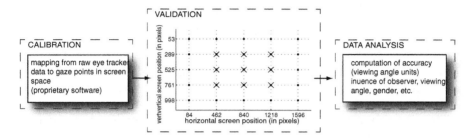

Fig. 1. The eye tracker accuracy measurement procedure. The validation module shows locations of the target points.

Calibration is preformed by the proprietary eye tracker software. We assume the calibration as a native feature of the eye tracker and do not test the accuracy of calibration separately. Multiple repetitions of the measurement for one observer provide an objective assessment of the calibration accuracy and take the calibration error into account in the overall results.

In *validation* phase participants look at the circle marker displayed for 2 s at 25 different positions (see Fig. 1). These positions, called the target points, act as known and imposed fixation points. The marker is moved between target points in random order. Smooth animation of the marker between target points allows for faster observer's fixation and reduces number of outliers. Additionally, the marker is minified when reaches its target position to focus observer's attention on a smaller area.

We record gaze direction captured by the eye tracker during the whole validation phase. Locations of the gaze points are transformed from screen coordinates to the accuracy units (see Sect. 3.2). In the *data analysis* phase, average accuracy, standard deviation and maximum error values are computed to express the accuracy and robustness of an eye tracker (see Sect. 5).

3.2   Accuracy Units

We express the *eye tracker accuracy* (ac) as the angle between the rays from the eye centre to the point captured by an eye tracker (the gaze point) and to the target point. We compute this angle as *arccos* of the dot product of the mentioned direction vectors. Both vectors $(\mathbf{v}_{gaze}$ and $\mathbf{v}_{tp})$ are calculated based on the distance from the eyes' plane to the screen $(hs$) as well as the distance from the centre of the screen (x_c, y_c) to gaze point position (x_{gaze}, y_{gaze}), and to target point position(x_{tp}, y_{tp}):

$$\begin{cases} x_{ps} = width/x_{res}, \\ y_{ps} = height/y_{res}, \\ \\ \mathbf{v}_{gaze} = \frac{(x_c - x_{gaze})*x_{ps}, (y_c - y_{gaze})*y_{ps}, hs}{|((x_c - x_{gaze})*x_{ps}, (y_c - y_{gaze})*y_{ps}, hs)|}, \\ \\ \mathbf{v}_{tp} = \frac{(x_c - x_{tp})*x_{ps}, (y_c - y_{tp})*y_{ps}, hs}{|((x_c - x_{tp})*x_{ps}, (y_c - y_{tp})*y_{ps}, hs)|}, \\ \\ ac = arccos(\mathbf{v}_{gaze} \bullet \mathbf{v}_{tp}), \end{cases} \qquad (2)$$

where (x_{res}, y_{res}) denotes screen resolution is pixels. The screen *width* and *height* are expressed in the same units as *hs*, all other values are expressed in the screen pixel coordinates.

The above formula provides more accurate calculation than the standard computation of the angular deviation from the centre of the screen (see equations in [17, Sect. 9.4]), which introduces distortions at edges of the screen. For example, the calculation precision of a 50 pixels shift between gaze point and target point at the screen corner for our eye tracker setup is 0.2674°, at horizontal and vertical edge: 0.1428° and 0.0634°, respectively.

3.3   Filtering

Filtering of the outliers is the most problematic issue in the eye tracker accuracy measurement. Eye trackers' vendors report the accuracy for the estimated fixation points computed by their proprietary software. This solution allows indeed reporting a lower accuracy error because the problematic data is filtered out. Commercial vendors in their eye tracking systems often calculate the accuracy with the use of pre-filtered data (e.g. with the proprietary fixation detection algorithms that are not disclosed [2]), what often artificially lowers the resulting error. However, these results do not reflect the accuracy in practical applications. For example, as we measured in a pilot study, SMI Experimental Center software [17] reports average accuracy of RED250 eye tracker as about 0.5° what is not consistent with our experimental evaluation (mean accuracy 1.66°, see Sect. 5) and practical insights.

We argue that a more useful accuracy estimation is based on the raw gaze data. We assume that a human is able to fixate at a point for some time constantly (2 s in our evaluation, ability to even 10 s fixation was reported [18]) and do not filter out any data. Obvious exception is skipping gaze points captured during displacement of the marker between target points. Also, we do not record data during blinks.

4   Experimental Evaluation

The main goal of our experimental evaluation was to measure the accuracy of two different eye trackers and analyse how human traits affect the collected results. We examine whether conditions like observation angle, observers' gender, fatigue, vision correction, measurement repetition affect the eye tracker accuracy.

4.1   Participants

Seven participants, age between 22 and 42 participated in our experiment (average of 33 years, 3 females and 4 males). Four participants had normal vision, one of them had corrected vision with contact lenses and two wore glasses. We asked each participant to repeat the experiment ten times at different times of the day. We conducted 70 measurement sessions per each eye tracker. They were completed within 15 days for RED250 and within 5 days for DIY eye tracker. Before each experiment, participants were asked to assess the fatigue of their vision in a 5-point Likert scale (1-excellent, 2-very good, 3-good, 4-tired, 5-very tired). No session took longer than 2 min for one participant to avoid fatigue. The participants were aware that accuracy of the eye tracker is tested, however they did not know details of the experiment.

4.2   Apparatus

Our experimental setup is presented in Fig. 2 (left). It consists of RED250 eye tracker controlled by the proprietary SMI iViewX (version 2.5) software running on dedicated PC. RED250 eye tracker is mounted under a 22" Dell E2210 LCD display with the screen dimensions 47.5×30 cm, and the native resolution 1680×1050 pixels (60 Hz). The same display was used for the assessment of the DIY eye tracker (presented in Fig. 2, right). DIY is controlled by the 3rd party ITU Gaze Tracker software (version 2.0 beta) developed at the IT University of Copenhagen [19] and running on the second PC (2.8 GHz Intel i7 930 CPU equipped with NVIDIA GeForce 480 GTI 512 MB graphics card and 8 GB of RAM, Windows 7 OS). This PC was also used to run our validation software which collects eye tracking data received from the external applications (SMI iView X or ITU Gaze Tracker), renders graphics and stores experiment results. The software was implemented in Matlab using Psychtoolbox [20] and additional mex files written in C++. It communicates with iViewX and ITU using the UDP (User Datagram Protocol) protocol.

Fig. 2. Left: Hardware setup used for the experiments, RED250 eye tracker is located under the display. Right: DIY eye tracker.

4.3   Stimuli and Procedure

Observers sat in the front of the display in 65 cm distance and used the chin-rest adopted from an ophthalmic slit lamp. Following ITU-R.REC.BT.500-1 recommendations [21], the experiment started with a training session in which observers could familiarise themselves with the task, interface, chin-rest, and the eye trackers. After that session, they could ask questions or start the main experiment.

The actual experiment started with a 9-point calibration controlled by iViewX or ITU software. This procedure took about 20 s and involved observation of the markers displayed in different areas of the screen. The data processing including computation of the calibration polynomial coefficients (see Sect. 2.2) was performed by the proprietary software. After calibration, the actual validation of eye tracker accuracy was performed based on procedure described in Sect. 3. The locations of the gaze points were still received from iView X or ITU, although the position of the target points and their rendering and display were imposed by our validation software.

Instead of measuring the accuracy separately for each eye, we average gaze position of left and right eye measured by RED250 eye tracker. The DIY eye trackers captures the position of only the left eye.

5   Results

The gaze points captured by RED250 and DIY eye trackers were transformed to the error accuracy units (see Sect. 3.2) and average error for the whole dataset was computed. We achieved an average error equal to 1.66° for RED250 and to 1.89° or DIY eye tracker (the detailed results, also for individual observers, are depicted in Table 1).

We express the average eye tracker error as the covariance ellipses (see Fig. 3). The direction of the radii of ellipse corresponds to the eigenvectors of the covariance matrix and their lengths to the square roots of the eigenvalues. It is assumed that an eye tracker has a good accuracy, if the distribution of the error has the circular shape corresponding to the normal distribution and the centre of this circle is located in (0,0). The ellipse radii should be as small as possible.

Average, Median and Maximum Error. The results are characterise by significant standard deviations (2.11° for RED250 and 2.37° for DIY) and large maximum accuracy error (39.87° and 43.89° for RED205 and DIY, respectively). It suggests that the gaze position cannot be estimated based on separate samples from eye tracker. Larger number of gaze points leads towards the normal distribution and improves the eye tracker accuracy.

Observers. Analysis of variance (ANOVA) shows strong dependence of the eye tracker accuracy on individual observers (see Fig. 4). For example, observer *eup* achieves significantly worse results than the average and this trend is visible

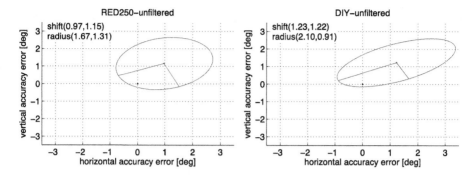

Fig. 3. Distribution of gaze directions around the target points averaged for the whole experimental dataset. The blue ellipses denote shift from (0,0) and co-variance values, for perfect distribution the circles with centre in (0,0) should be plotted. (Color figure online)

Table 1. Average accuracy of RED250 and DIY eye trackers in degrees of viewing angle, higher values mean lower accuracy.

Observer	Eye tracker accuracy error [°]							
	Horizontally		Vertically		Both directions			
	Mean	Std	Mean	Std	Mean	Median	Std	Max
RED250								
eup	1.49	2.91	2.11	2.90	2.89	1.98	3.75	37.47
rdm	0.78	0.76	0.91	0.93	1.30	1.13	1.01	38.34
bba	1.39	2.61	1.47	1.29	2.24	1.69	2.65	38.36
klu	0.85	1.42	1.00	0.95	1.42	1.14	1.56	37.48
sla	0.87	1.15	0.84	0.71	1.29	1.12	1.22	37.47
ant	0.64	0.89	0.89	0.93	1.18	1.01	1.15	39.87
pfo	0.75	0.67	0.79	0.67	1.18	1.05	0.76	28.77
all observers	0.97	1.75	1.15	1.48	**1.66**	1.24	2.11	39.87
DIY								
eup	2.84	3.15	1.74	2.27	3.52	2.71	3.59	43.89
rdm	0.87	1.20	1.66	1.24	1.98	1.78	1.54	16.76
bba	0.94	0.93	1.11	0.85	1.56	1.40	1.06	19.23
klu	0.55	0.68	0.70	0.57	0.96	0.82	0.78	18.17
sla	0.81	0.94	1.36	0.97	1.74	1.67	1.06	15.95
ant	1.70	3.62	1.24	1.62	2.23	1.26	3.83	28.75
pfo	0.70	0.78	0.71	0.63	1.07	0.89	0.87	13.39
all observers	1.23	2.15	1.22	1.37	**1.89**	1.34	2.37	43.89

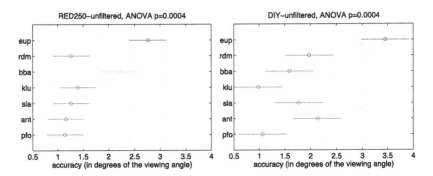

Fig. 4. Multiple comparison of the eye tracker accuracy computed individually for each observer. The mean accuracy values (circles) with confidence intervals (horizontal lines) denote significant difference between some observers for both RED250 and DIY eye trackers. Most of the observers achieved comparable results (the red lines). (Color figure online)

for both eye trackers. On the contrary, most of observers (5 for RED250 and 6 for DIY) achieved comparable results. It suggests that averaging the results for observers is acceptable but some individuals are not able to work with eye tracker due to some technical obstacles (e.g. thick spectacle frames or too strong makeup) or psychophysiological inability to stable focus eyes on a specific object for a longer time.

Viewing Angle. We measured how the viewing angle affects the eye tracker accuracy. The target points were divided into two groups: 16 exterior points and 9 interior points (see Fig. 1). As can be seen in Fig. 5, the accuracy for exterior points is lower than for the interior points. For RED250, the mean accuracy equals to 1.45° and 1.77° for interior and exterior points respectively (ANOVA reveals significant difference $p<0.05$). For DIY, the significant difference is not revealed, but the mean accuracy error for exterior points (1.92°) is higher than for interior points (1.83°).

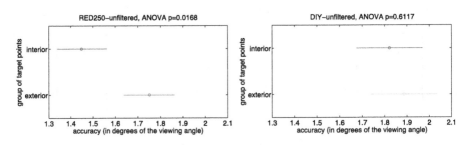

Fig. 5. Multiple comparison of the eye tracker accuracy computed for interior and exterior target points.

Other Factors. ANOVA analysis did not reveal dependence of the mean results on observers' visual fatigue ($p = 0.9025$ for RED250, $p = 0.2716$ for DIY), wearing glasses ($p = 0.9392$ for RED250, $p = 0.5519$ for DIY) and gender ($p = 0.5634$ for RED250, 0.3691 for DIY).

The accuracy strongly differs between repetitions. For example in various repetitions, observer *pfo* achieved values between $0.99°$ and $1.4°$ for RED250, and between $0.52°$ and $2.58°$ for DIY, regardless of the degree of fatigue (see Fig. 6).

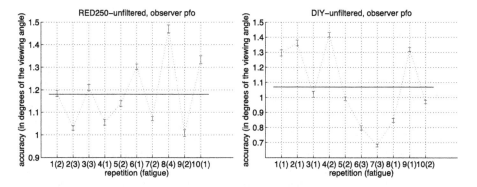

Fig. 6. Average accuracy for one observer computed for individual experiment repetitions. The values in brackets denote visual fatigue declared by the observer before each session (higher value means larger visual fatigue). The error bars depict the standard error of mean. The black horizontal line determines the mean accuracy for all repetitions.

6   Conclusions and Future Work

Both eye trackers used in the experiments have a low average accuracy close to 60 pixels on a 22" screen observed from 65 cm distance. Moreover, variance of results and maximum errors are large and show low robustness of the devices. The accuracy depends on a particular observer and significantly differs between the eye tracking sessions, which means that we cannot expect stable results for the observer in subsequent uses of the device. The visual fatigue does not affect the results in a systematic way. The accuracy does not depend on observer's gender and type of vision correction. On the contrary, the viewing angle is important and the eye tracker accuracy can differ for interior and exterior screen areas.

Eye tracking accuracy inevitably determines possible applications of the eye tracking devices [22–25]. As we assessed two representative eye tracking devices (RED250 and DIY), we argue that further development of eye tracking techniques is desirable to improve this accuracy. The major challenge lies not only in designing better hardware but also in better understanding of the visual fixation mechanism. For example an interesting approach has been proposed in the GDOT technique [26,27], in which fixation direction is determined not only based on the captured gaze direction but also using the information of location and movement of the objects in the scene.

Acknowledgement. The project was funded by the Polish National Science Centre (grant number DEC-2013/09/B/ST6/02270).

References

1. TOBII: Accuracy and precision test method for remote eye trackers (version 2.1). Technical report, TOBII (2012)
2. SMI: RED250 Technical Specification. SensoMotoric Instruments GmbH (2009)
3. Mantiuk, R., Kowalik, M., Nowosielski, A., Bazyluk, B.: Do-it-yourself eye tracker: low-cost pupil-based eye tracker for computer graphics applications. In: Schoeffmann, K., Merialdo, B., Hauptmann, A.G., Ngo, C.-W., Andreopoulos, Y., Breiteneder, C. (eds.) MMM 2012. LNCS, vol. 7131, pp. 115–125. Springer, Heidelberg (2012). doi:10.1007/978-3-642-27355-1_13
4. Reinhard, E., Khan, E.A., Akyuz, A.O., Johnson, G.: Color Imaging. Fundamentals and Applications. A K Peters, Wellesley (2008)
5. Charman, W.N.: Optics of the eye. In: Fundamentals, Techniques and Design, vol. 1, 2nd edn. McGraw-Hill, New York (1995)
6. Robinson, D.A.: The mechanics of human saccadic eye movement. J. Physiol. **174**, 245–264 (1964)
7. Rayner, K.: Eye movements in reading and information processing: 20 years of research. Psychol. Bull. **124**, 372–395 (1998)
8. Robinson, D.A.: The oculomotor control system: a review. Proc. IEEE **56**, 1032–1049 (1968)
9. Duchowski, A.T., Pelfrey, B., House, D.H., Wang, R.: Measuring gaze depth with an eye tracker during stereoscopic display. In: Proceedings of Symposium on Applied Perception in Graphics and Visualization, France, APGV 2011 (2011)
10. Salvucci, D.D., Goldberg, J.H.: Identifying fixations and saccades in eye-tracking protocols. In: Proceedings of the 2000 Symposium on Eye Tracking Research & Applications (ETRA), New York, pp. 71–78 (2000)
11. Duchowski, A.T.: Eye Tracking Methodology: Theory and Practice, 2nd edn. Springer, London (2007)
12. Blignaut, P.: Fixation identification: the optimum threshold for a dispersion algorithm. Attention Percept. Psychophys. **71**, 881–895 (2009)
13. Shic, F., Scassellati, B., Chawarska, K.: The incomplete fixation measure. In: Proceedings of the 2008 Symposium on Eye Tracking Research & #38; Applications, ETRA 2008, pp. 111–114. ACM, New York (2008)
14. Hoffman, J.E., Subramaniam, B.: The role of visual attention in saccadic eye movements. Attention Percept. Psychophys. **57**, 787–795 (1995)
15. Morimoto, C., Koons, D., Amir, A., Flickner, M., Zhai, S.: Keeping an eye for HCI. In: Proceedings of the XII Symposium on Computer Graphics and Image Processing, pp. 171–176 (1999)
16. Morimoto, C.H., Mimica, M.: Eye gaze tracking techniques for interactive applications. Comput. Vis. Image Underst. **98**, 4–24 (2005)
17. SMI: Experiment Center 2. Manual, version 2.4. SensoMotoric Instruments (2010)
18. Ditchburn, R.W.: The function of small saccades. Vis. Res. **16**, 271–272 (1980)

19. ITU: ITU Gaze Tracker software. IT University of Copenhagen, ITU GazeGroup (2009). http://www.gazegroup.org/home
20. Kleiner, M., Brainard, D., Pelli, D.: What's new in Psychtoolbox-3? A free cross-platform toolkit for Psychophysics with Matlab and GNU/Octave. Max Planck Institute for Biological Cybernetics (2008). http://psychtoolbox.org
21. ITU-R.REC.BT.500-11: Methodology for the subjective assessment of the quality for television pictures (2002)
22. Duchowski, A.T.: A breadth-first survey of eye-tracking applications. Behav. Res. Meth. Instrum. Comput. **34**, 455–470 (2002)
23. Mantiuk, R., Bazyluk, B., Tomaszewska, A.: Gaze-dependent depth-of-field effect rendering in virtual environments. In: Ma, M., Fradinho Oliveira, M., Madeiras Pereira, J. (eds.) SGDA 2011. LNCS, vol. 6944, pp. 1–12. Springer, Heidelberg (2011). doi:10.1007/978-3-642-23834-5_1
24. Mantiuk, R., Janus, S.: Gaze-dependent ambient occlusion. In: Bebis, G., Boyle, R., Parvin, B., Koracin, D., Fowlkes, C., Wang, S., Choi, M.-H., Mantler, S., Schulze, J., Acevedo, D., Mueller, K., Papka, M. (eds.) ISVC 2012. LNCS, vol. 7431, pp. 523–532. Springer, Heidelberg (2012). doi:10.1007/978-3-642-33179-4_50
25. Mantiuk, R., Markowski, M.: Gaze-dependent tone mapping. In: Kamel, M., Campilho, A. (eds.) ICIAR 2013. LNCS, vol. 7950, pp. 426–433. Springer, Heidelberg (2013). doi:10.1007/978-3-642-39094-4_48
26. Mantiuk, R., Bazyluk, B., Mantiuk, R.K.: Gaze-driven object tracking for real time rendering. Comput. Graph. Forum **32**, 163–173 (2013)
27. Bazyluk, B., Mantiuk, R.: Gaze-driven object tracking based on optical flow estimation. In: Chmielewski, L.J., Kozera, R., Shin, B.-S., Wojciechowski, K. (eds.) ICCVG 2014. LNCS, vol. 8671, pp. 84–91. Springer, Heidelberg (2014). doi:10.1007/978-3-319-11331-9_11

Mouth Features Extraction for Emotion Analysis

Robert Staniucha$^{(\boxtimes)}$ and Adam Wojciechowski

Institute of Information Technology, Lodz University of Technology,
Wólczańska 215, 90-924 Lodz, Poland
800671@edu.p.lodz.pl, adam.wojciechowski@p.lodz.pl

Abstract. Face emotions analysis is one of the fundamental techniques
that might be exploited in a natural human-computer interaction process
and thus is, one of the most studied topics in current computer vision
literature. In consequence face features extraction is an indispensable
element of the face emotion analysis as it influences decision making performance. The paper concentrates on mouth features extraction, which
next to eye region features becomes one of the most representative face
regions in the context of emotions retrieval. In the paper original, gradient based, mouth features extraction method was presented. Its high
performance (exceeding 90 % for selected features) was also verified for
a subset of the Yale images database.

1 Related Work

From a biologically point of view, facial expressions are generated by contractions
of facial muscles, which cause face features temporal deformations. The most
significant changes relate to eyelids, brews eyes, nose, lips or skin wrinkles. The
intensity of facial expression can be measured by geometric deformation of facial
features or analysis of facial texture, i.e. density of face appearing wrinkles.

Among anthropologically justified face core landmarks forming unquestion-
able framework for face features extraction researches [16] proposed 11 points:
pronasale, alare (left and right), subnasale, chelion (left and right), endocan-
thion (left and right), exocanthion (left and right) and sellion (Fig. 1). Then a
few minor landmarks can be evaluated.

All face features extraction methods, presented in available literature, can be
classified into two main groups: appearance based and geometric based methods [19]. Although the appearance based approach seems to be currently the
most popular, the geometric based methods seems to be recently neglected but
still promising. While this approach seems to be very well studied [8,33], according to Pali [20], one the most obvious aspects that can be still improved, within
face features extraction, there are dimensionality reduction, features extraction
techniques and features subset selection. Thus, the geometrical approach can
almost automatically reduce the space dimension of the problem and behave
more reliably in difficult scenarios where pose (in-plane and out of plane face
rotations) and illumination are not controlled [5].

Face features detection is a difficult task, because the quantity of local face
image structures, classical corner detectors are useless without considering their

© Springer International Publishing AG 2017
S. Kobayashi et al. (eds.), *Hard and Soft Computing for Artificial Intelligence, Multimedia and Security*, Advances in Intelligent Systems and Computing 534, DOI 10.1007/978-3-319-48429-7_21

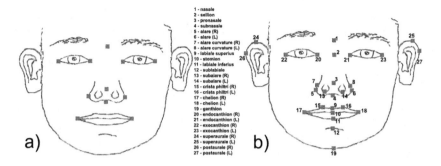

Fig. 1. (a) Set of 11 anthropological landmarks (small square) used for face identification and face expression analysis (b) Minor landmarks (small square) to improve accuracy

context. Researches estimating face inherent features and edges date back to well known Kanade work [11] and were further intensively developed (i.e. [3]). Authors attempted to reconstruct horizontal and vertical lines applying Laplacian operator and evaluating horizontal and vertical integral image projection achieve effectiveness of about 75 % on dedicated, self-prepared subset of a popular faces data source.

Castrillon [5], Castille [4] and Yang [31] suggested Viola-Jones object detection algorithm for coarse face parts (i.e.: eyes, mouth, nose) localization, but it did not localize the exact face landmarks and required further, more detailed features detectors. Even Panning [21] and Wang et al. [29] approaches, based on measuring distances between face regions, detected with originally elaborated features or Lienhart [17] extended Haar-like features, required consequent holistic extensive well trained classification system.

The process of face landmark detection should be made by an appropriate in-image face detection. Then extracted landmarks can be used then for classify facial expression. The localization of face and general face features extraction can be performed in various ways. Common solutions encompass deploying of a sequence of Haar-like features according to Viola-Jones algorithm [4,28], eigenspaces [22,27] also referring to 3D model [32], skin color segmentation [25], statistical methods [2,23] or active contour or shape models [12,14].

Some authors [9,10] used local binary patterns (LBP) idea for face image features extraction. Their extension: Local Direction Patterns (LDP) and locally assembled binary (LAB) features [30] seems to be quite efficient for face edges detection and partly inspired authors of this paper to analyze a gradient distribution.

If dedicated mouth features are further required, within face region analysis should be performed. Kim [13] and Chien [6] analyzed grid-based and coordinate-based lips features (width/height of outer/inner lips edges) for Korean language words recognition support. Matthews [18] exploited AAM and ASM for mouth visual shape description for lips reading. Shen [24] and Lewis [15] analyzed color space for lip features retrieving. He et al. [7] proposed modified Biologically

Inspired Model of face features extraction improving the SVM classification of face smile. Su [26] suggested geometrical and Gabor filter retrieved face features fusion can improve facial expression recognition.

Aforementioned approaches, although well studied and developed, lack of generic simplicity which lies in mouth lines detection. Described method robustly extracts simplified lips edges by means of originally elaborated gradient-based approach which can be subsequently interpreted. The solution provides a representative set of features for further mouth classification and consequent facial emotions recognition. It has also been tested on a subset of Yale faces database [1].

2   Mouth Features Extraction Method

To detect mouth shape features we need to find human face and relative localization of the mouth. Next, we can try extract mouth shape from it. Aggregated mouth features extraction process can be completed within subsequent steps:

1. Finding face,
2. Finding mouth on the face,
3. Mouth segmentation,
4. Features extraction.

In image face localization can be performed by means of Haar-like features method [17], but it will not be described here, because it is a part of another problem. Localization of mouth within the face region is a similar problem and can be completed with an analogical set of Haar-like features. That is why further, core mouth features analysis, assumes that face image is cropped to mouth with some border, as shown in Fig. 2

a)   b)   c)

Fig. 2. Example of mouth images: (a) happy, (b) neutral, (c) sad, for subject no. 1

Mouth segmentation is then performed in several steps. These can be described as:

1. Gradient calculation,
2. Resulting image normalization,
3. Resulting image filtering,
4. Resulting image thresholding,
5. Noise removal.

To retrieve information from mouth images, at the first step **gradient** should be calculated. Gradient of an image is calculated with formula 1.

$$\nabla f = \begin{bmatrix} g_x \\ g_y \end{bmatrix} = \begin{bmatrix} \frac{\partial f}{\partial x} \\ \frac{\partial f}{\partial y} \end{bmatrix} \tag{1}$$

During the experiments, it was noticed that for descent mouth retrieval there is no need to calculate the whole gradient, but only its vertical part. Using both dimensions of gradient does not give noticeable results improvement, thus, in the segmentation process there was used only vertical component.

To calculate a vertical gradient g_y of image A we can use simple filter, as it is shown in Eq. 2.

$$g_y = \frac{\partial f}{\partial y} = \begin{bmatrix} -1 \\ 0 \\ 1 \end{bmatrix} * A \tag{2}$$

Additionally, extended 3×3 matrix (Eq. 3) was used to reduce noise, which introduces additional column on the left and on the right side of pixel position.

$$\begin{bmatrix} -1 \\ 0 \\ 1 \end{bmatrix} \Rightarrow \begin{bmatrix} -1 & -1 & -1 \\ 0 & 0 & 0 \\ 1 & 1 & 1 \end{bmatrix} \tag{3}$$

a)   b)   c)

Fig. 3. Example gradient for exemplary mouth images: (a) happy, (b) neutral, (c) sad, for subject no. 1

The first step results in images with vertical gradient calculated over the whole their area. Additionally, gradient values were squared in order to level up output and remove useless noise (Fig. 3).

The next image processing step was the gradient **normalization**. MIN-MAX type normalization was used to adjust gradient to image full spectrum of brightness.

In the next step gradient image filtering was considered as to extract edges from it. Various filters was tested to extract shapes from images. The best results were obtained by filters represented in Fig. 4. Corresponding outputs obtained with matrix B are presented in Fig. 5.

After the filtering step a process of **thresholding** was applied to extract shape of lips. It was done by simple cut off image value below certain threshold of pixel brightness. Results were verified for a few threshold values, but overall the best was achieved with 250. Example results obtained with different values of threshold are shown in Fig. 6

$$A = \begin{bmatrix} -1 & -2 & -1 \\ 0 & 0 & 0 \\ 1 & 2 & 1 \end{bmatrix} \quad B = \begin{bmatrix} -3 & -10 & -3 \\ 0 & 0 & 0 \\ 3 & 10 & 3 \end{bmatrix} \quad C = \begin{bmatrix} -1 & -1 & -1 \\ 2 & 2 & 2 \\ -1 & -1 & -1 \end{bmatrix}$$

Fig. 4. Filter matrices used for edge extraction

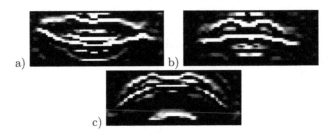

Fig. 5. Exemplary outputs obtained with matrix B from Fig. 4: (a) happy, (b) normal, (c) sad, for subject no. 1

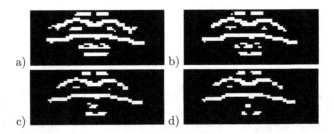

Fig. 6. Example results obtained with different threshold values: (a) 100, (b) 150, (c) 200 and (d) 250, for subject no. 1

The subsequent step of the proposed method is the noise reduction. It was achieved by morphological operations performed on image. The best noise reduction was obtained for closing, which is a combination of erosion and dilation. In result shape of mouth was closed as it is shown in Fig. 7.

The last stage concerned segmented image features extraction. Mouth corners were selected as initially considered features. They were found by the most extreme edges in all directions: down, left, right and up (Fig. 8).

Edge point c_e is defined as the farthest point in specified direction, as is shown in Eq. 4.

$$c_e = (x, y) \text{ where } e = \{\text{B} - bottom, \text{L} - left, \text{R} - right, \text{U} - topmost\} \quad (4)$$

In case of the left edge c_L, it is the most distant left point of the largest contour area. c_R can be defined analogously, but for the right edge. More difficult is to find the bottom and topmost edge points: c_B and c_U, because often the contour

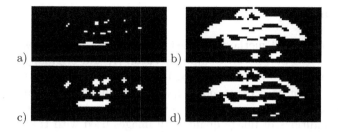

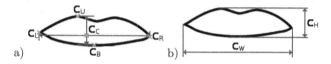

Fig. 7. Example results with different noise reduction: (a) erosion, (b) dilatation, (c) opening, (d) closing, for subject no. 12

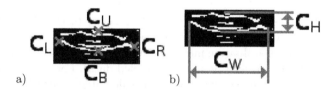

Fig. 8. Edge points on corners: (a) positions, (b) sizes, for subject no. 1

is composed of several separated pieces, so we need to check if the contour size is sufficiently large in relation to the image size. Calculation of the shape size was simply done by following each white pixel connected to initially located edge points (c_L and c_R). In case of contour inconsistence, averaged extreme values (bottom and topmost), retrieved from two separate edges, anchored independently from left and right corners were calculated.

The farthest edges are marked with white and light grey dashed lines. Brighter color is used to mark part of shape with features determining edges (Fig. 9).

Fig. 9. Feature extraction corners: (a) positions, (b) sizes, for subject no. 1

The collected information was sufficient to easily determine the next two parameters: c_W and c_H. The first one determines the maximum distance stretching horizontally the mouth area, between c_L and c_R, on the average height between them. Similarly value of c_H was determined.

With a **cross-section** c_C of width c_W and height c_H, it is possible to interpret shape of the the mouth. Cross-section is always defined, because interpreted shape corners c_e are also determined.

3   Evaluation

To evaluate presented method a subset of the Yale's face database was used. It comes from UC San Diego Computer Vision [1] and originally contains 165 grayscale images of 15 people, sized 320×243 pixels. Each subject had several images in different face expressions, quality and light conditions, but for all people those conditions were the same. This allowed us to test developed method in various conditions of image quality. All images used with this database was manually trimmed to mouth region with size of 75×30 pixels.

Each image was marked by expert as shown in Fig. 10: shape of mouth was built out of 6 points and it was marked by expert with white line, maximum width and maximum height. The selected elements have been measured and noted as m_e, i.e. width as m_W, the same way as the computed values c_W from authors' method.

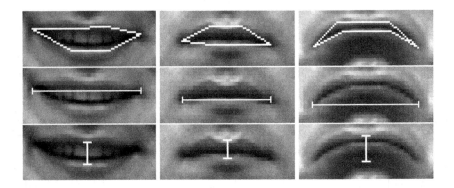

Fig. 10. Example labeled dataset with marked lines: shape, width and height

Exemplary results are shown in Fig. 11. In analogy to the reference images, maximum widths and heights are marked over the image.

Mouth key features evaluation method results c_e were subsequently compared with reference, expert marked points m_e (labeled data). For i-th image, each of key features c_e estimation accuracy ACC was calculated according to Eq. 5. For cross-section feature c_C estimation accuracy ACC was calculated according to Eq. 8, where distance is Manhattan distance between points and distance$_{MAX}$ is the maximum possible distance on the image.

$$ACC_i(c_e) = \frac{\min\{c_e.x; m_e.x\} \times 100\,\%}{\max\{c_e.x; m_e.x\}} \text{ where } e = \{L, R\} \qquad (5)$$

$$ACC_i(c_e) = \frac{\min\{c_e.y; m_e.y\} \times 100\,\%}{\max\{c_e.y; m_e.y\}} \text{ where } e = \{B, U\} \qquad (6)$$

$$ACC_i(c_e) = \frac{\min\{c_e; m_e\} \times 100\,\%}{\max\{c_e; m_e\}} \text{ where } e = \{W, H\} \qquad (7)$$

$$ACC_i(c_c) = 100\% - \frac{(\|c_c.x - m_c.x\| + \|c_c.y - m_c.y\|) \times 100\%}{75 + 30} \qquad (8)$$

Table 1 shown selected features positioning accuracy results are presented. The results were calculated as averaged value of selected ($n = 66$) Yale database images individually estimated accuracies $ACC_i(c_e)$ (Eq. 9).

$$ACC(c_e) = \frac{\sum_i ACC_i(c_e)}{n}$$
$$i \in \{1, 2, \ldots, n\}, \ e = \{L, R, B, U, W, H, C\} \qquad (9)$$

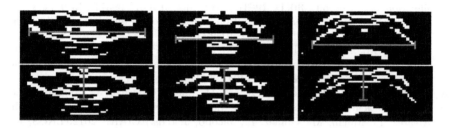

Fig. 11. Example results with marked with gray lines: width and height

Table 1. Mouth features recognition results

Measurement	Size		Position				
	$ACC(e_W)$	$ACC(e_H)$	$ACC(e_L)$	$ACC(e_R)$	$ACC(e_U)$	$ACC(e_B)$	$ACC(e_C)$
Result [%]	92.78	72.77	81.42	95.53	43.59	88.29	86.18
Std dev. [%]	7.05	17.93	18.4	3.84	25.5	13.13	9.4

The highest results were recorded for left (81.42 %), bottom (88.29 %) and right (95.53 %) part of mouth features detection. More problematic was upper part, what is closely related to variations in the received gradient of mouths. Resulting emotion characteristic determinants: mouth width (92.78 %), height (72.77 %) and cross-section point (86.18 %) revealed also high evaluation accuracy. Weak upper part of the mouth features estimation has a negative effectiveness of mouth shape determination, but associated mouth shape emotion analysis can be further successfully carried out to powerful classification algorithms.

4   Conclusion

This paper proposed a method to extract some facial features from mouth image to expression recognition. The experiments were performed on common database, which is often used in research methods dealing with face images. As shown in the previous section, it gives a generally good results, especially in horizontal measurement and should be sufficient for the right description of many facial expressions. In the future work it will be to collect other facial features and combine them together to better understand and identify face poses.

References

1. Belhumeur, P.N., Hespanha, J.P., Kriegman, D.J.: Eigenfaces vs. fisherfaces: recognition using class specific linear projection **19**(7), 711–720 (1997)
2. Berbar, M.A.: Three robust features extraction approaches for facial gender classification. Vis. Comput. **30**(1), 19–31 (2013). doi:10.1007/s00371-013-0774-8
3. Brunelli, R., Poggio, T.: Face recognition: features versus templates. IEEE Trans. Pattern Anal. Mach. Intell. **10**, 1042–1052 (1993)
4. Castrilln, M., Dniz, O., Hernndez, D., Lorenzo, J.: A comparison of face and facial feature detectors based on the violajones general object detection framework. Mach. Vis. Appl. **22**(3), 481–494 (2011)
5. Castrillón-Santana, M., Hernández-Sosa, D., Lorenzo-Navarro, J.: Combining face and facial feature detectors for face detection performance improvement. In: Alvarez, L., Mejail, M., Gomez, L., Jacobo, J. (eds.) CIARP 2012. LNCS, vol. 7441, pp. 82–89. Springer, Heidelberg (2012). doi:10.1007/978-3-642-33275-3_10
6. Chien, S.-I., Choi, I.: Face and facial landmarks location based on log-polar mapping. In: Lee, S.-W., Bülthoff, H.H., Poggio, T. (eds.) BMCV 2000. LNCS, vol. 1811, pp. 379–386. Springer, Heidelberg (2000). doi:10.1007/3-540-45482-9_38
7. He, C., Mao, H., Jin, L.: Realistic smile expression recognition using biologically inspired features. In: Wang, D., Reynolds, M. (eds.) AI 2011. LNCS (LNAI), vol. 7106, pp. 590–599. Springer, Heidelberg (2011). doi:10.1007/978-3-642-25832-9_60
8. Hjelmås, E., Low, B.K.: Face detection: a survey. Computer Vis. Image Underst. **83**(3), 236–274 (2001)
9. Hussain, A., Khan, M.S., Nazir, M., Iqbal, M.A.: Survey of various feature extraction and classification techniques for facial expression recognition. In: Proceedings of the 11th WSEAS International Conference on Electronics, Hardware, Wireless and Optical Communications, and Proceedings of the 11th WSEAS International Conference on Signal Processing, Robotics and Automation, and Proceedings of the 4th WSEAS International Conference on Nanotechnology, pp. 138–142 (2012)
10. Jabid, T., Kabir, M.H., Chae, O.: Robust facial expression recognition based on local directional pattern. ETRI J. **32**(5), 784–794 (2010)
11. Kanade, T.: Computer Recognition of Human Faces, vol. 47. Birkhäuser, Basel (1977)
12. Kass, M., Witkin, A., Terzepulos, D.: Snakes: active contour models. In: First IEEE International Conference on Computer Vision, pp. 259–268 (1987)
13. Kim, Y.K., Lim, J.G., Kim, M.H.: Comparison of lip image feature extraction methods for improvement of isolated word recognition rate, August 2015. http://dx.doi.org/10.14257/astl.2015.107.14
14. Lee, Y.H., Kim, C.G., Kim, Y., Whangbo, T.K.: Facial landmarks detection using improved active shape model on android platform. Multimedia Tools Appl. **74**(20), 8821–8830 (2013). doi:10.1007/s11042-013-1565-y
15. Lewis, T.W., Powers, D.M.W.: Lip feature extraction using red exclusion. In: Selected Papers from the Pan-Sydney Workshop on Visualisation, VIP 2000, vol. 2. pp. 61–67. Australian Computer Society Inc., Darlinghurst (2001). http://dl.acm.org/citation.cfm?id=563752.563761

16. Liang, S., Wu, J., Weinberg, S.M., Shapiro, L.G.: Improved detection of landmarks on 3d human face data. In: 2013 35th Annual International Conference of the IEEE Engineering in Medicine and Biology Society (EMBC), pp. 6482–6485, July 2013. doi:10.1109/EMBC.2013.6611039
17. Lienhart, R., Maydt, J.: An extended set of haar-like features for rapid object detection. In: Proceedings of the 2002 International Conference on Image Processing, vol. 1, p. I-900 (2002)
18. Matthews, I., Cootes, T., Bangham, J., Cox, S., Harvey, R.: Extraction of visual features for lipreading. IEEE Trans. Pattern Anal. Machine Intell. **24**(2), 198–213 (2002). doi:10.1109/34.982900
19. Mishra, S., Dhole, A.: A survey on facial expression recognition techniques. Int. J. Sci. Res. **4**(4), 1247–1250 (2015)
20. Pali, V., Goswami, S., Bhaiya, L.: An extensive survey on feature extraction techniques for facial image processing. In: Sixth International Conference on Computational Intelligence and Communication Networks, pp. 142–148 (2014)
21. Panning, A., Al-Hamadi, A.K., Niese, R., Michaelis, B.: Facial expression recognition based on haar-like feature detection. Pattern Recogn. Image Anal. **18**(3), 447–452 (2008)
22. Pentland, A., Moghaddam, B., Starner, T.: View-based and modular eigenspaces for face recognition. In: IEEE Conference on Computer Vision and Pattern Recognition, Seattle, NA, USA, pp. 84–91 (1994)
23. Schneiderman, H., Kanade, T.: A statistical method for 3d object detection applied to faces and cars. In: Proceedings of the IEEE Conference on Computer Vision and Pattern Recognition, vol. 1, pp. 746–751. IEEE (2000)
24. Shen, X.G., Wu, W.: An algorithm of lips secondary positioning and feature extraction based on ycbcr color space. In: International Conference on Advances in Mechanical Engineering and Industrial Informatics. Atlantis Press (2015). http://dx.doi.org/10.2991/ameii-15.2015.271
25. Sobottka, K., Pitas, I.: A novel method for automatic face segmentation, facial feature extraction and tracking. Sig. Process. Image Commun. **12**(3), 263–281 (1998)
26. Su, C., Deng, J., Yang, Y., Wang, G.: Expression recognition methods based on feature fusion. In: Yao, Y., Sun, R., Poggio, T., Liu, J., Zhong, N., Huang, J. (eds.) BI 2010. LNCS (LNAI), vol. 6334, pp. 346–356. Springer, Heidelberg (2010). doi:10.1007/978-3-642-15314-3_33
27. Turk, M., Pentland, A.: Eigenfaces for recognition. J. Cogn. Neurosci. **3**(1), 71–86 (1991)
28. Viola, P., Jones, M.: Robust real-time face detection. Int. J. Comput. Vis. **57**, 137–154 (2004)
29. Wang, Q., Zhao, C., Yang, J.: Robust facial feature location on gray intensity face. In: Wada, T., Huang, F., Lin, S. (eds.) PSIVT 2009. LNCS, vol. 5414, pp. 542–549. Springer, Heidelberg (2009). doi:10.1007/978-3-540-92957-4_47
30. Yan, S., Shan, S., Chen, X., Gao, W.: Locally assembled binary (lab) feature with feature-centric cascade for fast and accurate face detection. In: IEEE Conference on Computer Vision and Pattern Recognition, CVPR 2008, pp. 1–7 (2008)
31. Yang, M.T., Cheng, Y.J., Shih, Y.C.: Facial expression recognition for learning status analysis (2011). http://dx.doi.org/10.1007/978-3-642-21619-0-18
32. Yang, W., Sun, C., Zheng, W., Ricanek, K.: Gender classification using 3D statistical models. Multimedia Tools Appl., March 2016. http://dx.doi.org/10.1007/s11042-016-3446-7
33. Zhang, C., Zhang, Z.: A survey of recent advances in face detection. Technical report, Microsoft Research (2010)

Sensitivity of Area–Perimeter Relation for Image Analysis and Image Segmentation Purposes

Dorota Oszutowska–Mazurek[1,2(✉)] and Przemysław Mazurek[3]

[1] Faculty of Motor Transport, Higher School of Technology
and Economics in Szczecin, Klonowica 14 St., 71244 Szczecin, Poland
adorotta@op.pl
[2] Department of Epidemiology and Management, Pomeranian Medical University,
Zolnierska 48 St., 71210 Szczecin, Poland
[3] Department of Signal Processing and Multimedia Engineering, West–Pomeranian
University of Technology, Szczecin, 26. Kwietnia 10 St., 71126 Szczecin, Poland
przemyslaw.mazurek@zut.edu.pl

Abstract. Image analysis with the use of fractal estimators is important for the description of grayscale images. The sensitivity of Area–Perimeter Relation (APR) using Brodatz texture database and Monte Carlo approach is evaluated in this paper. Obtained APR curve is approximated using polynomial and two parameters of polynomial are applied as discrimination parameters. A few techniques for the evaluation of APR are applied. The results show the possibility of the discrimination using single or two polynomial parameters even for a few textures. The quality of discrimination (separation between textures classes) could be improved if larger window analysis sizes is applied.

Keywords: Fractals · Textures analysis · Area–Perimeter Relation

1 Introduction

Fractals are essential for the description of different object [7], like image textures, that is important in image analysis. Different fractal estimators are available, and classical estimators return single value of fractal dimension that corresponds to the complexity of analyzed object [13,14]. Such estimators could be applied for the analysis of synthetic objects, generated with the use of mathematical formulas. It means that fractal dimension is not a parameter of the scale. Natural objects are more complex and the fractal dimension is the parameter of scale, so multiple values for the description of objects are necessary. Estimators that return multiple value descriptors are multifractal estimators [4,16]. Multifractal analysis is better fitted for the typical image analysis tasks, because few parameters are required to distinguish between different textures.

There are numerous fractal descriptors but some of them require square or rectangular area of analysis. Irregular areas could be analyzed using variogram, Slit–Island Method (SIM) or Area–Perimeter Relation (APR). SIM is very poor

S. Kobayashi et al. (eds.), *Hard and Soft Computing for Artificial Intelligence, Multimedia and Security*, Advances in Intelligent Systems and Computing 534, DOI 10.1007/978-3-319-48429-7_22

approach, because obtained values are very sensitive to content, so comparison of very similar textures is possible only. Disadvantages of SIM are removed by APR and this estimator is much more interesting in practical applications.

1.1 Related Works

There are numerous fractal based estimators and the most notable is Box–Counting. Area–Perimeter Relation is proposed in [9]. Basic method – Slit Island Method is proposed in [8]. Multifractal analysis is possible using variogram [11,18] and lacunarity [12,15] directly, but the majority of methods could be adapted for multifractal analysis [10].

1.2 Content and Contribution of the Paper

Area–Perimeter Relation is briefly described in Sect. 2. Sensitivity analysis based on Monte Carlo approach is proposed in Sect. 3. Results and discussion are provided in Sects. 4 and 5 respectively. Final conclusion and further work is provided in Sect. 6.

The analysis of APR sensitivity allows the assessment of this method, but dedicated technique of analysis should be proposed. It is based on the applications of Brodatz texture database. Random size and position areas are analyzed using APR and the dependence on number of pixels is obtained.

SIM is very sensitive method and results are not reliable. Some of the examples are shown in [9] and another analyzes are provided in [5,6] for example.

2 Area–Perimeter Relation

This algorithm uses SIM approach for the layer cutting of object. SIM assumes analysis of 3D object that is sliced and transformed to binary images. APR is applied to the analysis of greyscale images that are sliced using threshold value. SIM method uses binary representation of layers using unipolar values. APR uses binary representation of layer using bipolar values and more carefully considers boundaries of objects.

The idea of image slicing is shown in Fig. 1.

Input image T is thresholded using the following formula:

$$X^T(x,y) = \begin{cases} -1 : I(x,y) < T \\ +1 : I(x,y) \geq T \end{cases}, \tag{1}$$

where x, y denotes coordinates of pixel and T is the threshold level.

This formula could be applied for example to the 8–bit greyscale images and there are $2^8 + 1$ of possible threshold levels, with two boundary cases. The first case corresponds to output image filled by -1 values and second to output image filled by $+1$.

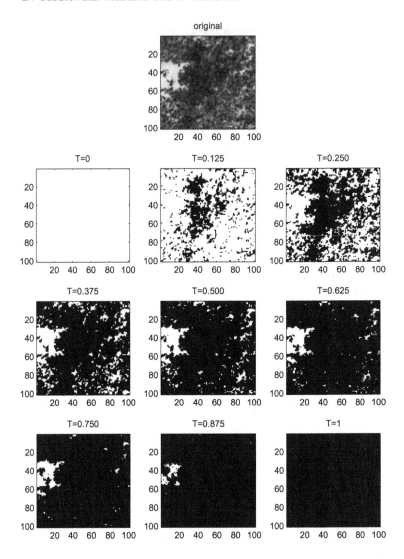

Fig. 1. Example square image and a few binary slices

APR uses computation of area A_i and perimeter P_i for every binary layer. Binary image requires assumption about technique of perimeter calculation, because perimeter is not unique. There are 4–way and 8–way neighborhood algorithms of perimeter calculations and modified 4–way is applied in APR. This modification is important for non–rectangular area of analysis.

The following formula is applied if background pixels B are omitted:

$$P^T = \frac{1}{2} \sum_{x,y} (C_1 + C_2 + C_3 + C_4) \tag{2}$$

$$C_1 = X^T(x,y) \wedge \overline{X^T(x,y-1)} \wedge \cdots$$
$$\cdots X^T(x,y) \neq B \wedge X^T(x,y-1) \neq B \tag{3}$$
$$C_2 = X^T(x,y) \wedge \overline{X^T(x,y+1)} \wedge \cdots$$
$$\cdots X^T(x,y) \neq B \wedge X^T(x,y+1) \neq B \tag{4}$$
$$C_3 = X^T(x,y) \wedge \overline{X^T(x-1,y)} \wedge \cdots$$
$$\cdots X^T(x,y) \neq B \wedge X^T(x-1,y) \neq B \tag{5}$$
$$C_4 = X^T(x,y) \wedge \overline{X^T(x+1,y)} \wedge \cdots$$
$$\cdots X^T(x,y) \neq B \wedge X^T(x+1,y) \neq B. \tag{6}$$

This formula forbids calculation of perimeter using edges between the object and background. Details about specific case of estimation are presented in [9]. The division by 2 is necessary for the fixing of result due to double counting of cliques.

The calculation of area is simple using counting of pixel with +1 value:

$$A^T = \sum_{x,y} \left(X^T(x,y) == +1 \right) \neq B. \tag{7}$$

Obtained area and perimeter pairs $\left(A^T, P^T \right)$ allows the visualization of values changes depending on threshold level. Obtained curve is not function, but relation.

The area is non–negative value and preserves the boundary conditions:

$$A^{T=0} = 0 \tag{8}$$

and

$$A^{T=max} = A_{max} \tag{9}$$

where A_{max} is the area of analysis. Additional shift $-A_{max}/2$ of area values gives a very important connection with the Ising model [2,3,17]. Such operation could be transformed to the following formula:

$$A_M^T = \sum_{x,y} X^T(x,y) \neq B, \tag{10}$$

where A_M is the magnetization, because bipolar values of X are used directly. The obtained curve is almost symmetrical and corresponds to the magnetization curve and preserves properties of II–type of critical phenomena [1]. The Curie point could be estimated, but in the context is not temperature based, so pseudo–Curie point naming is applied. Exemplary magnetization curve is shown in Fig. 2 for the previous image.

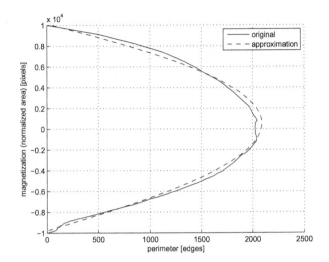

Fig. 2. Magnetization curve – Area–Perimeter Relation for Fig. 1

Different parameters of obtained curve could be considered. Polynomial fitting could be applied for the approximation and description of curve using polynomial weight coefficient. Such approximation is well fitted to the shape of observed curves, but another approaches are possible also. The zero order weight is the approximated pseudo–Curie point.

3   Sensitivity Analysis

The testing is based on Brodatz texture database without contrast scaling. This database contains numerous types of textures and is typical texture database used in different tests related to image analysis and segmentation.

Monte Carlo approach is applied for the sensitivity analysis and such approach allow the testing different properties of algorithms. The position of window is random, also. Estimated parameters depend on the position, due to small sample size, and number of pixel.

Polynomial approximation of APR is assumed in this paper and the following formula is used:

$$A(P) = a_4 P^4 + a_3 P^3 + a_2 P^2 + a_1 P + a_0, \tag{11}$$

where a_0 is the approximated pseudo–Curie point. Five parameters are obtained but lower number of parameters could be applied in image analysis applications.

There are 112 textures used in a test applied to four window sizes. The resolution of images is 640×640 pixels and window sizes are: 50×50, 100×100, 150×150 and 200×200. There are 400 samples drawn from image with random position.

4   Results

The first test uses single image with 100×100 nonoverlapping window position (Fig. 3). This sample image was selected randomly from database. There are 9 of window positions.

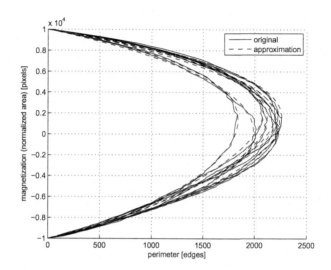

Fig. 3. Measurement of APR and corresponding approximation curves

The second test shows mean values of a_0 and a_1 polynomial coefficients. The mean value is obtained for 400 random positions of window (Fig. 5).

Two next examples show ranges (Figs. 5 and 6) obtained from 400 random positions of window. There are 10 randomly selected textures. There are available a few different textures in typical image analysis or segmentation applications. The Brodatz texture database applied in tests contains 112 of images, so it is possible to find similarity between textures using texture descriptors with a few parameters.

The range is defined by minimal and maximal value for a_0 and a_1 parameters. The lack of overlapping between rectangles means that two particular textures could be discriminated using a_0 and a_1 parameters.

5   Discussion

The similarity of results (similar APR curves) could be obtained for the same texture and results are shown in Fig. 3. This example is not a proof, but shows the possibility of the application of APR for image analysis and segmentation purposes.

244   D. Oszutowska–Mazurek and P. Mazurek

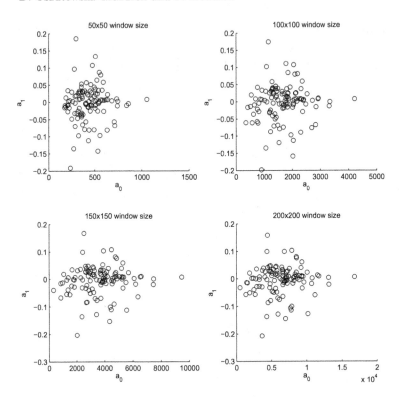

Fig. 4. Mean values of two polynomial coefficients - a_0 and a_1

The second test shows difference between two selected polynomial coefficients in mean sense (Fig. 4). The mean positions are dispersed so both coefficients could be applied together as a parameters for the discrimination between textures. In some cases differences are not high and higher density of position is visible in the center part of figures. It means that both parameters in mean sense are similar for two or more textures and the discrimination is difficult.

Two last tests (Figs. 5 and 6) shows a typical cases where a few textures are available. The difference between textures is well visible and some textures overlap (harder to discriminate), but in numerous cases two textures could be simply discriminated using two or single polynomial parameter. The larger size of the window improves discrimination, moreover. Range rectangles are less dispersed and the distance between them is larger. It is related to the scale of basic features in particular textures.

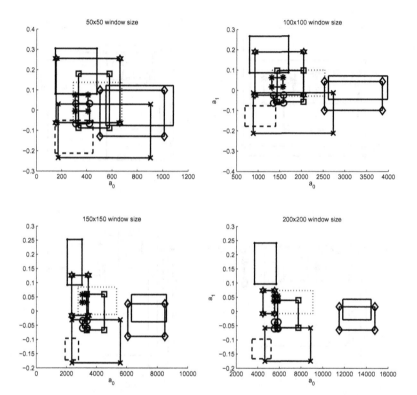

Fig. 5. Example no.1 - Range of two polynomial coefficients - a_0 and a_1 for ten random textures

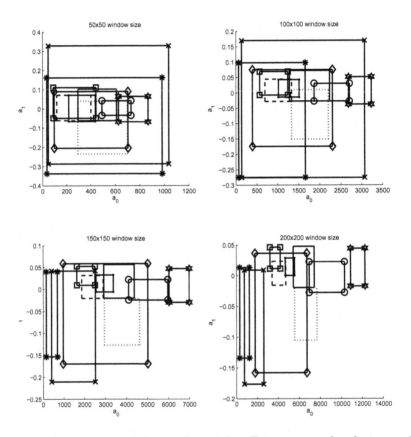

Fig. 6. Example no.2 - Range of two polynomial coefficients - a_0 and a_1 for ten random textures

6   Conclusions and Further Work

The importance of fractal/multifractal descriptors for the image analysis application is high, but they could be applied together with another (non–fractal) descriptors for the best performance. Fractal descriptors could be applied for fractal objects, but non fractal objects could be processed also, what is important feature.

The analysis of APR uses polynomial with two important parameters and alternative approaches for APR curve analysis are also interesting. One of the possible solutions is the analysis of pseudo–Curie region only, instead of overall curve.

There are five parameters, obtained during the approximation, and the number of them is estimated using test and trial method. The reduced number of parameters gives reduced quality of results. Only two parameters are important for the analysis, and the influence of other parameters is low.

The proposed method of texture discrimination could be applied for irregular areas or objects, that is important advantage of APR.

References

1. Binney, J., Dowrick, N., Fisher, A., Newman, M.: The Theory of Critical Phenomena. An Introduction to the Renormalization Group. Clarendon Press, Oxford (1992)
2. Cipra, B.: An introduction to the Ising model. Am. Math. Mon. **94**(10), 937–959 (1987)
3. Glauber, R.: Time-dependent statistics of the Ising model. J. Math. Phys. A **20**, 1299–1303 (1963)
4. Harte, D.: Multifractals: Theory and Applications. Chapman & Hall/CRC, Boca Raton (2001)
5. Huang, Z., Tian, J., Wang, Z.: A study of the slit island analysis as a method for measuring fractal dimension of fractured surface. Scripta Metall. Mater. **24**, 967–972 (1990)
6. Lu, C.: On the validity of the slit islands analysis in the measure of fractal dimension of fracture surfaces. Int. J. Fract. **69**, 77–80 (1995)
7. Mandelbrot, B.: The Fractal Geometry of the Nature. W.H. Freeman and Company, San Francisco (1983)
8. Mandelbrot, B., Passoja, D., Paullay, A.: Fractal character of fracture surfaces of metals. Nature **308**, 721–722 (1984)
9. Mazurek, P., Oszutowska-Mazurek, D.: From slit-island method to ising model - analysis of grayscale images. Intl. J. Appl. Math. Comput. Sci. **24**(1), 49–63 (2014)
10. Oszutowska-Mazurek, D., Mazurek, P., Sycz, K., Waker-Wójciuk, G.: Adaptive windowed threshold for box counting algorithm in cytoscreening applications. In: Advances in Intelligent Systems and Computing, vol. 233, pp. 3–12. Springer, Heidelberg (2014)
11. Oszutowska-Mazurek, D., Mazurek, P., Sycz, K., Wójciuk, G.W.: Variogram based estimator of fractal dimension for the analysis of cell nuclei from the papanicolaou smears. In: Advances in Intelligent Systems and Computing, vol. 184, pp. 47–54. Springer, Heidelberg (2013)

12. Oszutowska-Mazurek, D., Mazurek, P., Sycz, K., Waker-Wójciuk, G.: Lacunarity based estimator for the analysis of cell nuclei from the papanicolaou smears. In: Chmielewski, L.J., Kozera, R., Shin, B.-S., Wojciechowski, K. (eds.) ICCVG 2014. LNCS, vol. 8671, pp. 486–493. Springer, Heidelberg (2014). doi:10.1007/978-3-319-11331-9_58
13. Peitgen, H., Jürgens, H., Saupe, D.: Fractals for the Classrooms, vol. 1. Springer, New York (1991)
14. Peitgen, H., Jürgens, H., Saupe, D.: Fractals for the Classrooms, vol. 2. Springer-Verlag (1992)
15. Plotnick, R., Gardner, R., Hargrove, W., Prestegaard, K., Perlmutter, M.: Lacunarity analysis: a general technique for the analysis of spatial patterns. Phys. Rev. E **53**(5), 5461–5468 (1996)
16. Seuront, L.: Fractals and Multifractals in Ecology and Aquatic Science. CRC Press, London (2010)
17. Skomski, R.: Simple Models of Magnetism. Oxford University Press, Oxford (2008)
18. Wen, R., Sinding-Larsen, R.: Uncertainty in fractal dimension estimated from power spectra and variogram. Math. Geol. **29**(6), 727–753 (1997)

Vocal Tract Resonance Analysis Using LTAS in the Context of the Singer's Level of Advancement

Edward Półrolniczak$^{(\boxtimes)}$ and Michał Kramarczyk

Faculty of Computer Science and Information Technology,
West Pomeranian University of Technology, Szczecin, Poland
{epolrolniczak,mkramarczyk}@wi.zut.edu.pl

Abstract. The article presents the results of signal analysis of the recorded singing voice samples. The analysis is performed towards the presence of the resonances in singing voices. To this end the LTAS (Long-Term Average Spectrum) have been estimated over the vocal samples. The LTAS has been then analysed to extract the valuable information to conclude about the quality of the singer's voices. These studies are a part of a broader research on singing voice signal analysis. The results may contribute to the development of the diagnostics tools of computer analysis of singer's and speaker's voices.

Keywords: Signal analysis · Long term average spectrum · Resonance · Voice analysis · Singing voice

1 Introduction

Singing is a special case of the use of voice. Effective singing depends on the proper use of the potential in which the nature equipped a human. The potential lies in advanced construction of vocal tract. The important elements of this vocal tract are resonators. Because of the resonators the singer's voice is able to reach distant places in the concert hall.

The subject of the analysis carried out in this article is singing voice. The term "singing voice" is understood here as the voice at the singing "mode". At this level of analysis of the human voice, singing is distinguished from speech. Some features of the singing voice differentiate it from the speech. These include: intonation, a specific formant construction [10], vibrato. Regardless of whether we are considering singing or speech, in general, we can talk about voice quality.

Literature defines the quality of the voice specifying attributes that allow to distinct the voices [12]. According to this: volume (loudness) or intonation (pitch) may be indicators for singing voice quality. However, the change of one parameter may affect a change in a second one and therefore may not objectively assess the quality.

The parameter, which, regardless of the volume or intonation, can "judge" about the quality of singing voice is resonance. Therefore this article focuses on

© Springer International Publishing AG 2017
S. Kobayashi et al. (eds.), *Hard and Soft Computing for Artificial Intelligence, Multimedia and Security*, Advances in Intelligent Systems and Computing 534, DOI 10.1007/978-3-319-48429-7_23

the idea of resonance, which is achieved using natural resonators being a part of the construction of human vocal tract.

The resonance can be analysed by using long-term average spectra ($LTAS$) and thus the goal of the study is to investigate resonance characteristics of singing voice using the above mentioned method.

Long-term average spectrum ($LTAS$) analysis provides the average frequency distribution of the sound energy in a continuous singing sample. The method gives an information on the voice quality.

A total of 20 singers participated in that study. They were divided into groups by advancement and gender. All the singers were healthy. They are active members of The Jan Szyrocki Memorial Choir of West Pomeranian University of Technology in Szczecin (Poland). The investigation focuses mainly on the first three spectral peaks in $LTAS$ calculated over samples of recorded sung vowels (exercise: a-e-i-o-u). Peaks 1, 2 and 3 can be changed significantly in the context of advancement or gender [12]. It might be possible to observe the relationships between the results and the vocal experience of the singers.

2   Literature Review

Reviewing the literature it is found that $LTAS$ is applied in studying differences between normal and pathological voices [3,4] and to analyse various vocal pathologies [5,14]. $LTAS$ method has been also used to study individual and gender, age and language related differences [1,7]. Characteristics of vocal expression of emotions [9] and acoustic differences between specific voice qualities [8] have also been studied with this method. Moreover, $LTAS$ has been used to investigate singing voice, in particular to investigate the differences between voice categories and various singing styles [11].

In [2] the authors have selected to the experiment five men with English as their native language. Each one, alone or in a group, performed the songs of country music. The authors of the experiment asked the singers to sing and to speak the national anthem and a country song of their choice. For each registered sample analysis of $LTAS$ spectrum was carried out. The course of plotted spectrum has shown that the main energy of the sound lies within the range of 1–4 kHz. During the experiment spectrum graphs of trained, professional classical singers have been also analysed for a comparison. In the case of professional classical singers the presence of a singing formant around 3 kHz has been observed. In contrast, in the graphs for the country singers no significant values around 3 kHz were observed, and therefore the presence of a singing formant has not been detected. Another analysis of the graphs of the spectrum has confirmed the hypothesis of lack of a singing formant in the country singers group.

There are very few studies on the differences of singing skills in the group of untrained singers. One of such studies can be found in [13]. The authors have evaluated the abilities of precise reproduction of note intervals, as well as ability to control the pitch and timbre quality. In this context, a SPR (Singing Power Ratio), calculated over the $LTAS$ spectrum, has been used. According to the authors, the quality of singing, expressed by SPR, reflects the strengthening or attenuation in the vocal tract of the harmonics generated by the source of the sound. The values of this ratio were measured for untrained singers, divided on talented and not talented singers to objectively analyse differences in voice quality. The $LTAS$ has been also analysed in that study. It allowed to draw a conclusions on the degree of talent. Higher harmonics were more dominant in the spectrum of more talented singers. An increase of energy in the range of 2500–4000 Hz was observed.

3   Data Acquisition Method

During the studies a database consisting of representative samples presenting the abilities of choir singers has been used [6]. The database has been created under the research project of West Pomeranian University of Technology: "Computerized methods of supporting the process of training choir voices" and is still developed. The database consists of recordings representing different singing exercises. The exercises allow to analyse such important features of the singing as: pitch, vibrato, tremolo, characteristic of vowels and consonants.

The recordings have been taken in an acoustically modified studio. To get rid of voice reflections and reverberation sound-absorbing foam panels with an appropriate geometric profile and acoustic parameters have been used. The recordings were done with use of the capacitor microphone AT 4050 connected to high quality audio interface. The audio interface was connected to the computer equipped with appropriate software. All the singers sang situated at the microphone within a distance of about 40 cm. Every time the recorded person was led by the sound of the initial note before the singing.

For the purpose of this study two of the exercises from the database was selected. One of the chosen exercise consists of five vowels "a e i o u" sung at the same pitch. The second exercise consists of only one sung vowel "a". Each exercise was repeated and recorded on the following pitches (starting from the highest available to the singer). All of the singers came from the same university choir. The singers were chosen as representative voices from the group of 28 recorded singers. The age of eight males and twelve females is ranging from 21 to 37. Their experience in the choral singing is ranging from 1 to 18 years. Each person was in good vocal and health condition.

Table 1. Advancement level of singers experts classification

Advancement classification by experts	Male singers	Female singers	Quantity of singers in groups
Beginner	4	3	7
Intermediate	3	5	8
Advanced	2	3	5

Before being recorded each singer was assigned by expert to one of the three groups of advancement (Table 1). Before the analysis of the recorded samples towards the resonances in singing it has to be stressed out once again that all of the analysed persons are active singers of the same choir so the expected results may not differ too much from each other. The differences would be probably more visible comparing singers to non-singers or amateur to professional singers. To do that another database has to be built.

4   Data Extraction for Analysis

After the singing samples were acquired the $LTAS$ spectrum has been estimated for each recorded sample. Over the signal a simple extraction of values of peaks was performed to obtain the values to be analysed. The values of peaks of $LTAS$ are then analysed to prove hypothetical dependencies. The analysis of the data should give the answer to the posed questions. The whole process is illustrated in Fig. 1.

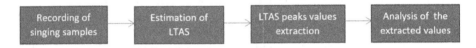

Fig. 1. Data extraction for the analysis

To estimate the $LTAS$ the signal has to be divided into L independent segments where each one contains N samples. For each segment a Fourier spectrum is calculated (by FFT in this case). Then the Power Spectra Distribution (PSD) function for each Fourier spectrum has to be calculated (Eq. 1).

$$PSD(k) = \frac{|X(k)|^2}{Ndt} \tag{1}$$

where:

$X(k)$ - signal spectrum,
dt - signal duration,
N - quantity of signal samples.

Taking mean value of each PSD the $LTAS_{dBHz}$ (Eq. 2) is calculated which is normalized $LTAS$ function. The $LTAS_{dBHz}$ better reflects the hearing curve and thus better corresponds to the hearing impressions of the human.

$$LTAS_{dBHz}(f) = 10log_{10}\frac{LTAS(f)}{P_0{}^2} \tag{2}$$

where:

P_0 - is the value of acoustic pressure understood as the threshold of human hearing at the frequency of 1 kHz, and the value of P_0 is $2 * 10^{(-5)}$ Pa. The unit of measure for $LTAS_{dB}$ is $\frac{dB}{Hz}$.

Concluding - $LTAS_{dBHz}$ represents the logarithmic power spectral density as a function of frequency, expressed in $\frac{dB}{Hz}$ relative to $2 * 10^{(-5)}$ Pa. In order to use proper notation and symbols, it is good to make a clear designation of the $LTAS$ of various types. The authors propose the following notation:

- $LTAS$ (written without special subscript symbols), $LTAS[dB]$ or $LTAS_{dB}$ means Long Term Average Spectrum in its general form,
- $LTAS_{dBHz}$ or $LTAS[\frac{dB}{Hz}]$ means normalized Long Term Average Spectrum form.

$LTAS$ can be the basis to calculate the Singing Power Ratio (SPR) [13] which mathematically is the ratio value of highest peak in the band 2–4 kHz to the value of the highest peak in the band 0–2 kHz. The SPR can be interpreted as the parameter representing the singing voice quality (how good singer is the analysed person). Higher SPR means better singer (better trained, more advanced in singing) but not all publications prove that assumption. It has to be also noticed that the SPR parameter, in contrast to $LTAS$, may be value depending on the content of the analysed samples (for example different kinds of the sung vowels). Thus the $LTAS$ analysis seems to be more reliable to conclude about voice signals. The $LTAS$ analysis is content independent. It points to the same conclusions irrespectively to the content of the analysed voice sample. It can be performed for example over the exercise containing a-e-i-o-u vowels what is more valuable as the content is more appropriate to reflect the real singing. In the next section the results of analysing of $LTAS$ will be presented, interpreted and discussed.

5   The Analysis and the Results

The data analysis focused on the first three spectral peaks in $LTAS$ calculated over samples of recorded sung vowels from exercise: a-e-i-o-u. The sounds were sung on the following pitches lying in the vocal range of the singer (the vocal range was tested before the recording). It was supposed that peaks 1, 2 and 3 may differ significantly in the context of advancement or gender. It was expected in the study that the relationship between the vocal experience of the singers and the peaks values would be observed.

The analysis performed over sound signals (singing samples) confirms the expectations.

The Fig. 2 shows the analysis for the female voices. In the figure the values of the first three peaks of the $LTAS$ have been visualised. Those peaks represent the values of the resonances found for the analysed singers in their vocal tracts. The found values of the peaks generally lie on the three curves. Those curves are

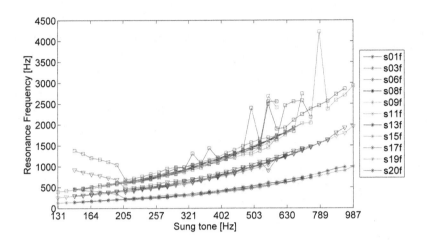

Fig. 2. *LTAS* analysis of the female voices

Table 2. Voices and advancement levels of the singers

Symbol of the singer	Voice	Advancement level
s01f	Alto	Beginner
s02m	Bass	Intermediate
s03f	Soprano	Intermediate
s04m	Bass	Intermediate
s05m	Bass	Beginner
s06f	Soprano	Intermediate
s07m	Bass	Advanced
s08f	Alto	Intermediate
s09f	Alto	Intermediate
s10m	Bass	Intermediate
s11f	Soprano	Advanced
s12m	Bass	Advanced
s13f	Soprano	Advanced
s14m	Bass	Beginner
s15f	Alto	Beginner
s16m	Bass	Beginner
s17f	Soprano	Advanced
s18m	Tenor	Beginner
s19f	Alto	Intermediate
s20f	Soprano	Beginner

going up slightly as the frequency of the sung sounds increases. Anyway some values lie outside the main courses and those values are the most interesting.

Studying the graph in Fig. 2 it can be found that female singer s03f, which is soprano singer (see Table 2), has some higher resonance values than other female singers in the group at the beginning of singing. After the inspection of the recorded sample it has been found that the singer (soprano second) had some vocal problems with such low frequencies. It is rather vocal range problem of the sample recording than the problem with the singer. The soprano singer shouldnt be recorded and shouldnt sing sounds lower than G^3 (at musical notation). It has to be recalled that for sopranos the regular scale is from c^1 (261.626 Hz) to a^2 (880.000 Hz) and up to d^3 (1174.659 Hz) for soloists. In that context the soprano s03f was singing much too low to her classification. The middle of the range 207.652 (G♯ /third octave) to 493.883 (c^2♯ /fifth octave) seems to be rather comfortable for all female voices in the group (both sopranos and altos).

Most of the problems are visible in the highest curve. Above the c^2♯ sound consistency of the resonances is becoming a problem. Some higher resonances than expected are visible at singers: S20f (soprano, beginner), s01f (alto, beginner), s09f (alto intermediate), s03f (soprano intermediate), s17f (soprano advanced). In case of singers: s01f (alto, beginner), s03f (soprano intermediate) and s09f (alto intermediate) the problems can be explained by the end of their scale. They should stop singing a little bit earlier (or just it proves the end of their abilities). The singer s20f has some accidental higher values in the context of resonance but generally all curves are pretty clean, the ranges are very wide. It can be seen that resonances of s20f are a little bit higher than the competitive s11f (soprano advanced) singer and that is because s20f is the beginner. The singer s17f was classified as advanced soprano but it seems she has some vocal problems and it looks like stressed or over-trained singer (taking into account her

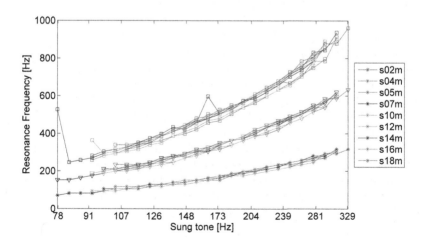

Fig. 3. *LTAS* analysis of the male voices

level of advancement). Generally beginners show lower or higher peaks values than mean in the groups.

The same analysis has been carried out for the male group. The values of the three highest peaks in the $LTAS$ spectrum for the consecutive sung notes is presented in Fig. 3.

Figure 3 shows that in the male group the curves are more consistent than in the female group. Such group will sound clean, coherently. Some accidental peaks are visible. If we take the most advanced singer s07m as a reference we can conclude about differences. The singers s04m (intermediate) and s10m (intermediate) have their resonances values generally a little below the reference singer resonances. They can sound a little under pitch. If one take into account the singers having their resonances higher than the reference singer it is clear that most of those singers are beginners. We can draw the conclusion that the reason is lack of voice training.

6   Conclusion

In this study a signal analysis has been carried out. The resonance values calculated over the $LTAS$ function for the group of singers have been estimated. The obtained results have been compared to the objective assessment of the level of advancement done by expert. The computer analysis performed over recorded samples of singing taken from 20 choral singers revealed some differences in their results. Those could be associated to the information about the level of advancement and the gender of the singers. The main differences were found on the ends of the singing ranges. It allows, despite minor differences in the middle of the obtained curves, to assess the level of advancement of the singers or reveal mistakes in the classification of the type of voice. The singers who cope better with singing at the ends of their singing scales are those more experienced or better trained.

The singers named as advanced showed stable courses of the curve of the resonance values measured for the consecutive sung notes. The beginners were the singers which do not cope well at the ends of their vocal range. For this group, there was also noticed a tendency to visible increases of the resonance values especially in the third level of the analysed resonances.

The performed analysis and investigation have shown the relationship between the expert assessment of the level of advancement and the obtained results. The method of signal analysis is clear and stable. It gives repeatable results and allows their valuable interpretation. This indicates the possibility of use of the $LTAS$ analysis in the systems for automatic assessment of singing quality and assessment of the level of advancement. The solution can be used to objectify the process of singers training.

References

1. Byrne, D., Dillon, H., Tran, K., Arlinger, S., Wilbraham, K., Cox, R., Hagerman, B., Hetu, R., Kei, J., Lui, C., et al.: An international comparison of long-term average speech spectra. J. Acoust. Soc. Am. **96**(4), 2108–2120 (1994)
2. Cleveland, T.F., Sundberg, J., Stone, R.: Long-term-average spectrum characteristics of country singers during speaking and singing. J. Voice **15**(1), 54–60 (2001)
3. Formby, C., Monsen, R.: Long-term average speech spectra for normal and hearing-impaired adolescents. J. Acoust. Soc. Am. **71**(1), 196–202 (1982)
4. Gauffin, J., Sundberg, J.: Clinical applications of acoustic voice analysis. part ii: acoustical analysis, results, and discussion. In: Speech Transmission Laboratory, Quarterly Progress and Status Report, vol. 2, pp. 39–43 (1977)
5. Hammarberg, B., Fritzell, B., Gaufin, J., Sundberg, J., Wedin, L.: Perceptual and acoustic correlates of abnormal voice qualities. Acta Oto-laryngol. **90**(1–6), 441–451 (1980)
6. Lazoryszczak, M., Półrolniczak, E.: Audio database for the as sessment of singing voice quality of choir members. Elektron. Konstrukcje Technol. Zastosowania **54**(3), 92–96 (2013)
7. Pavlovic, C.V., Rossi, M., Espesser, R.: Statistical distributions of speech for various languages. J. Acoust. Soc. Am. **88**(S1), S176 (1990)
8. Pittam, J.: Discrimination of five voice qualities and prediction to perceptual ratings. Phonetica **44**(1), 38–49 (1987)
9. Pittam, J., Gallois, C., Callan, V.: The long-term spectrum and perceived emotion. Speech Commun. **9**(3), 177–187 (1990)
10. Polrolniczak, E., Kramarczyk, M.: Formant analysis in assessment of the quality of choral singers. In: Signal Processing: Algorithms, Architectures, Arrangements, and Applications (SPA 2013), pp. 200–204, September 2013
11. Rossing, T.D., Sundberg, J., Ternström, S.: Acoustic comparison of voice use in solo and choir singing. J. Acoust. Soc. Am. **79**(6), 1975–1981 (1986)
12. Titze, I.: Principles of voice production (englewood cliffs, nj). Inc: Prentice Hall, ISBN 0-13-717893-X (1994)
13. Watts, C., Barnes-Burroughs, K., Estis, J., Blanton, D.: The singing power ratio as an objective measure of singing voice quality in untrained talented and nontalented singers. J. Voice **20**(1), 82–88 (2006)
14. Wendler, J., Doherty, E.T., Hollien, H.: Voice classification by means of long-term speech spectra. Folia Phoniatr. Logop. **32**(1), 51–60 (1980)

Parallel Facial Recognition System Based on 2DHMM

Janusz Bobulski[✉]

Institute of Computer and Information Science, Czestochowa University
of Technology, 73 Dabrowskiego Str., Czestochowa, Poland
januszb@icis.pcz.pl.pl

Abstract. The constantly growing amount of digital data more and more often requires applying increasingly efficient systems for processing them. Increase the performance of individual processors has reached its upper limit therefore we need to build multiprocessor systems. To exploit the potential of such systems, it is necessary to use parallel computing, i.e. creating computer systems based on parallel programming. In practice, most often their used adjusts the parallelization of the data processes, or regulates the parallelization of the query tasks. The system of face recognition that requires high computational power is one of potential application of the computations parallelization, especially for large database sizes. The aim of the research was to develop a parallel system of face recognition based on two-dimensional hidden Markov models. The results show that compared to sequential calculations, the best results were obtained for parallelization of tasks, and acceleration for training mode was 3.3 and for test mode - 2.8.

Keywords: Parallelization · Face recognition · 2D hidden Markov models · HMM · Parallel task

1 Introduction

The system of face recognition that requires high computational power is one of potential application of the computations parallelization, especially for large database sizes. It is forcing producers of computers to the prospecting of solutions which will be able to satisfy current and prospective needs. The evolution of parallel architectures has an intense influence on all computer systems, starting from smartphones until computing systems of the great scale. In particular, the multi-core became a main solution enabling to sustain the law Moore'a that is of exponential increasing the productivity of the computer what is being carried out above all by increasing the number of cores. A computational productivity and a demand for the electric energy are important factors affecting direction of the development of designs of parallel processors [1, 2]. So computational new architectures are being planned, of which characteristics are taking the liberty of getting the greater computing power at keeping or lowering the demand to the electric power. Increasing the frequency of the clocking was one of directions of the development [3]. Unfortunately, the technology based on silicon in this respect achieved ones of limits. Increasing the number of cores was another trend

© Springer International Publishing AG 2017
S. Kobayashi et al. (eds.), *Hard and Soft Computing for Artificial Intelligence, Multimedia and Security*, Advances in Intelligent Systems and Computing 534, DOI 10.1007/978-3-319-48429-7_24

of the development of processors of the general-purpose. Here a problem with the access to shared sources and the time and the manner of the transport appeared between cores. Functioning of processors producers show that future computational architectures in their nature will be solutions hybrid, finding application in very wide range. Hybrid, and at the same time heterogeneous systems, will be based on integration of two main kinds of components, multi-core processors of the general-purpose and dedicated, massively parallel computational accelerators. The virtue multi-core processors of the general-purpose is the fact that the substantial amount of cores in the uniform integrated circuit lets for increasing productivities is without raising frequencies of the clocking of the core, and consequently, without increasing the generation of the heat, what is being transferred on smaller power consumption. However, dedicated massively parallel computational accelerators, so as GPU overtook CPU processors in the productivity of the floating-point calculation. Applying two basic computational components in hybrid architectures [2, 4] in the form of CPU and computational accelerators is enjoying considerable influence to the productivity growth of numerical calculations. Traditional processors perform well best in carrying programs out with numerous branches, in which relatively it is necessary to process the little amount of data according to the complicated algorithm. However computational accelerators let the high performance getting indeed for the part of application, because are designed in order productively to carry the same set of the operation out on the large amount of data.

2 Parallel Computing

Contemporary processors of the general-purpose are ensuring excellent abilities for the parallel processing, letting for creating the application which the productivity of calculations will bring closer oneself to the peak performance of the arrangement. However exploiting the full potential of these processors constitutes the greatest challenge for programmers. The most popular methods of code parallelization are parallel data processing and task parallel processing [5, 6].

The model of data parallel processing is defining calculations as the sequence of instructions marked to a lot of elements put in the memory. The index space connected with the model of the workmanship is defining threads and the way of connecting data with them. The real model of parallel processed data assumes copying one to one between the weft and the element in the memory, on which kernel may parallel be made. Two forms of the realization of the model of data parallel processing are possible: (i) open form, programmer is defining the total number of executed parallel threads as well as a manner of the division into groups is describing by them, (ii) secret form, programmer he determines only a total number of executed parallel threads, however a programming environment is managing the division into groups.

The model of task parallel processing is establishing, that single authority kernel is being made irrespective of the index space. This attempt is analogous to the process of making kernel on the computational individual, where every group of threads contains only one thread. In this model the programmer is defining levels of the parallelism through: using vector supported data types by the device, performing many tasks simultaneously and workmanship kernels of other computational programming models, drawn up on the base [7].

Both recalled techniques parallelization of code were used for the parallel implementation of 2DHMM.

3   Evaluation of the Effectiveness of Parallel Processing

The evaluation of the productivity of parallel programs requires applying a few certificates what allows for efficiency analysis of complex problems. It is possible to rank among most often used metrics [8, 9]:

- execution time of calculation T_{obl}
- theoretical maximum performance R_{max};
- real performance R;
- speed up the S_p
- the efficiency E_p
- granularity G;
- memory bandwidth and I/O bus B.

The execution time of the parallel program is a basic parameter of the evaluation of the productivity of calculations. This time is being measured from the moment of commencing calculations all the way to the moment, which all performed calculations will end in. In the parallel algorithm it is possible to accept the lead time of calculations:

$$T_{obl} = T_s + T_p + T_c + T_i \qquad (1)$$

where T_s is a execution time of the sequential part of the code, T_p is a execution time of the parallel part, T_c - time of the communication, and T_i it is a time of the idleness.

The theoretical maximum performance multiprocessor system can be determined according to the equation:

$$R_{max} = f \cdot N \cdot I \qquad (2)$$

where f - frequency of the clocking of the processor, N - number of cores, I - the product of the number of elements of the vector and numbers of operations performed on one element of the vector.

The real performance is used to compare the get productivity with theoretical possibilities of the computer system. It is expressed with number of floating-point operations for a second (flop/s). A F_h number of indeed performed floating-point operations determines the real performance in the given time. This value, determined with formula:

$$R = \frac{F_h}{T_{obl}} \qquad (3)$$

and may never exceed the maximum theoretical performance of the computer.

Speedup S of the parallel algorithm on p processors, at the permanent size of the considered problem it is possible to define as:

$$S_p = \frac{T_1}{T_p} \tag{4}$$

where T_1 is a execution time of the algorithm executed on one individual processor ($p = 1$), T_p is a execution time of the parallel algorithm carried out on p processors.

Effectiveness E of parallel system determines the degree to which were used computational units:

$$E_p = \frac{S_p}{p} \tag{5}$$

The granularity of the algorithm is useful when selecting a parallel algorithm for a particular architecture. The granularity may be defined as the ratio of computation time T_{ref} to the time of communication T_{kom}, necessary for the implementation of the problem:

$$G = \frac{T_{obl}}{T_{kom}} \tag{6}$$

Memory bandwidth and I/O bus is an important parameter to judge performance of parallel algorithms. It may play a significant role in the case of applications that require large amounts of memory reference.

$$B_M = f \cdot M_i \cdot 2 \tag{7}$$

where f – frequency clock memory [Hz], M_i - the width of the memory interface [B], 2 – double way transfer.

4   2DHMM

There are two fundamental problems of interest that must be solved for 2DHMM, in order that be useful in face recognition applications. These problems are efficiently compute $P(O|\lambda)$ and estimate model parameters $\lambda = (A, B, \pi)$. Solutions to these problems are algorithms forward, backward and Baum-Welch [10].

Forward algorithm

$$\alpha_{t+1}(i,j,k) = \left[\sum_{l=1}^{N^2} \alpha_t(i,j,k) a_{ijl} \right] b_{ij}(o_{t+1}) \tag{8}$$

$$P(O|\lambda) = \sum_{i,j,k} \alpha_T(i,j,k) \tag{9}$$

Backward algorithm

$$\beta_t(i,j,k) = \sum_{l=1}^{N^2} a_{ijl} b_{ij}(o_{t+1}) \beta_{t+1}(i,j,k) \tag{10}$$

Baum-Welch algorithm

$$
\begin{aligned}
\xi_t(i,j,l) &= \frac{\alpha_t(i,j,k) a_{ijl} b_{ij}(o_{t+1}) \beta_{t+1}(i,j,k)}{P(O|\lambda)} \\
&= \frac{\alpha_t(i,j,k) a_{ijl} b_{ij}(o_{t+1}) \beta_{t+1}(i,j,k)}{\sum_{k=1}^{K} \sum_{l=1}^{N^2} \alpha_t(i,j,k) a_{ijl} b_{ij}(o_{t+1}) \beta_{t+1}(i,j,k)}
\end{aligned} \tag{11}
$$

5   Parallel Realization of 2DHMM Face Recognition System

For parallel realization of face recognition system based on 2DHMM we used the most popular methods of code parallelization are parallel data processing and task parallel processing. *Parfor* loop is used to implement data parallel processing, and execution of task parallel processing is shown in Figs. 1 and 2.

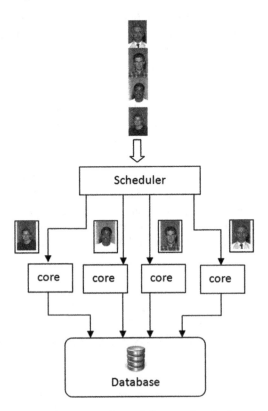

Fig. 1. Scheme of task parallel computing for learning

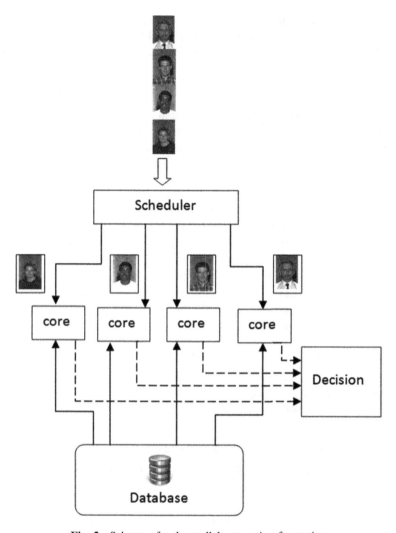

Fig. 2. Scheme of task parallel computing for testing

6 Experiment

The aim of the research was to develop a parallel system of face recognition based on two-dimensional hidden Markov models (2DHMM). The procedure of the person' identification used for feature extraction wavelet transform in this system. The feature vector obtained from this transformation is utilized for training and testing the system. In this method, for identification purposes, both 2D and 3D images of the face were exploited. The detailed description of this sequential method of the face recognition one may find in the article [10]. Because of long computation times in this method we decided to apply the parallel processing. The analysis of algorithms of the systems using 2DHMM allowed uses the parallel processing. The study of learning mode of the system allowed for the application the parallelization of data processing and the parallelization of tasks. While in test mode it was possible to uses only the parallelization of tasks. One task was the processing (training or testing) of one face image. For research we used the face base UMB-DB [11]. The experiment carried on the processor Intel i5 3.3 GHz with 4 cores and 4 threads, and the results of the experiment are presented in Table 1.

The results show that compared to sequential calculations, the best results were obtained for parallelization of tasks. In the learning mode and the testing mode with the use of a 2D image, we got a low speedup of calculations, because the source data was much less. However for 3D images, 3.2 speedup was get for the learning mode, and 2.7

Table 1. Comparison of processing time

Type of processing	Type of face image	Time of learning [s]	Time of testing [s]
Sequential	2D	126	117
Sequential	3D	1145	1090
Sequential	2D+3D	1262	1282
DPP	2D	102	1632
DPP	3D	762	2117
DPP	2D+3D	1082	3776
TPP	2D	122	98
TPP	3D	359	401
TPP	2D+3D	386	451

DPP – data parallel processing.
TPP – task parallel processing.

Table 2. Speedup

Type of processing	Type of face image	Learning	Testing
DPP	2D	1.24	0.07
DPP	3D	1.50	0.51
DPP	2D+3D	1.17	0.34
TPP	2D	1.03	1.19
TPP	3D	3.19	2.72
TPP	2D+3D	3.27	2.84

Table 3. Efficiency

Type of processing	Type of face image	Learning	Testing
DPP	2D	0.31	0.02
DPP	3D	0.38	0.13
DPP	2D+3D	0.29	0.09
TPP	2D	0.26	0.30
TPP	3D	0.80	0.68
TPP	2D+3D	0.82	0.71

for the testing mode. In case of using for the identification both of image, 2D as well as 3D, speedup for training mode and test mode was respectively 3.3 and 2.8.

To evaluate the efficiency of the implementation of parallel algorithms for learning and testing HMM used in computation time, acceleration and efficiency. The results are shown in Tables 2 and 3.

7 Conclusion

The use of data parallel processing with *parfor* loop did not give significant speedup calculations. This is due to the structure of training and testing algorithms of HMM, that they have in their structure the operations that do not allow parallelization of the calculation process.

In conclusion, when we are creating a biometric system, using a face image of individuals and it is based on 2DHMM, it is worth to use parallel processing with application parallelization of tasks. In such a system, each processor independently processes one object, i.e. one person. The result of applying a parallel structure system is a three-fold reduction in computation time.

References

1. Culler, D.E., Singh, J.P., Gupta, A.: Parallel Computer Architecture – A Hardware/Software Approach. Morgan Kaufmann Publishers (1999)
2. Kurzak, J., Bader, D., Dongara, J.: Scientific Computing with Multicore and Accelerators. Chapman & Hall/CRC Computer and Information Science Series (2010)
3. Vetter, J.: Keeneland: bringing heterogeneous gpu computing to the computational science community. Comput. Sci. Eng. **13**, 90–95 (2011)
4. Kurowski, K., Back, W., Dubitzky, W., Gulyas, L., Kampis, G., Manowski, M., Szemes, G., Swain, M.: Complex system symulation with QosCosGrid. In: 9th International Conference on Computational Science, pp. 387–396 (2009)
5. Blaziewicz, M., Brandt, S., Kierzynka, M., Kurowski, K., Ludwiczak, B., Tao, J., Weglarz, J.: CaKernel – a parallel application programming framework for heterogenous computing architectures. Sci. Programm. **19**(4), 185–197 (2011)
6. Czech, Z.: Wprowadzenie do obliczeń równoległych, Wydawnictwa Naukowe PWN (2010)
7. Wyrzykowski, R., Rojek, K., Szustak, L.: Model-driven adaptation of double-precision matrix multiplication to the Cell processor architecture. Parallel Comput. **38**(4), 260–276 (2012)
8. Hockney, R.W.: The Science of Computer Benchmarking. SIAM, Philadelphia (1995)
9. Wyrzykowski, R.: Klastry komputerów PC I architektury wielordzeniowe: budowa i wykorzystanie. Exit, Warszawa (2009)
10. Bobulski, J.: 2DHMM-based face recognition method, image processing and communications challenges 7. Adv. Intell. Syst. Comput. **389**, 11–18 (2016)
11. Colombo, A., Cusano, C., Schettini, R.: Umb-db: a database of partially occluded 3d faces. In: Proceedings of ICCV 2011 Workshops 1, pp. 2113–2119 (2011)

System of Acoustic Assistance in Spatial Orientation for the Blind

Mariusz Kubanek[1]([✉]), Filip Depta[2], and Dorota Smorawa[1]

[1] Institute of Computer and Information Science, Czestochowa University
of Technology, Dabrowskiego Street 73, 42-201 Czestochowa, Poland
{mariusz.kubanek,dorota.smorawa}@icis.pcz.pl
[2] Faculty of Mechanical Engineering and Computer Science,
Armii Krajowej Street 21, 42-201 Czestochowa, Poland
deptafilip@gmail.com

Abstract. The technological development provides better and more mobile devices to aid disabled people. All work conducted to improve the life of disabled people constitute an important part of science. The paper describes the issues related to anatomy and physiology of sight and hearing organs, imagination abilities of the blind and assisting devices for those people. The authors of this work have developed a prototype of an electronic device which navigates a blind person by means of sound signals. Sounds are meant to provide the blind with a simplified map of the object depth in their path. What makes the work innovative is the Kinect sensor applied to scan the space in front of the user as well as the set of algorithms designed to learn and generate acoustic space, which also take into account the tilt of the head. The carried out experiments indicate the correct interpretation of the sound signals being modelled. The tests conducted on the people prove the developed concept to be highly efficient.

Keywords: Bioinformatics · Bioengineering · The blind · Acoustic space

1 Introduction

The system of acoustic assistance in spatial orientation for the blind is a proto-type device intended to facilitate blind people moving indoors and outdoors. The sound signals they can hear contain information about the distance to obstacles, located in various places of the distance map, intercepted by the used sensor - Xbox Kinect.

A blind person is able to orient themselves in the space by receiving a series of sounds which give them the information about the place they are in and the distance to an obstacle. The brain of a blind person which is taught proper interpretation of signals based on provided information should start automatically process sound signals so they reflect the environment as they could really see it.

S. Kobayashi et al. (eds.), *Hard and Soft Computing for Artificial Intelligence, Multimedia and Security*, Advances in Intelligent Systems and Computing 534, DOI 10.1007/978-3-319-48429-7_25

Distance map of the space in front of the blind person, intercepted with the Xbox Kinect sensor, is processed by the miniature computer (Raspberry Pi 1 Model B+) which recreates the appropriate sound signals (which were generated earlier and saved in files) with the adequate volume for the left and right channel. The recreated signals are emitted via insert earphones. The energy bank with a capacity of 8400 mAh, supply voltage of 5 V and two outputs with the maximum power supply of 1.0 ampere and 2.1 ampere power the sensor and the computer. The operating system, software and audio files used by the computer, are saved on the Micro SDHC card of 8 GB of tenth grade speed. The sensor is attached to a bicycle helmet (being the only stable attachment to the head) by means of a special TV handle designed for the Xbox Kinect sensor.

1.1 Spatial Hearing

A man is able to localise a sound source with high precision. Spatial hearing is possible due to a number of phenomena such as sound volume difference between one and the other ear, inter-aural differences of time and finally spatial filtering [1].

The difference in sound intensity between one and the other ear is not much relevant in locating the source of the sound if it is lower than 2000 Hz and 3000 Hz, because below these frequencies the difference is too small, which is due to the fact that the head is unable to damp low tones. However, if the sound is a tone with a frequency greater than specified, the intensity difference in both ears becomes useful. Such sound frequency can be effectively damped by the head. Location of the sound source based on the intensity difference is called duplex theory. This phenomenon works with the high tones, but not with complex and low sounds localized by means of inter-aural time difference [1].

The inter-aural time difference is the phenomenon of a phase shift of sound between the two ears when a sound reaches one ear before it reaches the other. Due to this phenomenon it is possible to determine the direction a sound under 1500 Hz comes from, or a complex sound (which may include high frequencies), a single sound (eg. crash, bang, slam, snap) or recurring one, where the recurrence frequency is under 600 Hz [1].

The above-described mechanisms are not able to determine the precise direction the sound comes from. These mechanisms are not able to indicate whether the sound comes from behind, from above, or from another direction. Hearing uses another mechanism. This phenomenon is called spatial filtering. Due to the shape of the ear (auricles) and the head, a spectrum of sound delivered to the inside of the ear is changed accordingly. It is possible to precisely determine the direction from which the sound is coming due to sound spectrum changes. The auricle affects only the sounds over 6000 Hz, but the head can affect frequencies from 500 Hz to 16 000 Hz [1,2].

Each person has individual characteristics of the filter. Functions describing the changes in spectral composition of sounds coming from different directions are called Head Related Transfer Function (HRTF). The auricle also helps to determine whether the sound is coming from the inside (e.g. earphones effect) or from the outside [1,3,4].

1.2 Imagination Abilities of the Blind

Undoubtedly, spatial imagination occurs in blind people. Apart from all the research there is evidence of such imagination in the form of sculptures and raised-line drawings, invented and made by the blind, effective use of the tactile maps and other situations which require spatial imagination. Here the visually-impaired are doing very well [5,6].

Moreover, spatial imagination also occurs in people who have never seen, never experienced any visual perception. Some claim that the blind have such imagination or in other words the ability to visualize, however, others do not agree with this statement and use other terms which consider to be more accurate [5].

The ability to create imaginary space model based on oral description [5,7] is an important ability for the blind when developing a system to enhance spatial orientation. Although the prototype does not use oral descriptions of space, the above-mentioned abilities indicate that blind people's imagination is sufficient to use the acoustic system, however, preliminary practice is necessary to make use of the device.

1.3 Overview of Devices and Assistive Software

There are many types of devices and software supporting blind people. These include Braille printers, Braille watches, talking watches, regular white canes, software designed for blind people in the form of keyboard overlays or independent programs - such as speech synthesizers for reading texts or special versions of the GPS software [8]. A specific group of devices supporting blind people are devices which facilitate spatial orientation these are GPS devices, specially designed for the blind, but also the obstacle detectors and imagers.

White Cane is probably the first of the devices used by blind people to assess the environment. The idea of cane was formed much earlier than their white colour. It was much later when they used white colour to make it more visible to other people. The white colour became a symbol to inform that a person using a cane is blind [9].

UltraCane is a device in the form of a white cane, but its handle is equipped with ultrasonic obstacle detectors. The user is informed about an obstacle detected (in the direction shown) by two vibrating button-shaped handle parts [9].

'K' Sonar is a small hand-held device which may also be mounted on a regular white cane, which increases its stability and allows to use both the white cane and the device simultaneously. In this configuration the user should hold the set directly by the handle of the device, and not by the cane handle [9].

The vOICe is a program or set of programs rather than a specific device. The program allows the user to build the imaging device for the blind. To build the device, a user needs a computer or a smartphone (the program is available in versions for Windows and Android), headphones and camera (webcam, low resolution) [9].

BrainPort® V100 is another imaging device which transmits the image to a blind person, not using sound or vibration, but using an array of electrodes placed on the tongue. Although tests of the device had lasted for many years it was launched on the market on June 19, 2015 in the United States. Currently the device is available in the United States, Hong Kong and the European Union [9].

2 Project of the Device Assisting Blind People

any more or less similar solutions. The device does not have to be an electronic replacement of the white cane but a kind of supplement, extension and improvement (because it has to be an imaging device) it should therefore be used together with a white cane. As the cerebral cortex tends to be neuroplastic there is a chance to develop the habit of using the device as a simplified replacement for the sight so the device can be used automatically and will not require much attention. The device should help 'seeing' distant objects (in greater distances it just detects an object rather than it's shape), passageways or shapes of rooms. It should also allow for non-contact detection of other people's position and presence, so as not to tap them with the white cane.

All the device components are installed on a bicycle helmet to move aside uncomfortable wire, which is easy to hook or pull.

2.1 Principle of Operation

As mentioned previously, the device has to define the depth map of the environment observed by the sensor. Depth map is fed to a blind person through a pair of stereo headphones and a few sounds (tones of specific frequencies). Installed software and software embedded directly into the computer Raspberry Pi 1 Model B+ execute the appropriate sound. Computer communicates with the sensor via the USB port and with the headphones via embedded sound card and a 3.5 mm jack.

2.2 The Depth Values Returned by the Kinect Sensor (Using *Libfreenect* Library)

Sensor returns depth value as an 11-bit, non-negative number. It follows that the minimum depth value read would be 0 and the maximum 2047 (with 11-bit depth). In fact, there is no zero value. When its optical axis is set perpendicular to the flat surface of an obstacle the sensor 'sees' it just at a minimum distance of about 57 cm and returns a value of 488. If the surface of the obstacle is set at an angle, it is possible to slightly reduce the distance. The maximum value (2047) is in the readings, but it does not signal the maximum distance at which the surface was detected, but there is no reading. This means that the value of 2047 is always present when the distance value could not be read, and so when the obstacle was too close, too far away, when the structural light was covered, when it was not properly reflected (for example, shiny surfaces) or the structure of the light was suppressed by ambient light.

As a preparation phase of the prototype measurements were made to compare the values returned by the sensor to the actual dimensions. All the measurement results are given in Table 1.

Table 1. The values returned by the Xbox Kinect sensor (model 1414; with *libfreenect* library), depending on the actual distance

Distance [m]	Kinect Pkt	Distance [m]	Kinect Pkt	Distance [m]	Kinect Pkt
0.57	488	1.20	802	2.90	972
0.59	505	1.25	814	3.10	978
0.61	525	1.30	825	3.30	985
0.63	544	1.35	833	3.50	993
0.65	561	1.40	844	3.70	998
0.67	576	1.45	852	3.90	1002
0.70	598	1.50	860	410	1007
0.73	619	1.55	868	4.30	1012
0.76	637	1.60	876	4.50	1015
0.79	653	1.70	888	4.80	1020
0.82	670	1.80	898	5.00	1023
0.86	689	1.90	909	5.50	1028
0.90	708	2.00	918	6.00	1035
0.94	723	2.10	926	6.50	1038
0.98	738	2.20	933	7.00	1043
1.02	751	2.30	942	8.00	1049
1.06	764	2.40	947	9.00	1054
1.10	776	2.50	953		
1.15	791	2.70	963		

A program, based on the sample program, available on the project *OpenKinect* website, was written to perform the measurements. The program displays the depth value read by the sensor in the center of the visual field. The preview of the entire captured depth map was made available to make the aiming easier. The distance of the sensor to an obstacles was measured with a tape measure. The distance value, having been read and saved, was shifted around by a certain distance. Although measurements were not very accurate but rather illustrative, they were accurate enough to compile the sensor's profile, which later allowed to execute the function transforming the readings, which customize the sound volume to the returned distance value. It is easy to notice that the characteristic is nonlinear. A graph - Fig. 1, based on the data presented in Table 1 was prepared to better illustrate the characteristics of the sensor.

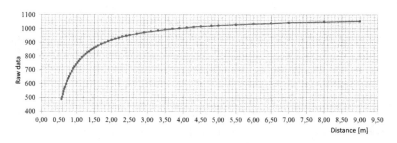

Fig. 1. A graph showing the relationship between the actual values of the distance and the raw values returned by the sensor

What can be clearly observed is high accuracy and great changes in raw values returned by the sensor, with a small change in distance if the obstacle is very close. The longer the distance, the smaller changes in the raw values are. Despite this fact, it is possible to use such a wide range because the reading is stable. Lack of precision in large distances is not a problem for the sensor we focused on in this work. The function *Mix_SetPosition(...)* from the *SDL_mixer* library (on which the developed program is based) takes an 8-bit parameter as a volume/distance value (and therefore the maximum value is 255 - silence). Therefore, it was necessary to convert the raw values to the useful range and a little more linear characteristics with the appropriate functions.

Figure 2 shows the graphs of functions converting raw values of distance, the 11-bit to 8-bit value. Different variants of the conversion function make it possible to adapt the device to more open space or smaller space, for example a room.

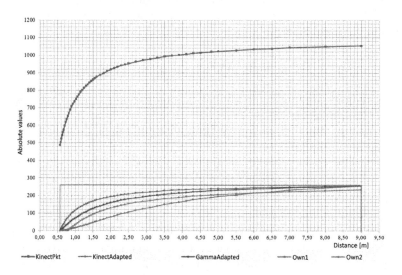

Fig. 2. Graphs of functions that transform raw data of distances (up to 9 m) into useful values within the range of 0–255

Any changes in the converting function are undesirable. The most universal function rather than converting functions are preferred due to the fact that the user's brain gets used to one function and will always be able to relate a specific volume to a specific distance (in order to facilitate this, a new reference sound of constant volume was added). This mechanism is not reliable but probable. It is also possible to subordinate the choice of conversion function to a variable factor, for example whether the person is indoors, or outdoors.

The chart labelled *KinectPkt* represents the raw values returned by the sensor (also shown in Fig. 1). Orange lines limit the chart field which includes the useful range of the actual distance, and converted distance values within the range of 0–255, returned by the sensor.

The chart labelled *KinectAdapted* is a graph indicating the function values with regard to the actual distance values:

$$d_2(KinectPkt) = \frac{KinectPkt - 488}{2.215} \tag{1}$$

This function merely scales the raw values so as to fit in the desired range. It can be useful in tight space (where the Kinect sensor is not suitable).

The chart labelled *Gamma Adapted* is a graph showing the function values with regard to the actual distance values:

$$d_3(KinectPkt) = \frac{(\frac{KinectPkt}{2048})^3 \cdot 36 \cdot 256 - 124}{4.44} \tag{2}$$

The function is based on another one - Gamma, which is arranged in the demonstrational source code of *libfreenect* library. Function has been transformed in such a way so that the returned values are within the desired range.

The chart labelled *Own1* is a graph showing the function values relative to the actual distance value:

$$d_4(KinectPkt) = \frac{2^{\frac{KinectPkt}{80}}}{36} - 1 \tag{3}$$

This is an empirical function. It is more suitable for open spaces rather than the function (1) or function (2).

The chart labelled *Own2* is a graph showing the function values relative to the actual distance value:

$$d_5(KinectPkt) = \frac{(\frac{KinectPkt}{2048})^3 \cdot 36 \cdot 256 - 300}{4.1} \tag{4}$$

This is also an empirical function. For the longest actual distance being read it does not produce complete silence. It also omits certain portion of the shortest distance detected by the sensor, where volume is maximal. Negative values must equal zero (workable under conditional instructions).

The above functions should be reduced by means of conditional instructions so as not to go beyond 0–255 range.

2.3   Transmission of Depth Map by Means of Sounds

Depth map is relayed to the user by means of sounds - tones with constant frequency. Increased tone volume means less distance from the surface of the obstacle. Distribution of volume/pitch dependent on the distance are provided by the conversion functions, described above.

The whole depth map returned by the sensor (640 × 480 px) is divided into smaller blocks, which will be called small fields in this paper. The division must be made due to the fact that it would be impossible or very difficult to communicate as much information through hearing. Another argument in favour of dividing the image is quite common areas where the sensor could not read the distance - such areas would introduce much confusion in the signal and could build the incorrect depth map in the user's mind.

The number of small fields determines the final resolution of depth map being transmitted. In the developed prototype the resolution of transmitted depth map is 5 columns by 6 lines (5 × 6 small fields), which generates 30 small fields (Fig. 3). Compared with a resolution of 640 × 480 px, which generates 307200 px it is not much. The resolution may be increased.

Fig. 3. On the left side - generated depth map for the sample scene. On the right side depth map of the same scene, but with a resolution reduced to 6 × 5 px. Scaling shown in the above illustration is made with a different method than the program which operates the device. Illustrations are only to present the perception difference of various resolutions and does not take into account the device dynamics

Coordinates of the transferred point of depth map are dependent on the lateralization of sound signal (changing in time) and the pitch height. Depth map is transmitted as a 'scan' of five columns from the left to the right and from the right to the left (the information contained in these two 'scans' is not the same - what is explained later). Each transmitted column takes a certain time - presently it is a long time, 200 ms, which is enough time to listen intently to the sounds. In one column there are 6 small fields (6 lines of depth map), but not all are transmitted collaterally.

The height of individual sounds determines the height the small fields represented by them are at - higher sound represents a small box above, lower sound represents a small box below. The sounds are played simultaneously (3 tones, not 6), and the volume of each corresponds to the distance provided by the small field. The scan from the left to the right is responsible for providing three upper lines (using three higher sounds). The scan from the right to the left is responsible for the transmission of three bottom lines (using three lower sounds). Graphical explanation of the above descriptions are shown in Fig. 4.

Each cycle of depth map transmittion begins with a brief sound, always with the same volume, frequency (100 Hz) and duration (300 ms), which does not belong to the depth map. It is designed to provide a reference volume when a person is in the quieter or in louder environment. It can also be a potential solution to the problem of the phenomenon of auditory adaptation or auditory fatigue (these phenomena are described in [10]).

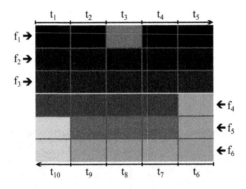

Fig. 4. The process of sound transmission of depth map (based on the map from Fig. 3). The illustration does not include pauses or reference sound. The arrows labelled fx indicate the direction of 'scan' and the sound, where f1 is the sound of the highest frequency, and f6 is the sound of the lowest frequency. The following times of the columns being transmitted are marked as t1–t10

The sound signal of the transferred depth map is also supplemented with pauses to distinguish the different phases of communication better. In the cycle of transmission there are: Step 1 - reference sound - 300 ms long; Step 2 - pause - 200 ms long; Step 3 - transfer of five columns from the three top rows - 200 ms long (per one column) 5 times, so 1000 ms long; Step 4 - pause - 200 ms long; Step 5 - transfer of five columns from the three bottom lines - 200 ms long (per one column) 5 times, so 1000 ms long; Step 6 - pause - 200 ms long.

With the described values the total duration of the cycle can be easily calculated, which equals - 2900 ms. It results in low frequency of communication, which is about 0345 Hz. It should be noted that before sounds from the next column are emitted, the depth map is updated. There is a partial refresh rate, which is 5 Hz (excluding pauses between 'scans').

These times are selected in such a way that beginners could hear all the sounds and learn to use them. The next stages of learning to use the device may be gradual shortening of the length of the various phases until the minimum time is reached.

Values of the small fields are calculated as follows: 1 - The value of each pixel in a small field is converted through a transition table which stores the converted distance values (with one of conversion function) when the distance value program is executed. This method was applied in the library freenect demonstration code. 2 - Values of all correctly read pixels are added up, and their number is counted. Pixels whose values are too big are rejected. This data is used to average the values. In addition, among the pixels the smallest indicated distance value is searched for. 3 - If the minimum indicated distance value within all pixels in a small field is greater than the determined limit (currently 1.2 m), and if there is at least one pixel whose value has been read correctly, the small field is assigned the average value of correctly read pixels. 4 - If the above condition is not fulfilled (if a near obstacle has been detected), the small field is assigned the value of the read minimum distance occurring within the small field.

It is possible to use an alternative method of the above algorithm, but probably it is less favourable. This method does not include point 2 and point 3 - values are not averaged, and the small field is always assigned the minimum value occurring within the area.

2.4 Learning Mode

Since the dependence of the height of view on the height of pitch is not natural, it is necessary to allow the user the opportunity to learn this ability. A special program has been executed (that runs directly on the device), which should help the blind learn to subconsciously relate looking high to the high pitch. Apart from the program, a person can also focus on learning to recognize sounds, but this is probably a much more difficult way to learn.

Learning mode operates in a similar way to an ordinary obstacle detector which detects an obstacle located exactly in the central area of the sensor's field of view (currently, this area has a size of 2×2 px) and indicates the minimum distance from the obstacle read out in this area with a single sound, the intensity of which increases when approaching an obstacle (the same conversion function should be used as in the normal mode).

The height of alerting sound is selected on the basis of the sensor tilt. The sensor inclination is read with a built-in accelerometer. If a person moves his head down, the distance to the obstacle is signalled with lower and lower sound (depending on the angle of inclination). If a person moves his head up, this distance is indicated with higher sounds. No sound is generated when the tilt angle exceeds the angle of the sensor's field of view in the normal mode, which is $43°$, so $21.5°$ down or $21.5°$ up. It is easy to calculate that one line (and thus a sound) accounts for about $7.17°$. Accelerometer data provided by the library *libfreenect* is returned with a resolution of $0.5°$.

The refresh rate is much higher than in the normal mode, and is $20\,\mathrm{Hz}$ (refreshes every 50 ms). Reference sound is played every 80 readings, so every 4 seconds. It takes as much time as in normal mode - 300 ms and pauses before and after take 200 ms each.

3   Functionality Testing

Since there was no access to a larger group of blind people, and for the sake of safety during testing the research on device functionality was conducted primarily on non-sight-impaired patients.

The people's task, who took part in the test, was to walk across a 20 m long room using our assistive technology. They had worn special glasses which covered their eyes tightly. There were 10 obstacles to navigate around, both horizontal and vertical. There were two modes used on the track 'no learning' and 'learning'. The study involved 120 people. Each participant took five different paths, each path had a different obstacle layout. Table 2 presents the results of all 120 people having walked five different paths in the 'no learning' mode. The task effectiveness was calculated for 120 people and took into account all walked round obstacles (of all obstacles on the track).

Table 2. The results of going through 5 different paths in 'no learning' mode

Kind of test	Track number to go				
	Track 1	Track 2	Track 3	Track 4	Track 5
The effectiveness [%]	54	63	66	65	68
The average time [s]	134	164	158	173	177

Table 3 shows the results of the effectiveness of achieving the objective by users on selected tracks, but this time in the 'learning' mode. This mode provides the built-in learning mode, which allows for better assimilation of generated sounds and improved sounds distinction when scanning the space. Each of the test subjects had the opportunity to learn the device only once. Such a requirement assumed to generate reliable results for all users of the system.

Table 3. The results of going through 5 different paths in 'learning' mode

Kind of test	Track number to go				
	Track 1	Track 2	Track 3	Track 4	Track 5
The effectiveness [%]	89	93	96	98	99
The average time [s]	84	103	112	107	121

Having analysed the results, it is easy to notice that the device requires learning. Each of the users felt more confident having learnt the device. Indisputably, the time results of walking across a 20-meter path are not satisfactory, but the more the participant learnt, the better the results were. As the consequence of obstacles to navigate around on the way, we got different times and different effectiveness. The biggest problems were caused by obstacles lying on the ground, which required a tilt of the head. Other obstacles (on the sides, and placed from the top on the head level) were almost correctly avoided after learning phase. We should consider the behaviour of the visually-impaired people. These tests will allow to determine the real features of the device assisting the blind.

4 Conclusion and Future Work

The acoustic system assisting blind people, or rather its prototype, is a device which still needs to be changed and developed. This work is only the beginning to create a much better device. The very Kinect sensor has a feature (does not work in sunlight), which will hamper the further development of the device, although on a large scale it has become an affordable 3D scanner for many people. Because of such high accessibility (especially low price) it was also used in this work.

The new approach is learning the height of signal based on the tilt of the head. If the device undergoes further refinements, we can expect better and more effective communication methods of depth maps. For this type of project the best outcome would be the constructive feedback from the visually impaired. What we currently pursue is checking the device with the blind people which would definitely give new insights in improving the prototype.

References

1. Sek, A., Skrodzka, E., Marszalkiewicz, M.: Psychoacoustics in a pill. Institute of Acoustic UAM (2000). (In Polish)
2. Pec, M., Bujacz, M., Strumio, P.: Personalized head related transfer function measurement and verification through sound localization resolution. In: 15th European Signal Processing Conference (EUSIPCO 2007), pp. 2326–2330 (2007)
3. Pec, M., Strumillo, P., Pelczynski, P., Bujacz, M.: The hearing images - support systems of blind people in the perception of the environment. Tech. Inf. Bull. Branch Board Lodz SEP (6), 6–11 (2006). (In Polish)
4. Hojan, E., Skrodzka, E.: Polish research on acoustical assistance for blind and visually handicapped persons. In: International Conference on Auditory Display, pp. 13–16 (2013)
5. Denis, M., Afonso, A., Picinali, L., Katz, B.F.G.: Blind people's spatial representations: learning indoor environments from virtual navigational experience. In: 11th European Congress of Psychology (2009)
6. Szubielska, M.: Imagination abilities of the blind children in the field of scanning and rotation of the shape of the touched objects. Ann. Psychol. **2**, 145–160 (2010)
7. Moore, B.: An Introduction to the Psychology of Hearing, 5th edn. Elsevier Academic Press (2003)
8. Tolman, B., et al.: The GPS toolkit - open source GPS software. In: 17th International Technical Meeting of the Satellite Division of the ION (ION GNSS 2004) (2004)
9. Velazquez, R.: Wearable assistive devices for the blind. In: Lay-Ekuakille, A., Mukhopadhyay, S.C. (eds.) Wearable and Autonomous Systems. LNEE, vol. 75, pp. 331–349. Springer, Heidelberg (2010)
10. Sliwinska-Kowalska, M.: Clinical Audiology. MEDITON Printing House (2005)

Software Technologies

Performance and Energy Efficiency in Distributed Computing

Andriy Luntovskyy$^{(\boxtimes)}$

BA Dresden University of Cooperative Education,
Hans-Grundig-Str. 25, 01307 Dresden, Germany
Andriy.Luntovskyy@ba-dresden.de

Abstract. Performance-to-energy models and tradeoffs in Distributed Computing (Clusters, Grids, and Clouds) are discussed. Performance models are examined. Energy optimization brings better PUE values for the data centers. Better tradeoff "Performance-to-energy" can be reached under use of the advanced "green" technologies as well as in the so called Internet of Things (IoT).

Keywords: Distributed computing · Clusters · Grids · Clouds · Performance models · PUE · On-board-controllers · IoT · Fog computing

1 Motivation

After 2011 the development of cluster and cloud computing methods was commonly triggered by the trend of "green IT" with increasing energy demand and prices. The computing centers were built more often in colder regions of Earth. So, for example, Google Data Centers achieve the PUE (Power Usage Effectiveness) of 1.12 due to further optimization of hardware, waste heat recycling systems and building construction features like improved air circulation, reuse of waste heat etc. This means that only 12 % of energy required for computing was used not by servers, but by other services like conditioning, energy distribution, lighting, surveillance systems etc. However, according to Uptime Institute 2012 Data Center Survey, the average PUE in the domain was about 1.89, which means a significant improvement on the side of Google, i.e.:

$$Max(PUE) \wedge [QoS \geq QoS_{min}] \wedge$$
$$[Costs \leq Costs_{max}] \tag{1}$$

where $Costs_{max}, QoS_{min}$ – the Cost and Quality of Service constraints.

2 Distributed Computing: Performance and Efficiency

2.1 Trends to Low-Cost and Low-Energy Computing Nodes

A new trend to low-cost and low-energy computing nodes based on cheap on-board microprocessors (RISC/ARM) is to consider nowadays as an serious alternative to expensive state-of-the art nodes within the up-to-date IoS (Internet of Services) [1, 2].

© Springer International Publishing AG 2017
S. Kobayashi et al. (eds.), *Hard and Soft Computing for Artificial Intelligence, Multimedia and Security*, Advances in Intelligent Systems and Computing 534, DOI 10.1007/978-3-319-48429-7_26

The deployment of low-cost and low-energy computing nodes such as Arduino, Raspberry Pi, Intel Edison means significant increasing of energy outcomes as well as technologically important new step towards of IoT (Internet of Things) [3, 4].

The scenarios for the so-called Fog Computing within IoT will steady gain in importance in mid-term. Instead of use of IoS with heavy-weighed processors and VMs agile and energy-efficient on-board microprocessors should be operated: see the view of future transfer from Clouds/IoS to the Fog Computing/IoT (Fig. 1). Surely the deployment of low-cost and low-energy computing nodes based on on-board micro-processors like Arduino, Raspberry Pi, Intel Edison should be defined via the impor-tance, priority and is strong dependent on appropriate resource use in the frame of a given math-log problem!

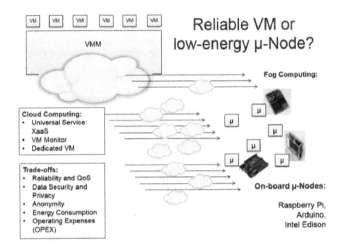

Fig. 1. Energy-efficient on-board computing nodes as a basis for distributed computing

2.2   Performance Models

First of them let us define the most important performance factors. The performance parameters of modern computers are as follows:

- Tact (Clock) Frequency, f;
- MIPS: Million Instructions per Second, Mega IPS;
- FLOPS: Floating Point Operations per Second.

System clock signal (System Clock) synchronizes the operation of multiple func-tional blocks within a CPU. The system tact is a periodical function based on Peirce-Function, NOR. Some examples about the performance of certain CPU models from last production years are given below (Table 1).

The most representative performance unit is FLOPS-value. The value acts as an integrated characteristic defined via the computer architecture properties, including RAM, bus and cache construction, assembler type like ARM/RISC/CISC, as well as compiler peculiarities. The following performance formula can be used [1, 2, 5]:

Table 1. Performance of certain selected CPU Models

#	CPU models (2006–2015)	Performance, MIPS	Tact frequency, GHz
1	AMD Athlon FX60	18.938	2,6
2	Intel Xeon Harpertown	9.368	3
3	ARM Cortex-A15	35.000	2,5
4	AMD FX-8150	108.890	3,6
5	Intel Core i7 2600K	128.300	3,4
6	AMD A12 Pro-8800B (2015)	More than 150.000	3,4

$$P = f \cdot n_1 \cdot I \cdot n_2 \tag{2}$$

where P - performance in GFLOPS, f – CPU tact frequency in GHz, n_1 – number of cores within a CPU, I – CPU instructions per tact, n_2 – number of CPUs per computing node. Let us consider the integral performance criterion FLOPS:

Example 1. Let's consider a 2-Socket-Server with CPU Intel X5675 (3,06 GHz, 6 cores, 4 instructions/tact):

Performance = 3,06 × 6 × 4 × 2 = 146,88 GFLOPS.

Example 2. We have a 2-Socket-Server with CPU Intel E5-2670 (2,6 GHz, 8 cores, 8 instructions/tact):

Performance = 2,6 × 8 × 8 × 2 = 332,8 GFLOPS.

2.3   Speedup and Effectiveness of Computing Process

Factors of speedup and effectiveness are computed as follows:

$$An = \frac{T_1}{T_n}, \quad E_n = \frac{A_n}{n} \cdot 100 \tag{3}$$

where T1 – computing time for a math-log problem with use of only one CPU, Tn – computing time of the parallelized on n processors or threads solution A_n – speedup factor, E_n – effectiveness for speedup on n CPUs.

An exemplarily sections' distribution by task parallelization and the influence of cluster communication exchanges by Message Passing between the processors or threads is depicted in Fig. 2. The computation time gain is possible only due to higher (p/s) – ratio within a parallelized task (a math-log problem). The time estimations are as follows, refer (4):

$$T = s \vee T = s + p \vee T = s + \frac{p}{n}$$

$$T = s + \frac{p}{n} + k \cdot n \tag{4}$$

$$e = 1 - p$$

where T – overall computing time, s – sequential part of a task (percentage), p – potentially parallelized part of a task (a math-log problem), i.e. on n threads or CPUs, e – part for sequential computing time, k – negative influence of communication by message passing between CPU/threads (this component can also be neglected, k = 0).

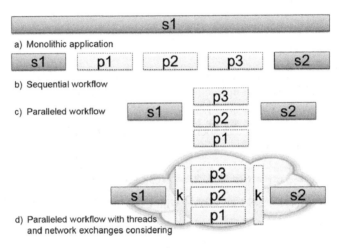

Fig. 2. Sections distribution by a math-log problem parallelization and the influence of cluster communication (exchanges) by message passing

2.4 Amdahl's Law

One of mostly appropriate and useful approximations for Speedup factor is the G.M. Amdahl's one (1967):

$$T = 1, \ 1 \equiv (1 - p) + p$$

$$A_n = \frac{1}{(1 - p) + p/n} \leq \frac{1}{1 - p}$$

$$A_{max} = \frac{1}{1 - p} \tag{5}$$

$$A_{n,k} = \frac{1}{(1 - p) + p/n + k \cdot n}$$

where p – potentially paralleled part of a math-log problem, n – number of available CPU/threads, k – negative influence of communication by message passing between CPU/threads (this component can also be neglected, k = 0).

Example 3. Let's consider a math-log problem with Toverall = 20 h; Tsec = 1 h (i.e. 5%), Tpar = 19 h (i.e. 95%); SpeedupMAX = 20.

Then by n = 10 processors (threads): p = 0,95, Speedup = 1/((1–0,95) + 0,95/10) = 1/(0,05 + 0,095) = 6,9 < SpeedupMAX.
On the other hand: n = 95 processors (threads) Speedup = 16,7 only!

So we can obtain the following graduated depiction of the speedup factor (Fig. 3). There are some critic points regarding to this realistic model: too pessimistic representation of the parallel computing status. But other models talk a lot also about the saturation effects, especially due to communication processes within a cluster between the processors (threads) and energy losses (in form redundant warm, waste heat).

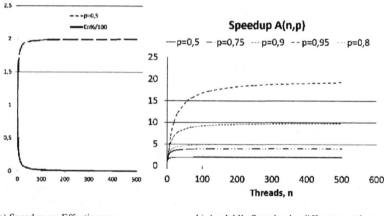

a) Speedup vs. Effectiveness   b) Amdahl's Speedup by different p-values

Fig. 3. Pessimistic Amdahl's model for the speedup factor depending on p = {0,5 … 0,95}: saturation effect, no more profit due to increasing of n – number of threads

2.5   Speedup Model Overview

Table 2 illustrates the set of integrated models and approximations of speedup factors are typically used for distributed (parallel) computing. The approximations of speedup factor [6, 7] are given on dependence of the criteria {n, p = 1–e, k}. There are the mostly used models and laws including Amdahl's, Barsis-Gustafson's, Moore's law (exponential model) and some further models.

Example 4. We would like to define herewith the value e (refer formula (4)), i.e. the normally unknown part for sequential computing time for a math-log problem, on the basis of the Karp-Flatt-Metric (1990), refer Table 2 (Pos. 9):

Number of CPU n = 100, measured speedup A = 10, 1/A = 0,1;
e = (0,1 – 0,01)/(1 – 0,01) = 0,09/0,99 = 0,0909;
e = 9,1%; it can be for parallelized p = 91%!!!
Number of CPU n = 100, measured speedup A = 25, 1/A = 0,04;
e = (0,04 – 0,01)/(1 – 0,01) = 0,03/0,99 = 0,0303;
e = 3,03%; it can be parallelized for p = 97%!!!
Number of CPU n = 100, Speedup A = 66, 1/A = 0,0151;
e = (0,0151 – 0,01)/(1 – 0,01) = 0,0051/0,99 = 0,0052;
e = 0,52%; it can be parallelized for p = 99,5%!!!

Considering (3), (4) and Table 2 we can obtain the next useful formula (6) for the criterion:

$$A_n > 1$$

$$e(A_n, n) = 1 - p = \frac{1/A_n - 1/n}{1 - 1/n} \tag{6}$$

$$p = \frac{1 - 1/A_n}{1 - 1/n} = \frac{A_n - 1}{A_n - A_n/n} = \frac{A_n - 1}{A_n - E_n/100}$$

Example 5. Let us consider the number of CPU should n = 100, Speedup A = 66, effectiveness E = 66 %, then the math-log problem can be parallelized for the $p-$ ratio: $p = (66 - 1)/(66 - 0.66) = 65/65.34 = 0,995$ *(cp. Example 4).*

2.6   Own Experiments

For the hardware basis (Fig. 4a) offered at Dresden University of Technology the following own results (Table 3) on speedup have been obtained. It was a voluminous experiment in Nov. 2006 aimed to simulation of signal power propagation of WLAN/WiMAX networks on complex 2D-environments (maps of the obstacles with given material features) [2, 8–11].

The simulation has been realized with use of CANDY Software and Web Services for SSL-access to MARS. The following results are obtained (Fig. 5, refer Table 3). These results can be approximated as follows, cp. Grosch's law:

$$An = \frac{T_1}{T_n} = n^\alpha, T_1 = 8021 \text{ sec}, \tag{7}$$

$$\alpha \approx 0,95$$

The following results are obtained (Fig. 5, refer Table 3).

Example 6. The new hardware basis from the same hand is called TAURUS Bull HPC-Cluster [2]. This cluster is more powerful than the formerly leading MARS and has nowadays the following features (Fig. 4b):

(a) Hardware basis: High Performance Computing
Cluster MARS SGI Altix 4700 @ TUD with 1024 cores
possesses the performance 13,1TFLOPS

(b) Up-to-date Hardware basis: TAURUS Bull
HPC-Cluster with 137TFLOPS

Fig. 4. Hardware basis: high performance computing at TUD

- Island 1: 4320 cores Intel E5-2690 (Sandy Bridge) 2.90 GHz.
- Island 2: 704 cores Intel E5-2450 (Sandy Bridge) 2.10 GHz as well as 88 NVidia Tesla K20x GPUs.
- Island 3: 2160 cores Intel X5660 (Westmere) 2.80 GHz.
- SMP (Symmetric Multiprocessing) nodes with 1 TB RAM
- 1 PB SAN disk storage.
- Bullx Linux 6.3, batch system Slurm.
- 137 TFLOPS total peak performance (without GPUs).

The so called SNP (Symmetric Multiprocessing) with large RAM capacities gains in its deployment nowadays more sympathizers than the NUMA (Non-Uniform Memory Access) with the offered unique address spaces as well as correspondingly the cache-coherent NUMAs. A performance comparison is given in Table 4. Herewith some world-wide known clusters and grids from the Top500 list [12] are referred in correspondence to the above mentioned performance of MARS and TAURUS systems. MARS performance is given as "a unit".

Example 7. A generalized graphical comparison of speedup factors is depicted in Fig. 6. The mostly used models are shown: a trivial one (3) as well as optimistic; by B.–G. (8), i.e. more realistic, and A. (5), i.e. a pessimistic one, refer Table 2: (3), (5), (8):

2.7 On-Board-Microcontrollers and Mini-PC

But none of the above mentioned computing systems is enough energy-efficient. The electricity consumption surrounds in just MWh area. Energy efficient solutions can be provided via small, low-cost and low-energy on-board processors. The electricity consumption surrounds in this case at most kWh area. Low-energy home intelligent nodes (3–10 W) for private Cloud-solutions, File Server, Web Server, Multimedia Home Centre etc. can be placed on the low-cost energy-efficient on-board microcontrollers like Arduino, Raspberry Pi or Intel Edison etc. as the trade-off solution. They offer a cheap alternative and symbolize step-by-step shift to the IoT. An appropriate

Table 2. Speedup model overview

Speedup factor $An = \frac{T_1}{T_n}$	Speedup model	Conventions	Title of an empirical model
1	$A_n = \sqrt{n}$	The type of math-log problem is not considered	Grosch's law (1965)
2	$A_n = n^b$	The type of math-log problem is not considered	Generalized Grosch's law $0,5 \leq b \leq 1$
3	$A_n = n$	The type of math-log problem is not considered	Proportional Amdahl law for $p = 1$ $s = 0$
4	$A_n = \log_2(n)$	The type of math-log problem is not considered	Logarithmic Law
5	$A_n = \frac{1}{(1-p)+p/n}$	$0,5 \leq p \leq 0,999\ldots$ $k = 0$	Amdahl's Law (1967)
6	$A_n = \frac{1}{(1-p)+p/n+k \cdot n}$	$0,5 \leq p \leq 0,999\ldots$ $k \approx 10^{-4}..10^{-5}$	Corrected Amdahl's model with inter-processor communication considering
7	$A_n = 2$ $n = 70\%/r\%$	The type of math-log problem is not considered, $r = 1\ldots2\%$ characterizes inter-processor communication losses	Empirical law "69–70–72" for CPU-number n, which provides double speedup of computing time
8	$A_n = (1-p) + p \cdot n$	$0,5 \leq p \leq 0,999\ldots$ $k = 0$	Barsis-Gustafson-Law (1988)
9	$A_n > 1$ $e(A_n, n) = 1 - p = \frac{1/A_n - 1/n}{1 - 1/n}$	$e = 1-p$ – the unknown part for sequential computing time; $0,5 \leq p \leq 0.999\ldots$ $k = 0$	Karp-Flatt-Metric (1990) for Amdahl's or Barsis-Gustafson-Law

solution will be the Raspberry Pi on-board-microcontroller (firstly deployed in 2011 in Cambridge, UK) with only credit card dimensions, in a pod like a matchbox as well with the following characteristics [13, 14]:

- ARM/RISC CPU; f=700900 MHz, RAM = 256 MB–1 GB.
- Extern storage like a SD-Card.
- Ports: USB 3.0, HDMI, LAN 10/100-Ethernet

Table 3. Computing time for a complex simulation task of WLAN/WiMAX propagation

Number of threads	Computing time, s	Speedup factor An = T1/Tn
1	8021	1
2	4163	1,9
5	1749	4,6
10	908	8,8
20	471	17,0
30	321	25.0
55	181	44,3
70	144	55,7

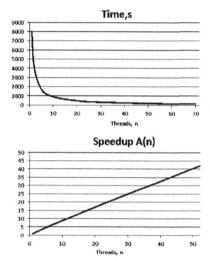

Fig. 5. Computing time and speedup factor in depending on threads number obtained on the multi-core high-performance computer MARS @ TU Dresden (Basis – CANDY Framework 2006)

- Low-cost, price depends on the models Pi A/B or Pi2 B.
- OS: Linux/BSD, RISC OS or Raspbian OS, as well as diversity on GPL-Software, in mid-term under Win 10.
- Protocol support: SSH/Samba.
- USB-Drive for about 0,5 – 3 TB.
- Low power, depends on models A/B: ca. 3,55 W.

Naturally there are a lot of scenarios on economical network nodes. The newest Raspberry Pi 2 Model B supports Windows 10 and acts as a mini-PC with 6 times CPU-performance due to tact frequency 900 MHz, RAM = 1 GB and quad-core architecture being oriented to the Windows Developer Program for IoT. Thus the most significant criterion for the choice of appropriate computing system is not only performance considering but also use of the combination of the further criteria as follows:

Table 4. Peak performance comparison (Status 2016)

Cluster or grid	Peak performance, PFLOPS	Multiplicity factor (regarding to "MARS units")
Tianhe-3 (this specifying is without awareness)	100	7692
Tianhe-253 (a supercomputer from Popular Republic China)	33,86	2605
Titan (USA-supercomputer)	17,59	1353
BOINC (grid system hosted at Berkeley University of California, USA)	9	692
SuperMuc (hosted at Leibniz-Institute in Munich)	6,8	523
Juqueen (hosted at Research Centre FZ Juelich, supported via IBM)	4,1	315
TAURUS (hosted at TU Dresden)	0,137	11
MARS (TU Dresden, taken into account via the authors of this paper for our own WLAN/WiMAX simulation experiments with signal power propagation in 2006)	0,013	1
Home desktop PC	0,0001	–

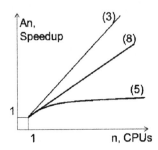

Fig. 6. Speedup comparison

- Reliability and QoS.
- Data security, anonymity and privacy.
- Energy minimum.
- Tiny OPEX.

Unfortunately there are no definitive answers on the given question (refer Fig. 1); the questions are today open!

Example 8. Herewith a small example addressing the discussed trade-offs. A "super-computer" with 64 cheap Raspberry Pi's und two Lego racks is depicted in Fig. 7. This low energy cluster (64 × 3,5 W, max 0,25 kW) is built with use of low-cost and energy-efficient on-board microcontrollers. The small but smart Raspberry Pi Cluster for parallel computing offers the following features [13, 14]:

Fig. 7. Energy efficient Raspberry Pi Cluster (source: http://www.pro-linux.de)

- DC Supplying per USB, 3,5 W/CPU 700 MHz;
- Energy efficient Cloud IaaS;
- SD-Card as extern disk drive;
- Low power Data transfer and exchange via LAN Ethernet;
- OS Raspbian.

3   Conclusions

Performance-to-energy tradeoffs in parallel computing are discussed. Performance models are examined. Energy optimization can be reached under use of the advanced technologies like IoT. The most significant criterion for the choice of appropriate computing system is not only performance considering but also use of the combination of the criteria, inter alia energy minimum [5–7].

By my opinion, the essential development stages of modern Clustering and Parallel Computing are as follows [8–10]:

1. Meta-computing, pioneer grid projects like GRID based on active involvement of the technologies from scientific areas to everyday life.
2. Convergence with Web technologies (e.g. BOINC) wide-spreading of grids.
3. Efforts to solving of wider range of problems: secured access, interoperability, resource discovery on the basis of deployment of standardized MW (middleware) like OGSA (Open Grid Services Architecture).
4. Wide-spread acceptance of grid services in the same way as delivering of water and electricity and, then, inset of the SOA approach (service-oriented architectures) via standardized Web Services deployment and composition (WS-BPEL, Business Process Execution Language).
5. Wide-spreading of Cloud Computing as a model for enabling ubiquitous, convenient, on-demand network access to a shared pool of configurable computing resources with essential measured services like XaaS and rapid elasticity.
6. Integration of grid services within high-available clouds (mostly PaaS) together with parallel clusters (IaaS) and capable network storages (RAIC, Redundant Array of Independent.
7. and, "last but not least", development of new energy-efficient grids, clusters and cloud services, within Smart Grid technology towards IoT and Fog Computing.
8. Acknowledgment

Acknowledgment. My graceful heart's acknowledgement to Dr. J. Spillner with ZHAW (Zurich, Switzerland) for the helpful and friendly support, the inspirations and co-operation in meaning construction while completing this work.

References

1. BOINC Grid. http://boinc.berkeley.edu/
2. HPC Clusters in Dresden, ZIH@TUD. http://zih.tu-dresden.de
3. Bonomi, F., Milito, R., Zhu, J., Addepalli, S.: Fog computing and its role in the internet of things, 15 p. CISCO Corp., CA, USA (2007)
4. Shelby, Z., Bormann, C.: 6LoWPAN: the wireless embedded Internet. In: EE Times (2011)
5. Performance of Grids and Clouds. http://cidse.engineering.asu.edu/seminar-performance-of-grids-and-clouds-may-20/
6. Gustafson, J.L.: Re-evaluating Amdahl's Law. Commun. ACM **31**(5), 532–533 (1988)
7. Karp, A.H., Flatt, H.P.: Measuring parallel processor performance. Commun. ACM **33**(5), 539–543 (1990). doi: 10.1145/78607.7861
8. Luntovskyy, A.O., Guetter, D.G., Melnyk, I.V.: Planung und Optimierung von Rechnernetzen: Methoden, Modelle, Tools für Entwurf, Diagnose und Management im Lebenszyklus von drahtgebundenen und drahtlosen Rechnernetzen, 411 p. Handbuch. – Springer, Vieweg + Teubner Verlag, Wiesbaden (2011) (in German)
9. Luntovskyy, A.O., Klymash, M.M., Semenko, A.I.: Distributed Services for Telecommunication Networks: Ubiquitous Computing and Cloud Technologies, Monograph, 368 p. Lvivska Politechnika, Lviv (2012) (in Ukrainian)
10. Luntovskyy, A.O., Klymash, M.M.: Data Security in Distributed Systems, Monograph, 464 p. Lvivska Politechnika, Lviv (2014) (in Ukrainian)
11. CANDY Framework and Online Platform. http://candy.inf.tu-dresden.de
12. Top500 list of supercomputers. http://www.top500.org
13. Richardson, M., Wallace, S.: Getting Started with Raspberry Pi, 175. p. O'Reilly Media, Sebastopol (2012). ISBN: 978-1-449-34421-4
14. Orsini, L.: How To Host a Website with Raspberry Pi. http://www.readwrite.com/

The Approach to Web Services Composition

Alexander Koval, Larisa Globa, and Rina Novogrudska[✉]

Information Telecommunication Networks Department,
Kyiv Polytechnic Institute, National Technical University of Ukraine,
Peremoga Avenue 37, Kiev 03056, Ukraine
avkovalgm@gmail.com, lgloba@its.kpi.ua, rinan@ukr.net

Abstract. The paper presents an approach to the composition of web services based on their meta descriptions. The process of web services sequence formation is depicted. Such web services sequence describes execution of certain user's task. The method of user's tasks scenario formation is proposed that allows to define dynamically an ordered sequence of web services required to run a specific user's tasks. The scenario formation for real user's task "Calculation of the strength for the power components of magnetic systems" is represented, approving the applicability and efficiency of proposed approach.

Keywords: Web services · Scenario · User's task · Meta description · Composition

1 Introduction

Nowadays, a large number of web services are provided over the web from companies such as Google, Microsoft and Amazon. By utilizing these services, end user is expected to create value added services to fulfil the requirement. This process is commonly known as web services composition. An analysis of the literature related to the service composition focuses that, frequently performed tasks throughout the composition are the discovery and selection of the candidate web services which are participating to generate the composition plan. Although, there is a close relationship among discovery, selection and composition, researchers treat them individually and independently.

Web services usually are used for different processes performance. Such process can be:

- General user's tasks,
- Calculation and computational tasks,
- Business processes.

Moreover, different users want to achieve the increased value of satisfaction from the resultant composition solution. Thus, the execution of user's tasks is closely correlated with web services execution. User's task is the conceptual notion. For the execution of user's task, the sequence of web services must be executed (for the web services that correlate to this user's task). Web services composition into such sequence should be optimal to provide correct execution of user's tasks.

S. Kobayashi et al. (eds.), *Hard and Soft Computing for Artificial Intelligence, Multimedia and Security*, Advances in Intelligent Systems and Computing 534, DOI 10.1007/978-3-319-48429-7_27

The main objective of the web services composition "on the fly" is the organization in the dynamics of such an ordered sequence of web services that allows to execute a specific end-user task. Thus, it is needed to develop the method of web services dynamic integration for a specific end-user task execution.

This paper describes a novel approach for web services composition given their meta descriptions that is based on the method of user's tasks scenario formation.

The paper is structured as follows: Sect. 2 provides the background and analyses of related works on web services composition. Section 3 gives main notions and their formal description as well as characteristic features of web services. In this section the system of meta descriptions for web services and user's tasks is suggested, such system forms the basis of web services composition. Section 4 describes the method of user's tasks scenario formation. Section 5 contains the description of the proposed method application for the scenario formation that represent the execution of real complex computational user's task in the subject domain "Strength of materials". Section 6 concludes the work with a summary and outlook on future work.

2   Backgrounds and Related Work

Web Services Composition (WSC) is a method to connect different web services that are used for creating high-level business architecture by compiling of web services in order to provide functionalities that are not available during design [1]. Consequently, there is a possibility to develop a new functionality by simply reusing of components that are already available, but unable to complete a task successfully on their own.

Static and Dynamic WSC are two major types of web services composition. Dynamic web service composition occurs when user or software application queries for a web services at runtime while static web services composition is performed at the compile time. Dynamic web services composition is less trivial and requires much more work compared to static web services composition.

Various authors classify other WSC approaches. In [2], Static and Dynamic Web Services Composition approaches are grouped into a single approach. This and other possible approaches for composition are listed below:

- Static and Dynamic Composition;
- Model Driven Service Composition;
- Declarative Service Composition;
- Automated and Manual Composition;
- Context-based Service Discovery and Composition.

The ability to select and compose heterogeneous web services over the web efficiently and effectively at runtime is an important step towards the development of the web services applications [3]. By utilizing web services, end user is able to create composite services to fulfil the requirement when single service unable to do it.

Most of the approaches related to the web services composition [4–7] realized the fact that the prerequisite tasks to generate the composition solution are the service discovery and service selection of the candidate web services stored in the service repository.

Semantic web services [8–11] provide an open, extensible, semantic framework for describing and publishing semantic content, improved interoperability, automated service composition, discovery and invocation, access to knowledge on the Internet [12].

[13, 14] describes discovery and composition issues of web services in the proposed extended SOA architecture. The concept of automated service discovery and composition process using Semantic Web services is presented.

In agent-based solutions for web services, composition agencies gather QoS data from agents, store, aggregate, and present it to agents [15]. Approach, which solves these problems, is introduced in [16]. Distinctive feature of this approach is utilization of SLA via WS-Agreement [17] during both workflow enactment and workflow analysis stages.

A composite service is a set of individual services that are effectively combined and reused in order to achieve a desired effect. Automatic service composition consists of four phases: Discovery, Selection, Planning, and Execution [18]. The first phase involves creating a plan, i.e., sequence of services in desired composition. The plan creation could be manual, semi-automatic, or automatic. The second phase embodies service discovery due to the plan. Planning and discovery are often combined into one step. After discovery of suitable services, the selection phase starts. It embodies a selection of the optimal composition from the available combinations of web services judging on nonfunctional properties like QoS properties. The final phase involves executing the services due to the plan. If some service is not available, another one takes its place. Concerning WS performance on engineering knowledge portals the most time-consuming is the phase of Discovery. As it is in need to choose from WS variety only those that are to be composed for complex engineering calculation performance.

In the case of solving the problem of web services composition aiming specific end-user's tasks execution it is in need to override these stages meaning. First it is needed to discover and select from web services set those web services that are used for described user's task execution. Then the user's task scenario is to be formed from this web services. Next the stage of execution is performed.

3   The Characteristic Features of Web Services

Figure 1 describes the correlation between notions: user's task (UT), scenario and web service (WS). Each user's task stage is associated with certain web service that execute this stage. The scenario is an ordered sequence of this web services, scenario presents the conceptual process of user's task execution.

Each of user's task has a specific set of attributes, describing its features and characteristics. As well, each WS can be described by some set of parameters. Such sets forms their meta descriptions. Specification of WS meta descriptions is based on Dublin Core standard and defines meta description as $MD = <A,D>$, where A is meta description attribute (represent important characteristic or parameter of task or WS) and D is data (represent the name of this parameter or characteristic for appropriate task or WS). Thus, each task and web service can be described by certain set of meta descriptions and each of such meta description has some meaning exactly for this task or WS. Such system of meta description allows to describe task or WS in terms of their

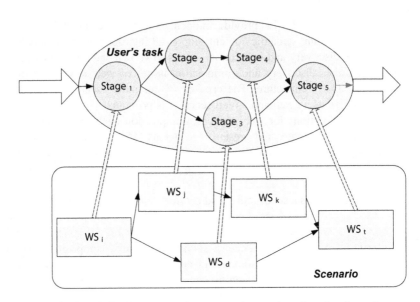

Fig. 1. Correlation between notions: scenario, user's task and web services

parameters meanings for appropriate subject domain (that is not depicted by WS-CDL or WS-BPEL).

The structure of user's tasks meta descriptions and web services meta descriptions is similar. Thus, the dynamic generation of web services sequence (the scenario) that corresponds to the certain user's task can be based on the comparison of such web services meta descriptions and user's task meta descriptions.

All meta descriptions corresponds to two types:

1. General meta descriptions – describe different general characteristics of user's tasks and web services.
2. Specific meta descriptions – describe different parameters and characteristics of user's tasks and web services that depend on UTs and WSs subject domain (they reflect the specific of subject domain and may indicate the services necessary to perform a certain user's task).

Table 1 presents the suggested system of meta description for user's tasks and web services.

All web services are stored in independent storage, while meta descriptions and their values are stored in database tables (Fig. 2).

User's tasks set is described as:

$$UT \ni UT_i, \ UT_i = \langle T_i, MD_{ip} \rangle, \tag{1}$$

where UT – the set of user's tasks;

UT_i – i-th user's task;

$T_i,$ – the title of i-th user's task;

Table 1. System of meta description for user's task and web services

Meta descriptions title	Meta descriptions description
Standard meta descriptions	
Title	Title of UT or WS
Identifier	Unique ID number that is considered as UT or WS number
Subject domain	Subject domain of UT or WS
Description	Text description of web service
Data	Data type, for the data that is used for UT or WS execution
Regulatory	Title of the normative document (GOST, ISO, Standard), guiding which user's task is executed
Address	The reference to web service location
Specific meta descriptions (as an example – specific meta descriptions for subject domain "Strength of materials")	
Parameters	Parameters that are used for engineering calculation task (user's task)
Loading	Loading types that are used for engineering calculation task (user's task)
Method	Method of calculation that is used within engineering calculation task (user's task) execution
Element	Construction element that is used in engineering calculation task (user's task)
Measurement area	The area of construction that is used in engineering calculation task (user's task)

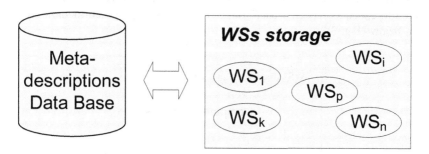

Fig. 2. Storage of WSs and meta descriptions

MD_{ip} – p-th meta description of i-th user's task.
Scenarios set is described as:

$$Sc \ni Sc_k, \qquad (2)$$

where Sc – the set of scenarios;

Sc_k – k-th scenario that is characterized by certain web services sequence.

Correlation between user's task and scenario (that corresponds to this user's task) is described by equivalence relation:

$$UT_k \sim Sc_k \tag{3}$$

Web services set is described as:

$$WS \ni WS_l, \ WS_l = \langle T_l, MD_{lq} \rangle, \tag{4}$$

where WS – the set of web services;

WS_l – l-th web service;

T_l, – the title of l-th web service;

MD_{lq} – q-th meta description of l-th web service.

Correlation between the scenario and web services is described by inclusion relation (WSs are included into the scenario):

$$Sc_k \supset (WS_l, \dots WS_d) \tag{5}$$

It is in need to develop the method that enables scenario formation based on comparison of web services meta descriptions values and provides the ability to perform real end-user's tasks by forming sequence of web services dynamically.

4 The Method of User's Tasks Scenario Formation

This section describes the steps of the proposed method of user's tasks scenario formation. This method allows to define dynamically an ordered sequence of web services required to execute a specific user's task. The stages of proposed method are represented below (Fig. 3).

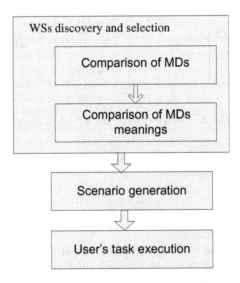

Fig. 3. Method of user's tasks scenario formation

Stage 1. WSs discovery and selection.
Stage 1.1. MD comparison.
On this stage it is in need to discard from the WSs set those WSs, for which all MD_{lq} are not equal to MD_{iv}. It is in need to compare the descriptions of WSs and to select those WSs which meta descriptions match user's task meta descriptions according to the rule:

$$MD_{ip} = \bigcup_l^m MD_{lq}. \tag{6}$$

whereby possible the fulfilment: $MD_{lq} \cap MD_{dq}$, where $i, j \in l, l = \overline{1,m}$.

As the result the subset of WSs set is received, such subset satisfies rule (6).

Stage 1.2. MD meaning comparison.
It is necessary to compare meanings of web services meta description - $MN(MD_{lq})$ to cut the subset of web services, that has common meta descriptions, but the meanings of this meta descriptions are not equal. Thus, on this stage the values of WSs meta descriptions (only for those WSs that were left in the set after stage 1.1) are compared to the values of user's task meta description:

$$Mn(MD_{lq}) < > Mn(MD_{iv}), \tag{7}$$

under the condition that $MD_{lq} = MD_{iv}$. The result can be one of the following options:

1 *Meta description value domains match* - $Mn(MD_{iv}) = Mn(MD_{lq})$. In this case, domain of the v-th meta description value of i-th user's task matches domain of q-th meta description value of l-th web service. Thus, specified web service is included in the user's task scenario.
2 *One meta description value domain includes another one:*
   2.1. $Mn(MD_{lq}) \subset Mn(MD_{iv})$. In this case, domain of the q-th meta description value of l-th web service includes domain of v-th meta description value of i-th user's task. Thus, specified web service is included in the user's task scenario.
   2.2. $Mn(MD_{iv}) \subset Mn(MD_{lq})$. In this case, domain of the v-th meta description value of i-th user's task includes domain of q-th meta description value of l-th web service. Thus, specified web service is included in the user's task scenario, but analysis of the meta description value for other web services continues in order to find such web service, which meta description value domain in combination with specified web service meta description value domain will form user's task meta description values domain: $Mn(MD_{lq}) \cup Mn(MD_{hd}) = Mn(MD_{iv})$.
3 *Meta description value domains overlap* - $Mn(MD_{iv}) \cap Mn(MD_{lq})$. In this case, domain of the v-th meta description value of i-th user's task overlaps domain of q-th meta description value of l-th web service. Thus, specified web service is included in the user's task scenario, but analysis of the meta description value for other web services continues in order to find such web service, which meta description value domain in combination with specified web service meta description value domain will form user's task meta description values domain: $Mn(MD_{lq}) \cup Mn(MD_{hd}) = Mn(MD_{iv})$.

4 *Meta description value domains don't overlap* – $Mn(MD_{lq}) \neq Mn(MD_{iv})$. In this case, domain of the v-th meta description value of i-th user's task matches domain of q-th meta description value of l-th web service. Thus, specified web service is not included in the user's task scenario.

Stage 2. Scenario generation.

This step can be regarded as an intermediate. At this stage, the sequence of the web services selected at the previous stage is formed into the scenario. Scenario formation involves redundancy and duplication verification. While the formation of WSs sequence is held it is in need to verify web services input and output data. If the result of any WS is the input data to another WS, then first WS should be placed before the second WS in the sequence.

Stage 3. User's task execution.

This stage is the stage of user's task direct execution. At this stage, the execution of all web services included into scenario is made using appropriate software.

5   The Estimation of the Web Services Composition Efficiency

Suggested approach was used for scenarios that represents the user's tasks in subject domain "Strength of materials" formation [19, 20]. To confirm the efficiency of the proposed method the test group of complex tasks for problem domain "Strength of materials" was selected (Table 2), which consists of five complex tasks. Each of complex tasks consists of several partial tasks (each partial tasks corresponds to certain web service).

Let us show the example of complex task from the test group formation using proposed method. This task is "Calculation of the strength for the power components of magnetic systems". The system of meta description for such user's task is shown in Table 3.

At the 1-st stage the analysis of WSs set that represents partial tasks was done. It was in need to choose from this set only those WSs that satisfies condition (6). Such WS are:

Table 2. Test group of complex tasks

№	The title of user's task	The amount of stages (WSs) in each user's task
1	Calculation of the strength for the power components of magnetic systems	56
2	Calculation of the strength for equipment and pipelines of nuclear power systems	28
3	ITER Structural Design Criteria for magnetic components	33
4	Magnet DDD 1.1–1.3. Magnet System Design Criteria	48
5	Calculation of the strength for the elements of equipment and pipelines of ship nuclear steam generating systems with a water reactors	18

Table 3. Meta description for Calculation of the strength for the power components of magnetic systems

Meta description title	Meta description value
Title	Calculation of the strength for the power components of magnetic systems
Identifier	1
Subject field	Strength of materials
Description	The calculation, carried out for the cylinder formed by the internal resistance of toroidal field coils
Data	Data, formulas
Regulatory	PNAEG-7-002-86
Address	http\\calc266\19.portal.ua
Parameters	$a_1 = 10$ – allowable pressure $E_2 = 52$ – average loading $E_1 = 48$ – basic loading $R = 5$ – winding radius $c = 10$ – the length of connection between construction body and winding $l = 15$ –critical length $D_m = < 10...12 > =$ maximal diameter $P_i = 18$ – dimension of construction body $\eta = 20$ – critical pressure $\phi = 5$ – allowable tension pressure $v = 28$ – winding length $d = 44$ – winding strength $e = 6$ – constant of static level $M = 12$ – general membrane strength
Loading	–
Methods	–
Element	Constructions shell, constructions corps
Measurement area	Cross section, length

WS_1 – for task "calculation of basic parameters";
WS_2 – for task "verifying calculation"
WS_3 – for task "strength calculation"
WS_4 – for task "critical constant calculation"
WS_5 – for task "resistance calculation"
WS_6 – for task "dimensioning of reactors coil corps with arched slit"
WS_7 – for task "calculation of reducing the strength of welds"
WS_8 – for task "calculation of flanges, rings and fasteners"
WS_9 – for task "calculation of static strength"
WS_{10} – for task "calculation of stability"
WS_{11} – for task "calculation of cyclic strength"
WS_{12} – for task "calculation of crack resistance"
WS_{13} – for task "calculation of voltage variation"

WS_{14} – for task "calculation of seismic impacts"
WS_{15} – for task "calculation of the vibration strength"
WS_{16} – for task "calculation of the limiting value of deformation"
WS_{17} – for task "calculation of the vibration stability"

At 2-nd stage, the comparison of WSs meta descriptions meanings were held according to the rule (7). Some WSs were discarded from the set formed on the first stage, as some of their MD are out of range of values with values of the same parameters of task. The following WSs were discarded:

WS_4 – for task "critical constant calculation"
WS_{10} – for task "calculation of stability"
WS_{11} – for task "calculation of cyclic strength"
WS_{13} – for task "calculation of voltage variation"
WS_{17} – for task "calculation of the vibration stability"

The estimation of the time required to develop scenarios for complex tasks was made. The estimation was made for static way of WSs composition and for their composition with the help of proposed method (Fig. 4).

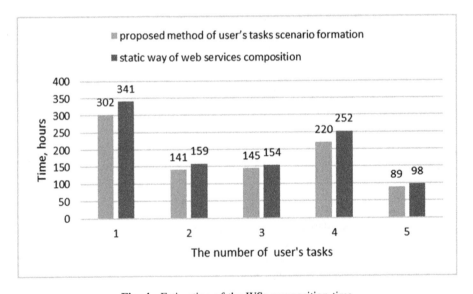

Fig. 4. Estimation of the WSs composition time

Time needed for services development with the help of static way WSs composition is 1004 h and with the help of proposed method is 897 h. Thus, the proposed method allowed to reduce the time for scenario for complex task development by 11 %.

6 Conclusions

The paper presents an approach to web services composition. The method of engineering tasks composition is proposed that allows to integrate web services into a scenario based on the web services meta descriptions comparison. The usage of such method makes it possible to execute real end-user's tasks based on dynamically formed sequence of web services.

Future work is aimed on implementation of suggested approach to different types of uses tasks. This will allow to approve its applicability and efficiency on real world scenarios. Developed software tool will be tested and verified on real world scenarios when engineering knowledge portals are developed. Quantitative evaluation of the proposed approach and tool efficiency will be obtained for different subject domains: the average time of web services composition, their correctness and quality will be validated.

References

1. Pukhkaiev, D., Kot, T., Globa, L., Schill, A.: A novel SLA-aware approach for web service composition. In: IEEE EUROCON, pp. 327–334 (2013)
2. Dustdar, S., Schreiner, W.: A survey on web services composition. Int. J. Web Grid Serv. **1** (1), 1–30 (2005)
3. Sheng, Q.Z., Qiao, X., Vasilakos, A.V., Szabo, C., Boume, S., Xu, X.: Web services composition: a decade's overview. Inf. Sci. **280**, 218–238 (2014)
4. Moghaddam, M., Davis, J.G.: Service selection in web service composition: a comparative review of existing approaches. In: Bouguettaya, A., Sheng, Q.Z., Daniel, F. (eds.) Web Services Foundations, pp. 321–346. Springer, New York (2014)
5. Shehu, U., Epiphaniou, G., Safdar, G.A.: A survey of QoS-aware web service composition techniques. Int. J. Comput. Appl. **89**(12), 10–17 (2014)
6. Upadhyaya, B.: Composing heterogeneous services from end users' perspective (2014)
7. McIlraith, S.A., Son, T.C., Zeng, H.: Semantic web services. IEEE Intell. Syst. **2**, 46–53 (2001)
8. Medjahed, B., Bouguettaya, A.: Service Composition for the Semantic Web. Springer, New York (2011)
9. Martin, D., Paolucci, M., McIlraith, S.A., Burstein, M., McDermott, D., McGuinness, D.L., Parsia, B., Payne, T.R., Sabou, M., Solanki, M., Srinivasan, N., Sycara, K.: Bringing semantics to web services: the OWL-S approach. In: Cardoso, J., Sheth, A.P. (eds.) SWSWPC 2004. LNCS, vol. 3387, pp. 26–42. Springer, Heidelberg (2005)
10. Miller, J., Verma, K., Rajasekaran, P., Sheth, A., Aggarwal, R., Sivashanmugam, K.: Wsdl-s: adding semantics to wsdl-white paper. LSDIS Lab, University of Georgia, Georgia, USA (2004)
11. De Oliveira Jr, F.G.A., de Oliveira, J.M.P.: QoS-based approach for dynamic web service composition. J. UCS **17**(5), 712–741 (2011)
12. Papazoglou, M.P., Van Den Heuvel, W.J.: Service oriented architectures: approaches, technologies and research issues. VLDB J. **16**(3), 389–415 (2007)

13. Hatzi, O., Vrakas, D., Nikolaidou, M., Bassiliades, N., Anagnostopoulos, D., Vlahavas, I.: An integrated approach to automated semantic web service composition through planning. IEEE Trans. Serv. Comput. **5**(3), 319–332 (2012)
14. Ngan, L.D., Kanagasabai, R.: Semantic web service discovery: stateof-the-art and research challenges. Pers. Ubiquit. Comput. **17**(8), 1741–1752 (2013)
15. Maximilien, E.M., Singh, M.P.: A framework and ontology for dynamic web services selection. IEEE Internet Comput. **8**, 84–93 (2004)
16. Blake, M.B., Cummings, D.J.: Workflow composition of service-level agreements. In: Proceedings of IEEE International Conference on Services Computing (SCC), pp. 138–145, July 2007
17. Andrieux, A., Czajkowski, K., Dan, A., Keahey, K., Ludwig, H., Nakata, T., Pruyne, J., Rofrano, J., Tuecke, S., Xu, M.: Web Services Agreement Specification (WS-Agreement) (2007). http://www.ogf.org/documents/GFD.107.pdf
18. Cardoso, J., Sheth, A.: Semantic Web Services, Processes and Applications. Springer, New York (2006)
19. Norms for the strength calculation for the power components of magnetic systems. – Kyiv: ISP – 73 (1984)
20. Norms for the strength calculation for equipment and pipelines of nuclear power systems – M.: Energoatomizdat, – 525 (1989)

Loop Nest Tiling for Image Processing and Communication Applications

Wlodzimierz Bielecki$^{(\boxtimes)}$ and Marek Palkowski

Faculty of Computer Science and Information Systems, West Pomeranian University
of Technology in Szczecin, Zolnierska 49, 71210 Szczecin, Poland
{wbielecki,mpalkowski}@wi.zut.edu.pl
http://www.wi.zut.edu.pl

Abstract. Loop nest tiling is one of the most important loop nest optimizations. This paper presents a practical framework for automatic tiling of affine loop nests to reduce time of application execution which is crucial for the quality of image processing and communication systems. Our framework is derived via a combination of the Polyhedral and Iteration Space Slicing models and uses the transitive closure of loop nest dependence graphs. To describe and implement the approach in the source-to-source TRACO compiler, loop dependences are presented in the form of tuple relations. We expose the applicability of the framework to generate tiled code for image analysis, encoding and communication program loop nests from the UTDSP Benchmark Suite. Experimental results demonstrate the speed-up of optimized tiled programs generated by means of the approach implemented in TRACO.

Keywords: Optimizing compilers · Tiling · Transitive closure · Data locality · Image processing · Communication

1 Introduction

Tiling is a very important iteration reordering transformation for improving data locality in software code. Tiling for improving locality groups loop statement instances in a loop iteration space into smaller blocks (tiles) allowing reuse when the block fits in local memory.

To our best knowledge, well-known tiling techniques are based on linear or affine transformations of program loop nests [7,11,13,15,25]. In paper [5], we presented a novel approach for generating tiled code for affine loop nests which is based on the transitive closure of a program dependence graph. The proposed approach allows producing tiled code even when there does not exist an affine transformation allowing for producing a fully permutable loop nest.

In this paper, we examine loop nest tiling algorithms for improving data locality in DSP applications. We consider loop nests of image analysis, encoding, and communication applications from the UTDSP benchmark suite [26]. Experimental results, illustrating the achieved speedup of tiled code generated are presented.

© Springer International Publishing AG 2017
S. Kobayashi et al. (eds.), *Hard and Soft Computing for Artificial Intelligence, Multimedia and Security*, Advances in Intelligent Systems and Computing 534, DOI 10.1007/978-3-319-48429-7_28

2   Image Processing and Communication Algoritmhs Implemented in the UTDSP Benchmark Suite

The UTDSP Benchmark Suite [26] was created in 1992 at the University of Toronto to evaluate the quality of code generated by a high-level language (such as C) compiler targeting a programmable digital signal processor (DSP). This evaluation was used to drive the development of specific compiler optimizations to improve the quality of generated code and to modify the architecture of the target processor to simplify compiler's task [19].

A C compiler is used to translate target applications into machine operations that can be executed by the model architecture. The C compiler generates sequential code and performs register allocation based on an instruction set and the number of registers defined for the model architecture. A post-optimizer is then used to exploit the DSP-specific features of the model architecture. The post-optimizer also exploits parallelism in sequential code and creates executable code that runs on the model architecture. The code is provided in multiple styles (usage of arrays versus pointers), but this is now irrelevant to modern compilers that produce the same performance level for all styles [19].

UTDSP benchmarks contain compute-intensive DSP kernels as well as applications composed of more complex algorithms and data structures [19]. DSP kernels are simply small code fragments that represent important calculations in DSP applications. Examples of DSP kernels are: fast Fourier transform, matrix multiply, finite and infinite impulse response and other filters.

DSP applications, on the other hand, are entire programs that would be executed on a DSP in a commercial product. Using the suite of benchmarks with the application-driven design methodology thus makes the resulting model architecture an ideal design for embedded DSP processors. Examples of DSP applications are: linear predictive coding, image compression, edge detection, histogram equalization, and an implementation of speech encoders.

In this paper, we demonstrate how to improve the locality of UTDSP benchmark code by means of the loop nest tiling algorithms based on the transitive closure of dependence graphs.

3   Loop Tiling Based on the Transitive Closure of Loop Nest Dependence Graphs

3.1   Background

In this paper, we deal with affine loop nests where, for given loop indices, lower and upper bounds as well as array subscripts and conditionals are affine functions of surrounding loop indices and possibly of structure parameters (defining loop index bounds), and the loop steps are known constants.

To implement tiling algorithms, we have chosen the dependence analysis proposed by Pugh and Wonnacott [22], where dependences are represented by dependence relations. A dependence relation is a tuple relation of the form *[input list]*→*[output list]*: *formula*, where *input list* and *output list* are the lists of variables and/or expressions used to describe input and output tuples, and *formula* describes the constraints imposed upon input and output lists and it is a Presburger formula built of constraints represented by algebraic expressions and using logical and existential operators [22].

Standard operations on relations and sets are used, such as intersection ($\cap$), union ($\cup$), difference (-), domain (dom R), range (ran R), relation application ($S' = R(S) : e' \in S'$ iff exists e s.t. $e \to e' \in R, e \in S$). In detail, the description of these operations is presented in [22,23].

$$R^+ = \{e \to e' : e \to e' \in R \vee \exists e'' \text{s.t. } e \to e'' \in R \wedge e'' \to e' \in R^+\}. \quad (1)$$

It describes which vertices e' in a dependence graph (represented by relation R) are connected directly or transitively with vertex e.

Transitive closure, R^*, is defined as follows [14]:

$$R^* = R^+ \cup I, \quad (2)$$

where I is the identity relation. It describes the same connections in the dependence graph (represented by R) that R^+ does plus connections of each vertex with itself.

For carrying out experiments, the ISL function `isl_map_transitive_closure` has been used for calculating transitive closure [23].

In sequential loops, the iteration i executes before j if i is *lexicographically less* than j, denoted as $i \prec j$, i.e., $i_1 < j_1 \vee \exists k \geq 1 : i_k < j_k \wedge i_t = j_t, for\ t < k$.

3.2   Loop Nest Tiling Based on the Transitive Closure of Dependence Graphs

In this paper, to generate valid tiled code, we apply the approach presented in paper [5], which is based on the transitive closure of dependence graphs.

Let us briefly remind the steps of that approach which envisages building the following sets:

- *TILE(**II**, B)* includes iterations belonging to a parametric tile, as follows
  $TILE(\mathbf{II}, \mathbf{B}) = \{[\mathbf{I}] \mid \mathbf{B}^*\mathbf{II} + \mathbf{LB} \leq \mathbf{I} \leq \min(\mathbf{B}^*(\mathbf{II} + 1) + \mathbf{LB} - 1, \mathbf{UB}) \wedge \mathbf{II} \geq 0\}$, where vectors **LB** and **UB** include the lower and upper loop index bounds of the original loop nest, respectively; diagonal matrix **B** defines the size of a rectangular original tile; elements of vector **I** represent the original loop nest iterations contained in the tile whose identifier is **II**; **1** is the vector whose all elements have value 1,[1]
- *TILE_LT* is the union of all the tiles whose identifiers are lexicographically less than that of *TILE(**II**, B)* : $TILE\_LT = \{[\mathbf{I}] \mid \exists\ \mathbf{II}'\ \text{s. t. } \mathbf{II}' \prec \mathbf{II} \wedge \mathbf{II} \geq 0$ AND $\mathbf{B}^*\mathbf{II} + \mathbf{LB} \leq \mathbf{UB} \wedge \mathbf{II}' \geq 0$ and $\mathbf{B}^*\mathbf{II}' + \mathbf{LB} \leq \mathbf{UB} \wedge \mathbf{I}$ in $TILE(\mathbf{II}', \mathbf{B})\}$,

---

[1] The notation $x \geq (\leq) y$ where x, y are two vectors in $\mathbb{Z}^n$ corresponds to the component-wise inequality, that is, $x \geq (\leq) y \Longleftrightarrow x_i \geq (\leq) y_i$, i=1,2,...,n.

- $TILE\_GT$ is the union of all the tiles whose identifiers are lexicographically greater than that of $TILE(\boldsymbol{II}, \boldsymbol{B})$: $TILE\_GT =\{[\boldsymbol{I}] \mid \exists\ \boldsymbol{II}$'s. t. $\boldsymbol{II'} \succ \boldsymbol{II} \wedge \boldsymbol{II} \geq 0$ AND $\boldsymbol{B*II}+\boldsymbol{LB} \leq \boldsymbol{UB} \wedge \boldsymbol{II'} \geq 0$ and $\boldsymbol{B*II'}+\boldsymbol{LB} \leq \boldsymbol{UB} \wedge \boldsymbol{I}$ in $TILE(\boldsymbol{II'}, \boldsymbol{B})\}$,[2]
- $TILE\_ITR = TILE - R^+(\ TILE\_GT\ )$, which does not include any invalid dependence target, i.e., it does not include any dependence target whose source is within set $TILE\_GT$,
- $TVLD\_LT = (R^+\ (TILE\_ITR) \cap TILE\_LT) - R^+(TILE\_GT)$ includes all the iterations that (i) belong to the tiles whose identifiers are lexicographically less than that of set $TILE\_ITR$, (ii) are the targets of the dependences whose sources are contained in set $TILE\_ITR$, and (iii) are not any target of a dependence whose source belong to set $TILE\_GT$,
- $TILE\_VLD = TILE\_ITR \cup TVLD\_LT$ defines target tiles,
- $TILE\_VLD\_EXT$ is built by means of inserting (i) into the first positions of the tuple of set $TILE\_VLD$ elements of vector $\boldsymbol{II}$: $ii_1, ii_2, ..., ii_d$; (ii) into the constraints of set $TILE\_VLD$ the constraints defining tile identifiers $\boldsymbol{II} \geq 0$ and $\boldsymbol{B*II}+\boldsymbol{LB} \leq \boldsymbol{UB}$.

Target code is generated by means of applying any code generator allowing for scanning elements of set $TILE\_VLD\_EXT$ in the lexicographic order, for example, CLooG [1] or the ISL code generator [12].

3.3   An Illustrative Example from the UTDSP Benchmark Suite

In this section, we present the usage of the algorithm presented in our paper [5] extended with the algorithm outlined in the paper [17] to optimize the following loop nest from the UTDSP edge detector code, *Edge_detect_1*.

```
for ( i=0;  i<N;  i++)
 for ( j=0;  j<N;  j++){
  temp1 = image_buffer1 [i][j];
  temp2 = image_buffer2 [i][j];
  temp3 = temp1 + temp2;
  image_buffer3 [i][j] = temp3 + T;
 }
```

First, applying the approach (variable privatization), presented in paper [17], we eliminate some loop nest dependences. The following relations describe the reminding dependences in that program.

```
R1 := [i, j, 13] -> [i, j, 18] : i >= 0 and i <= 15+N and j >= 0
and j <= 15 + N;
R2 := [i, j, 15] -> [i, j, 18] : i >= 0 and i <= 15+N and j >= 0
and j <= 15 + N;
R3 := [N] -> [i, j, 18] -> [i, j, 20] : i >= 0 and i <= 15 + N
and j >= 0 and j <= 15 + N.
```

<hr>

[2] "$\prec$" and "$\succ$" denote the lexicographical relation operators for two vectors.

Below, we demonstrate how tiled code can be generated for the loop nest example. First, we define a rectangular parametric tile of the size 16x16 as follows.

TILE := [ii, jj, N] -> { [i, j, v] : i >= 16ii and i >= 0 and
i <= 15 + N and i <= 15 + 16ii and j >= 16jj and j >= 0 and
j <= 15 + N and j <= 15 + 16jj and ii >= 0 and jj >= 0 and
N >= -15 and (v = 13 or v = 15 or v = 18 or v = 20).

Next, we calculate the following sets:

TILE_LT := [ii, jj, N] -> { [i, j, v] : i >= 16ii and i >= 0
and i <= 15+ N and i <= 15 + 16ii and j >= 0 and j <= -1 + 16jj
and N >= -15 + 16jj and (v = 13 or v = 15 or v = 18 or v = 20)};

TILE_GT := [ii, jj, N] -> { [i, j, v] : i >= 16ii and i >= 0
and i <= 15 + N and i <= 15 + 16ii and j >= 16 + 16jj and
j <= 15 + N   and (v = 13 or v = 15 or v = 18 or v = 20)};

TILE_ITR := [N, ii, jj] -> { [i, j, v] : ii >= 0 and
jj >= 0 and i >= 16ii and i <= 15 + 16ii and i <= 15 + N and
j >= 16jjand j <= 15 + 16jj and j <= 15 + N and (v = 13 or
v = 15 or v = 18 or v = 20)};

TVLD_LT := [N, ii, jj] -> { [i, j, v] : 1 = 0 };
TILE_VLD := TILE_ITR;

TILE_VLD_ext := [N] -> { [i0, i1, i2, i3, v] : i0 >= 0 and i1 >= 0
and i2 >= 16i0 and i2 <= 15 + 16i0 and i2 <= 15 + N and i3 >= 16i1
and i3 <= 15 + 16i1 and i3 <= 15 + N and (v = 13 or v = 15 or
v = 18 or v = 20).

Applying CLooG [1] to set *TILE_VLD_EXT,* we generated the following tiled code.

```
for (c0 = 0; c0 <= (N − 1) / 16; c0++)
  for (c1 = 0; c1 <= (N − 1) / 16; c1++)
    for (c2 = 16 * c0; c2 <= min(N − 1, 16 * c0 + 15); c2++)
      for (c3 = 16 * c1; c3 <= min(N−1, 16 * c1 + 15); c3++) {
        temp1=image_buffer1[c2][c3];
        temp2=image_buffer2[c2][c3];
        temp3=temp1+temp2;
        image_buffer3[c2][c3]=temp3+T;
      }
```

4   Related Work

The literature presents a wide number of approaches to optimize and/or parallelize UTDSP programs by means of manual, semi-automatic and automatic techniques. Murray et al. presents code transformation and instruction set extension in paper [16]. Real-time partitioned scheduling on multi-core systems with local and global memories for UTDSP is outlined by Chang et al. in [8]. Pouchet et al. analyse iterative optimization in the polyhedral model [20] and performance distribution of affine schedules [21] for UTDSP programs. In paper [4], we presented an approach to parallelize UTDSP by means of Iteration Space Slicing [2] which is also based on the transitive closure of dependence graph, but without applying any loop nest tiling.

Well-known automatic code locality improvement algorithms are based on the polyhedral model, which provides an abstraction to perform high-level transformations such as loop-nest optimization on affine loop nests. The powerful polyhedral source to source tools: Pluto [7], POCC and PTile [18] transform C programs to expose parallelism and improve data locality simultaneously. The core transformation framework mainly works by finding affine transformations for efficient loop nest tiling and fusion, but not limited to those.

The polyhedral model approach includes the following three steps [7,9,10]:

1. program analysis aimed at translating source codes to their polyhedral representation with data dependence analysis,
2. program transformations with the aim of code optimization,
3. code generation.

All above three steps are available in the approach examined in this paper. But there exists the following difference in the second step, which is not represented by a set of affine functions [3]. Our approach does not use any set of affine functions [3], instead it applies the transitive closure of a program dependence graph to transform the original loop nest.

Some limitations of affine transformation are considered in papers [2,14,24]. The main drawback of the affine transformation framework is that there does not always exist solutions to the time-partition constraints. To enable loop nest tiling, our approach needs only the transitive closure (exact or approximated) of a dependence graph, so it lacks the drawbacks inherent to affine transformations.

5   Experimental Study

Tiling based on transitive closure has been implemented in the optimizing compiler TRACO, public available at the website traco.sourceforge.net. To extract dependences, the Petit tool [22] was used. Petit was able to find dependences in 32 UTDSP loop nests. We have excluded from those programs fourteen loop nests of depth 1 (tiling improves data locality only for loop nests of depth 2 and more)

Table 1. Execution times of the original and tiled programs from the UTDSP benchmark suite

Loop nest	Parameters	Times of program execution			
		Original	Tiled [16]	Tiled [32]	Tiled [64]
Matrix_multiply_1	N = 2000	1,245	0,839	0,722	1,059
	N = 3000	4,738	3,340	3,191	3,559
	N = 4000	7,054	6,132	6,006	6,412
Edge_detect_1	N = 1000	0,192	0,159	0,144	0,138
	N = 2000	0,472	0,364	0,334	0,402
	N = 3000	2,514	1,853	1,766	1,939
Edge_detect_3	N = 500, K=10	1,644	1,631	1,596	1,693
	N = 1000, K=10	4,203	3,916	3,759	3,999
	N = 1500, K=10	8,698	6,873	6,291	6,359
Compress_1	N = 1024	1,433	1,372	1,319	1,329
	N = 1536	2,217	1,912	1,843	1,885
	N = 2048	2,991	2,329	2,138	2,434

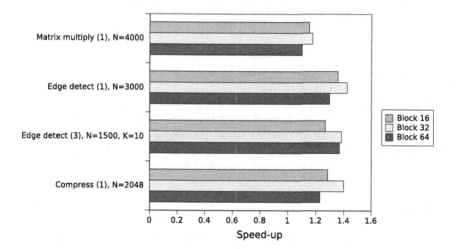

Fig. 1. Speedup of the studied UTDSP loop nests

and five fully sequential loop nests. So, we generated tiled code only for 13 UTDSP loop nests[3].

To examine generated loop nest speed-up, we chosen four arbitrary nested loop nests: one loop nest from the *Matrix Multiply* programs, two loop nests from the *Edge_detect* application, and one from the *Compress* application. For each

---

[3] Source codes with tiled version are available at the TRACO repository http://sourceforge.net/p/traco/code/HEAD/tree/trunk/examples/utdsp/.

examined loop nests, the exact transitive closure of a corresponding dependence graph was calculated and tiled code generated. To check the performance of tiled code, times of original and tiled code executions were examined for various sizes of problem presented in Table 1. Experiments were carried with the Intel Core i5-4670 processor (3.4 Ghz CPU, 6 MB Cache), 8 GB RAM and Ubuntu Linux. The programs were compiled by means of GCC 4.8.4 and the -O3 flag. The tiled codes were analysed with 16, 32 and 64 tile side. Table 1 presents execution times of the studied programs. Figure 1 illustrates positive speed-up $(S > 1)$, achieved for those programs, in a graphical way. Speed-up is the ratio of the original program execution time to the tiled program execution time. Analysing the results received, we may conclude that tiling, based on the transitive closure of dependence graphs, allows us to increase the locality of the examined programs. This in consequence leads to improving code performance.

6   Conclusion

In this paper, we demonstrated that the approach, based on the transitive closure of dependence graphs, is able to generate tiled code for image processing and communication applications. Experimental results confirm that tiling, applied to examined programs, improves code locality that allows us to achieve positive speed-up of those programs.

The main merit of applying transitive closure to tile loop nests is that it does not require full permutability of loops in a nest which is required by the Affine Transformation Framework [15, 25].

In the future, we are going to examine other benchmarks of image analysis, video encoding and communication using the presented approach. We also plan to improve code locality for digital signal processing programs including algorithms for extraction of synchronization-free parallelism and parallelism based on time partitions (free scheduling) presented in papers [2, 6].

References

1. Bastoul, C.: Code generation in the polyhedral model is easier than you think. In: PACT 2013 IEEE International Conference on Parallel Architecture and Compilation Techniques, pp. 7–16, Juan-les-Pins, September 2004
2. Beletska, A., Bielecki, W., Cohen, A., Palkowski, M., Siedlecki, K.: Coarse-grained loop parallelization: iteration space slicing vs affine transformations. Parallel Comput. **37**, 479–497 (2011)
3. Benabderrahmane, M.-W., Pouchet, L.-N., Cohen, A., Bastoul, C.: The polyhedral model is more widely applicable than you think. In: Gupta, R. (ed.) CC 2010. LNCS, vol. 6011, pp. 283–303. Springer, Heidelberg (2010). doi:10.1007/978-3-642-11970-5_16
4. Bielecki, W., Palkowski, M.: Coarse-grained loop parallelization for image processing and communication applications. In: Choraś, R.S. (ed.) Image Processing and Communications Challenges 2, pp. 307–314. Springer, Heidelberg (2010)

5. Bielecki, W., Palkowski, M.: Perfectly nested loop tiling transformations based on the transitive closure of the program dependence graph. In: Wiliński, A., Fray, I., Pejaś, J. (eds.) Soft Computing in Computer and Information Science. AISC, vol. 342, pp. 309–320. Springer, Heidelberg (2015). doi:10.1007/978-3-319-15147-2_26

6. Bielecki, W., Palkowski, M., Klimek, T.: Free scheduling for statement instances of parameterized arbitrarily nested affine loops. Parallel Comput. **38**(9), 518–532 (2012)

7. Bondhugula, U., Hartono, A., Ramanujam, J., Sadayappan, P.: A practical automatic polyhedral parallelizer and locality optimizer. SIGPLAN Not. **43**(6), 101–113 (2008)

8. Chang, C.W., Chen, J.J., Kuo, T.W., Falk, H.: Real-time partitioned scheduling on multi-core systems with local and global memories. In: 2013 18th Asia and South Pacific Design Automation Conference (ASP-DAC), pp. 467–472, January 2013

9. Feautrier, P.: Some efficient solutions to the affine scheduling problem: I. one-dimensional time. Int. J. Parallel Program. **21**(5), 313–348 (1992)

10. Feautrier, P.: Some efficient solutions to the affine scheduling problem: II. multi-dimensional time. Int. J. Parallel Program. **21**(5), 389–420 (1992)

11. Griebl, M.: Automatic parallelization of loop programs for distributed memory architectures (2004)

12. Grosser, T., Verdoolaege, S., Cohen, A.: Polyhedral AST generation is more than scanning polyhedra. ACM Trans. Program. Lang. Syst. **37**(4), 12:1–12:50 (2015)

13. Irigoin, F., Triolet, R.: Supernode partitioning. In: Proceedings of the 15th ACM SIGPLAN-SIGACT Symposium on Principles of Programming Languages, pp. 319–329, POPL 1988, NY, USA. ACM, New York (1988)

14. Kelly, W., Maslov, V., Pugh, W., Rosser, E., Shpeisman, T., Wonnacott, D.: The omega library interface guide. Technical report, College Park, MD, USA (1995)

15. Lim, A., Cheong, G.I., Lam, M.S.: An affine partitioning algorithm to maximize parallelism and minimize communication. In: Proceedings of the 13th ACM SIGARCH International Conference on Supercomputing, pp. 228–237. ACM Press (1999)

16. Murray, A.C., Bennett, R.V., Franke, B., Topham, N.: Code transformation and instruction set extension. ACM Trans. Embed. Comput. Syst. **8**(4), 26:1–26:31 (2009)

17. Palkowski, M.: Impact of variable privatization on extracting synchronization-free slices for multi-core computers. In: Keller, R., Kramer, D., Weiss, J.-P. (eds.) Facing the Multicore-Challenge III. LNCS, vol. 7686, pp. 72–83. Springer, Heidelberg (2013). doi:10.1007/978-3-642-35893-7_7

18. Park, E., Pouchet, L.N., Cavazos, J., Cohen, A., Sadayappan, P.: Predictive modeling in a polyhedral optimization space. In: 9th IEEE/ACM International Symposium on Code Generation and Optimization (CGO 2011), pp. 119–129. IEEE Computer Society press, Chamonix, France, April 2011

19. en Peng, S.H.: UTDSP: A VLIW programmable DSP processor. Thesis, University of Toronto (1999)

20. Pouchet, L.N., Bastoul, C., Cohen, A., Vasilache, N.: Iterative optimization in the polyhedral model: Part i, one-dimensional time and part ii, multi-dimensional time. In: International Symposium on Code Generation and Optimization (CGO 2007), pp. 144–156, March 2007

21. Pouchet, L.N., Bastoul, C., Cavazos, J., Cohen, A.: A note on the performance distribution of affine schedules. In: 2nd Workshop on Statistical and Machine Learning Approaches to Architectures and Compilation (SMART 2008), Göteborg, Sweden, January 2008

22. Pugh, W., Wonnacott, D.: An exact method for analysis of value-based array data dependences. In: Banerjee, U., Gelernter, D., Nicolau, A., Padua, D. (eds.) LCPC 1993. LNCS, vol. 768, pp. 546–566. Springer, Heidelberg (1994). doi:10. 1007/3-540-57659-2_31

23. Verdoolaege, S.: Integer set library - manual. Technical report (2011). www.kotnet. org/~skimo//isl/manual.pdf

24. Wonnacott, D.G., Strout, M.M., Wonnacott, D.G., Strout, M.M.: On the scalability of loop tiling techniques. In: Proceedings of the 3rd International Workshop on Polyhedral Compilation Techniques (IMPACT) (2013)

25. Xue, J.: On tiling as a loop transformation. Parallel Process. Lett. **7**(4), 409–424 (1997)

26. UTDSP benchmark suite. http://www.eecg.toronto.edu/~corinna/DSP/infrastruc ture/UTDSP.html

Tile Merging Technique to Generate Valid Tiled Code by Means of the Transitive Closure of a Dependence Graph

Włodzimierz Bielecki and Piotr Skotnicki$^{(\boxtimes)}$

Faculty of Computer Science, West Pomeranian University of Technology,
ul. Żołnierska 49, 71-210 Szczecin, Poland
{wbielecki,pskotnicki}@wi.zut.edu.pl
http://www.wi.zut.edu.pl

Abstract. The paper presents a novel approach for generation of parallel tiled code of arbitrarily nested loops whose original inter-tile dependence graphs contain cycles. We demonstrate that the problem of cyclic dependences can be reduced to the problem of finding strongly connected components of an inter-tile dependence graph, and then solved by merging tiles within each component. Our technique is derived via a combination of the Polyhedral Model and Iteration Space Slicing frameworks. The effectiveness and efficiency of generated code is verified by means of well-known linear algebra kernels from the PolyBench benchmark suite.

Keywords: Optimizing compilers · Tiling · Transitive closure · Parallel computing · Dependence graph · Locality enhancement

1 Introduction

Tiling [6] is a very important loop nest optimizing technique for locality enhancement and coarse-grained parallelism extraction, significantly improving program performance. The method involves aggregating statement instances, accessing a common address space, into smaller blocks (referred to as *tiles*), therefore allowing for an efficient use of hardware registers and cache memory. However, complex dependence patterns, resulting from the order of memory accesses in subsequent loop nest iterations, limit the range of possible reorderings and thus pose challenges to finding a valid code transformation. In particular, a tiling scheme must ensure that tiles can be executed atomically without violating data dependences, which is possible only if a corresponding inter-tile dependence graph is cycle-free. Various tiling-enabling strategies, from memory-expensive array expansion, to advanced transformations of iteration spaces, result in different amount and shape of tiles, thereby impacting the efficiency of generated code.

In this paper, we propose a tile merging algorithm to form a cycle-free inter-tile dependence graph. The approach involves partitioning an inter-tile dependence graph into strongly connected components, found by means of the application of the transitive closure of a dependence graph. Tiles, belonging to the same

© Springer International Publishing AG 2017
S. Kobayashi et al. (eds.), *Hard and Soft Computing for Artificial Intelligence, Multimedia and Security*, Advances in Intelligent Systems and Computing 534, DOI 10.1007/978-3-319-48429-7_29

component, are merged together, allowing for generation of valid resulting tiles, i.e., such that a corresponding inter-tile dependence graph is acyclic. Additionally, we demonstrate how to construct the free schedule of resulting tiles using an inter-tile dependence relation. Our approach is derived via a combination of the *Polyhedral Model* [4,5] and *Iteration Space Slicing* [11] frameworks.

The main merit of the presented approach, in comparison with well-known tiling techniques, is that it neither considers any affine transformations, nor it requires full permutability of loops to generate tiled code.

2   Background

In this paper, we deal with affine loop nests where lower and upper bounds, array subscripts, and conditionals are affine functions of surrounding loop indices and symbolic constants, and the loop steps are known constants.

The presented algorithm is based on a combination of the Polyhedral Model [4,5] and Iteration Space Slicing frameworks [11]. Let us remind that this approach includes the following steps: (i) program analysis aimed at translating high level codes to their polyhedral representation and providing a data dependence analysis based on this model, (ii) program transformation with the aim of improving program locality and/or parallelization, (iii) code generation.

A loop nest is *perfectly nested* if all its statements are surrounded by the same loops; otherwise the loop nest is *imperfectly nested.*

A *statement instance* S[I] is a particular execution of a loop statement S for a given iteration I. Two statement instances S1[I] and S2[J] are *dependent* if both access the same memory location and if at least one access is a write. S1[I] and S2[J] are called the source and the target of a dependence, respectively, provided that S1[I] is executed before S2[J]. The sequential ordering of statement instances, denoted S1[I] $\prec$ S2[J], is induced by the original execution ordering of iteration vectors, or by the textual ordering of statements if $I = J$.

An iteration vector can be represented by a k-integer tuple of loop indices in the $\mathbb{Z}^k$ iteration space. Consequently, a *dependence relation* is a mapping from tuples to tuples of the form { [*source*] $\rightarrow$ [*target*] | *constraints* }, where *source* defines dependence sources, *target* defines dependence targets, and *constraints* is a *Presburger formula* (built of affine equalities and inequalities, logical and existential operators) that imposes constraints on the variables and/or expressions within *source* and *target* tuples.

A dependence relation is a mathematical representation of a directed *data dependence graph* whose vertices correspond to loop statement instances while edges connect dependent instances.

It is often convenient to group related elements of a set – like instances of the same statement in particular – using a *named integer tuple* [13]. Typically, a name associated with a tuple is the same as the label of a corresponding statement (if the tuple represents an iteration vector), or as the name of a variable (if the tuple represents memory offsets in subsequent array dimensions).

For manipulating sets and relations, we use common operations, such as union ($\cup$), difference ($-$), composition ($\circ$), domain of a relation ($\mathrm{domain}(R)$), range of a relation ($\mathrm{range}(R)$).

The positive transitive closure of a given relation R, R^+, is defined as follows:

$$R^+ = \{\, [e] \to [e'] \mid [e] \to [e'] \in R \vee \exists e'' : [e] \to [e''] \in R \wedge [e''] \to [e'] \in R^+ \,\}. \quad (1)$$

It describes which vertices e' in a dependence graph (represented by relation R) are connected directly or transitively with vertex e.

The transitive closure, R^*, is defined as below:

$$R^* = R^+ \cup I, \quad (2)$$

where I is the identity relation. It describes the same connections in a dependence graph (represented by R) that R^+ does plus connections of each vertex with itself.

The application of relation R to set S, resulting in the range of relation R with domain S, is defined as follows:

$$R(S) = \{\, [e'] \mid \exists e \in S : [e] \to [e'] \in R \,\}. \quad (3)$$

The algorithm, presented in this paper, requires an exact representation of dependences and consequently an exact dependence analysis which detects a dependence if and only if it actually exists. The relation can be computed as a union of flow, anti- and output dependences, according to the formula [13]:

$$R = \left((RA^{-1} \circ WA) \cup (WA^{-1} \circ RA) \cup (WA^{-1} \circ WA) \right) \cap (S \prec S), \quad (4)$$

where RA/WA are read/write *access relations*, respectively, mapping an iteration vector to one or more memory locations, and S is the original *schedule* represented with a relation which maps an iteration vector of a statement to a corresponding multidimensional timestamp (i.e., a discrete time when the statement instance has to be executed). $S \prec S$ denotes a strict partial order of statement instances, computed as: $S^{-1} \circ (\{\, [e] \to [e'] \mid e \prec e' \,\} \circ S)$.

3   Extraction of Strongly Connected Components

The main idea of the presented approach is to merge all the tiles involved in each strongly connected component of a possibly unbounded/parameterized inter-tile dependence graph to get one resulting tile. Let us briefly remind the basic definitions related to graph theory.

Definition 1. A *strongly connected component* (SCC) of a directed graph is a maximal subset of vertices such that there is a directed path from each vertex to all others in the same component.

Given a dependence graph $G = (V, R)$, where V is a set of vertices and R is a Presburger relation capturing dependences between vertices thus modeling directed edges of the graph, we exploit the above definition to form the following relation allowing us to find the pairs of vertices sharing the same strongly connected component:

$$T\_CYCLE = \{ [e] \rightarrow [e'] \mid [e] \rightarrow [e'] \in R^+ \wedge [e'] \rightarrow [e] \in R^+ \wedge e \prec e' \}, \quad (5)$$

where R^+ is the positive transitive closure of relation R, and e, e' denote vertices.

We will refer to each strongly connected component using the lexicographically smallest vertex it includes. For this reason, relation T_CYCLE is constrained by the lexicographical inequality $(e \prec e')$ which enforces that each input tuple is lexicographically less than its target. Such a condition guarantees that the lexicographically smallest vertex of each component will be contained only in the domain of the relation. Therefore, the set of all SCC representatives can be computed as follows:

$$REPR = V - \text{range}(T\_CYCLE). \quad (6)$$

Eventually, we restore vertices included in a strongly connected component represented by $e \in REPR$, using the below formula:

$$SCC(e) = T\_CYCLE^*(\{[e]\}). \quad (7)$$

It is worth to note that set $REPR$ includes representatives of all strongly connected components, hence $T\_CYCLE^*(REPR) = V$.

4   Tiling Algorithm

In this section, we first present a tiling idea and finally a formal tiling algorithm. For that purpose, let us consider the following illustrative example:

Example 1

```
for (i = 0; i < 4; ++i) {
    for (j = 0; j <= i; ++j) {
S1:    A[i] = A[i] + A[j] * 5;
    }
S2: A[i] = A[i] + 1;
    }
```

The loop nest contains two statements – S1 and S2. As the first step, we translate the source code of the loop nest into its polyhedral representation using the *Polyhedral Extraction Tool* [14] which returns the following output:

$$\bigcup_{i=1}^{2} LD_i := \{ S1[i, j] \mid i \leq 3 \wedge 0 \leq j \leq i \} \cup \{ S2[i] \mid 0 \leq i \leq 3 \},$$

$$S := \{\, S1[i,j] \rightarrow [i,0,j]\,\} \cup \{\, S2[i] \rightarrow [i,1,0]\,\},$$

$$RA := \{\, S1[i,j] \rightarrow A[i] \mid 0 \le i \le 3 \wedge 0 \le j \le i\,\}$$
$$\cup \{\, S1[i,j] \rightarrow A[j] \mid 0 \le i \le 3 \wedge 0 \le j \le i\,\}$$
$$\cup \{\, S2[i] \rightarrow A[i] \mid 0 \le i \le 3\,\},$$

$$WA := \{\, S1[i,j] \rightarrow A[i] \mid 0 \le i \le 3 \wedge 0 \le j \le i\,\}$$
$$\cup \{\, S2[i] \rightarrow A[i] \mid 0 \le i \le 3\,\}.$$

Initially, we tile the iteration space of each statement of the loop nest domain $(LD_i, i = 1, 2)$ using rectangular blocks of the size 2×2. From a mathematical perspective, those original rectangular tiles can be represented by a parametric set $TILE_i, i = 1, 2$, that defines the instances of statement $Si, i = 1, 2$, included in a tile identified by the values of symbolic constants:

$$TILE_1 := [ii, jj] \rightarrow \{\, S1[i,j] \mid 2ii \le i \le 3 \wedge i \le 1 + 2ii \wedge j \ge 2jj$$
$$\wedge\ 0 \le j \le i \wedge j \le 1 + 2jj\,\},$$

$$TILE_2 := [ii] \rightarrow \{\, S2[i] \mid 0 \le ii \le 1 \wedge 2ii \le i \le 1 + 2ii\,\}.$$

With each set $TILE_i, i = 1, 2$, we associate another set, $II\_SET_i, i = 1, 2$, that includes the identifiers of corresponding tiles:

$$II\_SET_1 := \{\, [ii, jj] \mid 0 \le ii \le 1 \wedge 0 \le jj \le 1\,\},$$

$$II\_SET_2 := \{\, [ii] \mid 0 \le ii \le 1\,\}.$$

Following (4), we carry out a data dependence analysis over the read (RA) and write (WA) memory accesses, which produces the following set of relations:

$$\bigcup_{i=1}^{2} \bigcup_{j=1}^{2} R_{i,j} := \{\, S2[i] \rightarrow S1[i', j = i] \mid i \ge 0 \wedge i < i' \le 3\,\}$$
$$\cup \{\, S1[i,j] \rightarrow S2[i' = i] \mid i \le 3 \wedge 0 \le j \le i\,\}$$
$$\cup \{\, S1[i,j] \rightarrow S1[i', j' = i] \mid 0 \le j \le i \wedge i < i' \le 3\,\}$$
$$\cup \{\, S1[i,j] \rightarrow S1[i' = i, j'] \mid i \le 3 \wedge j \ge 0 \wedge j < j' \le i\,\}.$$

The dependences and iteration spaces of the working example, tiled with rectangular blocks of the size 2×2, are presented in Fig. 1a. As can be seen from the figure, a dependence source can belong to one iteration space while its corresponding dependence target can reside in a distinct space, for example, $S1[0,0] \rightarrow S2[0]$. To make computations on sets and relations feasible, and be able to handle separate statements uniformly, all the sets and relations that are subject to further processing need to be *normalized*. A normalization procedure involves extending corresponding tuples with extra dimensions, that will (i) make tuple sizes to be equal, (ii) allow us to unambiguously identify which statement a given normalized tuple refers to. In practice, to make sets and/or relations to be normalized, we can apply a scheduling function S computed by

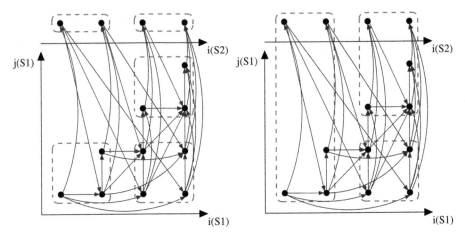

(a) Iteration space with initial 2x2 tiles.   (b) Merged tiles of the working example.

Fig. 1. Dependences and the tiled iteration space before and after merging.

the Polyhedral Extraction Tool [14] to tuples of these sets and/or relations. Consequently, we extend the parameter space of tile identifiers by introducing an additional symbolic constant for each fixed element of the normalized tuple of set $TILE_i$, $i = 1, 2$, followed by equating the corresponding dimensions.

That is, by applying schedule S to the tuples of sets $TILE_1$, $TILE_2$, and of relation R, and normalizing the parameter space of tile identifiers, we obtain:

$$TILE := [ii, kk, jj] \rightarrow \{ [i, k = 0, j] \mid kk = 0 \wedge 2ii \leq i \leq 3 \wedge i \leq 1 + 2ii$$
$$\wedge j \geq 2jj \wedge 0 \leq j \leq i \wedge j \leq 1 + 2jj \}$$
$$\cup [ii, kk, jj] \rightarrow \{ [i, k = 1, j = 0] \mid kk = 1 \wedge jj = 0 \wedge 0 \leq ii \leq 1$$
$$\wedge 2ii \leq i \leq 1 + 2ii \},$$

$$II\_SET := \{ [ii, kk = 0, jj] \mid 0 \leq ii \leq 1 \wedge 0 \leq jj \leq 1 \}$$
$$\cup \{ [ii, kk = 1, jj = 0] \mid 0 \leq ii \leq 1 \},$$

$$R := \{ [i, 0, j] \rightarrow [i', 0, i] \mid 0 \leq j \leq i \wedge i < i' \leq 3 \}$$
$$\cup \{ [i, 0, j] \rightarrow [i, 0, j'] \mid i \leq 3 \wedge j \geq 0 \wedge j < j' \leq i \}$$
$$\cup \{ [i, 1, 0] \rightarrow [i', 0, i] \mid i \geq 0 \wedge i < i' \leq 3 \}$$
$$\cup \{ [i, 0, j] \rightarrow [i, 1, 0] \mid i \leq 3 \wedge 0 \leq j \leq i \}.$$

We are now interested in constructing a relation, R_TILE, describing inter-tile dependences: a dependence between tiles exists iff there exists a data dependence such that its source originates from a statement instance within one tile and targets a statement instance included in another tile. For this purpose, we adapt the idea presented in paper [9] to form the following relation:

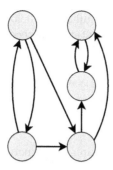

Fig. 2. Inter-tile dependence graph for the working example.

$$R\_TILE := \{\, [ii, kk, jj] \to [ii', kk', jj'] \mid kk \geq 0 \wedge 0 \leq jj \leq ii \wedge ii \leq ii' \leq 1$$
$$\wedge \; -kk - ii' < kk' \leq 1 - kk \wedge kk' \leq 1 + ii - ii' \wedge jj' \geq ii - kk'$$
$$\wedge \; 0 \leq jj' \leq 1 - jj \wedge jj' \leq 1 - kk' \wedge jj' \leq ii \,\}.$$

The inter-tile dependences can be illustrated by means of a directed graph whose vertices represent tiles and edges indicate the existence of at least one dependence which originates from a statement instance contained in one tile, and targets a statement instance contained in another tile. Figure 2 presents the inter-tile dependence graph, connecting dependent tiles of the working example. As we can see, the inter-tile dependence graph is cyclic. As a consequence, it is not possible to construct a scheduling function that would allow for atomic execution of the formed tiles without violating data dependences among statement instances in the graph represented with relation R_TILE. We would therefore like to eliminate all cycles within relation R_TILE.

Let us note that the inter-tile dependence graph is composed of a set of vertices II_SET, connected by edges described by relation R_TILE. Following this observation, we apply (5) to find all cyclic relations:

$$T\_CYCLE := \{\, [ii, kk = 0, jj = ii] \to [ii' = ii, kk' = 1, jj' = 0] \mid 0 \leq ii \leq 1 \,\}.$$

Based on the outcome, we conclude that there exist two pairs of tiles involved in cycles, $\{\, [1, 0, 1] \to [1, 1, 0], [0, 0, 0] \to [0, 1, 0] \,\}$. This allows us to extract the set of representatives of strongly connected components. Subtracting the range of relation T_CYCLE from II_SET (6), we obtain the following set:

$$II\_SET\_M := \{\, [ii, kk = 0, jj] \mid 0 \leq ii \leq 1 \wedge 0 \leq jj \leq 1 \,\}.$$

Enumerating the elements of set II_SET_M, we find the 3 non-empty strongly connected components, represented by tiles with identifiers $\{\, [1, 0, 1],$ $[1, 0, 0], [0, 0, 0] \,\}$. Eventually, for each SCC, we merge all its tiles into one tile. That is, we construct a new set of tiles, $TILE\_M$, that for given values of symbolic constants (ii, kk, jj), denoting an SCC representative, defines the statement instances of all the original rectangular tiles belonging to the same strongly connected component as that representative in accordance with (7):

$$TILE\_M := [ii, kk, jj] \rightarrow \{ [i, k = 0, j] \mid kk = 0 \wedge ii \leq 1 \wedge 2ii \leq i \leq 1 + 2ii$$
$$\wedge \; j \geq 2jj \wedge 0 \leq j \leq i \wedge j \leq 1 + 2jj \}$$
$$\cup \; [ii, kk, jj] \rightarrow \{ [i, k = 1, j = 0] \mid kk = 0 \wedge jj = ii$$
$$\wedge \; 0 \leq ii \leq 1 \wedge 2ii \leq i \leq 1 + 2ii \}.$$

A new inter-tile dependence graph, constructed based on relations connecting merged tiles, can be described by the following relation:

$$R\_TILE\_M := \{ [ii, kk = 0, jj = 0] \rightarrow [ii' = 1, kk' = 0, jj' = ii] \mid 0 \leq ii \leq 1 \}.$$

Let us highlight that the graph represented by R_TILE_M does not contain any cycles.

Figure 1b illustrates the final tiling scheme produced by the presented idea for the working example. If relation R_TILE_M is lexicographically-forward (i.e., $\forall x \rightarrow y \in R\_TILE\_M, y - x \succ 0$), we can generate code scanning tile identifiers and their iteration points in the lexicographic order. Utilizing the *Integer Set Library* [12], the following tiled code is produced:

```
for (ii = 0; ii <= 1; ii += 1)
   for (jj = 0; jj <= ii; jj += 1)
      for (i = 2 * ii; i <= 2 * ii + 1; i += 1) {
         for (j = 2 * jj; j <= min(2 * jj + 1, i); j += 1)
            A[i] = (A[i] + (A[j] * 5));
         if (jj == ii)
            A[i] = (A[i] + 1);
      }
```

Algorithm 1 provides a formal description of the presented idea and details the procedure for finding the free schedule for tile execution.

5   Results of Experiments

In order to evaluate the effectiveness and efficiency of code generated by means of the presented approach, the algorithm has been implemented into the *TC* optimizing compiler[1]. TC is a source-to-source compiler which utilizes a state-of-the-art polyhedral compilation toolchain, including the Polyhedral Extraction Tool [14] for extracting the polyhedral model of transformed loops, and the Integer Set Library [12] for performing dependence analysis, manipulating integer sets and relations, and generating output code.

Experiments were conducted on a highly-parallel machine (2x Intel Xeon E5-2699 v3 clocked at 2.3 GHz, 32 KB L1 data cache, 256 KB L2 cache, 256 GB

---

[1] http://tc-optimizer.sourceforge.net.

Algorithm 1. Tile merging for arbitrarily nested parameterized affine loops.

Input: Arbitrarily nested affine loops.
Output: Parallel tiled code.

1. Transform the loop nest into its polyhedral representation including: an iteration space, access relations, and global schedule S.
2. For each i, $i = 1, 2, ..., q$, and d_i, where q is the number of loop nest statements, and d_i is the number of loops surrounding statement Si, form the following vectors, matrix and sets:
   - vector I_i whose elements are original loop indices $i_1, i_2, ..., i_{d_i}$;
   - vector II_i whose elements $ii_1, ii_2, ..., ii_{d_i}$ define the identifier of a tile;
   - vectors LB_i and UB_i whose elements are lower $lb_1, ..., lb_{d_i}$ and upper $ub_1, ..., ub_{d_i}$ bounds of indices $i_1, i_2, ..., i_{d_i}$ of the enclosing loops, respectively;
   - vector 1_i whose all d_i elements are equal to the value 1;
   - vector 0_i whose all d_i elements are equal to the value 0;
   - diagonal matrix B_i whose diagonal elements are constants $b_1, b_2, ..., b_{d_i}$ defining a single tile size;
   - set $TILE_i$ including the iterations belonging to a parametric tile defined with parameters $ii_1, ii_2, ..., ii_{d_i}$ as follows:
     $TILE_i(II_i) = [II_i] \rightarrow \{ [I_i] \mid B_i * II_i + LB_i \leq I_i \leq \min(B_i * (II_i + 1_i) + LB_i - 1_i, UB_i) \land II_i \geq 0_i \}$;
   - set II_SET_i including the identifiers of corresponding tiles:
     $II\_SET_i = \{ [II_i] \mid II_i \geq 0_i \land B_i * II_i + LB_i \leq UB_i \}$.
3. Carry out a dependence analysis to produce a set of relations $R_{i,j}, i, j = 1, 2, ..., q$ describing all the dependences present in this loop nest.
4. Normalize the tuples of relations $R_{i,j}$ and sets $TILE_i$, $i, j = 1, 2, ..., q$ by applying schedule S, received in step 1. Introduce a new dimension in the normalized parameter space for each fixed element of the tuples, equal to the value of that element, and group the identifiers in sets II_SET_i.
5. Calculate a union of all normalized sets $TILE_i$, $i = 1, 2, ..., q$ and denote the result as $TILE$.
6. Calculate a union of all normalized sets II_SET_i, $i = 1, 2, ..., q$ and denote the result as II_SET.
7. Calculate a union of all normalized relations $R_{i,j}, i, j = 1, 2, ..., q$ and denote the result as R.
8. Form a relation R_TILE representing inter-tile dependences as follows:
   $R\_TILE = \{ [II] \rightarrow [JJ] \mid II, JJ \in II\_SET \land II \neq JJ \land \exists I, J : I \in TILE(II) \land J \in TILE(JJ) \land J \in R(I) \}$.
9. Form a relation T_CYCLE associating identifiers of cyclically-dependent tiles:
   $T\_CYCLE = \{ [II] \rightarrow [JJ] \mid II \in R\_TILE^+(JJ) \land JJ \in R\_TILE^+(II) \land II \prec JJ \}$.
10. Compute the set of representatives of strongly connected components:
    $II\_SET\_M = II\_SET - range(T\_CYCLE)$.
11. Form set $TILE\_M$, comprising statement instances sharing the same SCC:
    $TILE\_M = [II] \rightarrow \{ [I] \mid \exists JJ : II \in II\_SET\_M \land JJ \in T\_CYCLE^*(II) \land I \in TILE(JJ) \}$.
12. Form relation R_TILE_M describing dependences between merged tiles:
    $R\_TILE\_M = \{ [II] \rightarrow [JJ] \mid II, JJ \in II\_SET\_M \land II \neq JJ \land \exists I, J : I \in TILE\_M(II) \land J \in TILE\_M(JJ) \land J \in R(I) \}$.
13. Extend relation R_TILE_M of the form $\{ [X] \rightarrow [Y] \mid constraints \}$ to relation R_TILE_M' of the form $\{ [k, X] \rightarrow [k + 1, Y] \mid constraints \land k \geq 0 \}$.
14. Calculate the union of tile ultimate dependence sources and independent tiles:
    $UDS\_IND = II\_SET\_M - range(R\_TILE\_M)$.
15. Form relation FS representing the free schedule according to paper [3]:
    $FS = \{ [II] \rightarrow [k, JJ] \mid II \in UDS\_IND \land (k, JJ) \in R\_TILE\_M'^*(\{[0, II]\}) \land \nexists k' > k : (k', JJ) \in R\_TILE\_M'^+(\{[0, II]\}) \}$.
16. Extend set $TILE\_M$ by inserting the symbolic constants (k and tile identifiers) into the first positions of the tuple of set $TILE\_M$:
    $TILE\_M\_EXT = \{ [k, II, I] \mid (k, II) \in range(FS) \land I \in TILE\_M(II) \}$.
17. Generate code enumerating statement instances of a single parametric tile, represented with set $TILE\_M\_EXT$, in the lexicographic order, by means of applying any code generator. Parallelize any loop scanning tile identifiers II.

RAM clocked at 2133 MHz). The code of both original and transformed loop nests was compiled under the Linux kernel 3.10.0 x86_64 by GCC 4.8.3 with O3 optimization enabled, and then executed by 1, 2, 4, 8, 16 and 32 threads in subsequent runs. The parallelism of tiled loop nests is represented by means of the OpenMP API [8] using the *omp parallel for* directive.

For the experiments, we have chosen kernels of the PolyBench/C 4.1 [10] benchmark suite. Out of the 30 loop nests included in PolyBench, TC has found 13 kernels for which a corresponding inter-tile dependence graph is acyclic: *2mm, 3mm, atax, bicg, correlation, covariance, gemm, gemver, gesummv, mvt, syr2k, syrk, trmm*. For the remaining 17 suites, due to the complexity of inter-tile relations posing challenges on computing their transitive closures, TC is able to transform only 6 parameterized loop nests, namely: *cholesky, floyd-warshall, jacobi-1d, jacobi-2d, seidel-2d, trisolv*. The kernels *jacobi-1d, jacobi-2d, seidel-2d*, are the stencil codes whose dependences introduce cycles between adjacent tiles in all dimensions except for a time axis; as a result, the tile merging technique does not form multidimensional tiles. For the kernels *adi, doitgen, fdtd-2d, gram-schmidt, lu, ludcmp, symm*, TC is able to generate code only for compile-time-known loop upper-bounds. For the kernels *deriche, durbin, heat-3d, nussinov*, ISL cannot compute the positive transitive closure of inter-tile relations, even for known problem sizes. Below, we discuss results collected for parameterized loop nests *cholesky* and *trisolv*. Both kernels were transformed using tiles of the initial size 64 (in each dimension). Parallel tiled code generated by TC can be found in the *results* directory of the TC project repository.

The kernel *trisolv* is an imperfectly nested loop of $O(N^2)$ time complexity which has a dependence pattern similar to the working example studied in this paper. In particular, tile merging combines each tile of the last statement of the outermost loop with an adjacent 2-D tile. As a consequence, the number of tiles is always reduced by $\lceil N/64 \rceil$, without affecting data locality noticeably. However, positive speed-up is observed when using at least 8 threads of execution.

The second examined kernel, *cholesky*, is an $O(N^3)$ loop nest with 4 statements located at different depths. Depending on the problem size, the amount of merged tiles accounts for approximately 4 % up to 9 % of all tiles.

Table 1 summarizes the impact of merging original tiles on the total number of tiles. Figures 3a and b visualize the speed-up of parallel tiled code gained over the serial execution time of the original kernels analyzed in this section, for different upper bounds of the loops.

6   Related Work

Tiling has been studied from various perspectives. In this section, we focus only on known to us other techniques attempting to solve the problem of cyclic graphs.

Paper [2] introduced the idea of *tiles correction* by means of the application of the transitive closure of a data dependence graph. Let us remind that this technique involves relocating statement instances between tiles so that a target of any dependence is always inserted into the same or lexicographically greater tile

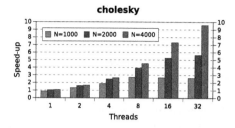

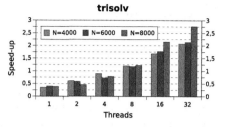

(a) Speed-up of the tiled *cholesky* kernel.   (b) Speed-up of the tiled *trisolv* kernel.

Fig. 3. Speed-up of the parallel tiled code over serial execution time.

Table 1. The impact of tile merging on the total number of tiles.

Kernel	Upper bounds	Number of tiles before merging	Number of tiles after merging
cholesky	N = 1000	1104	936
	N = 2000	7072	6480
	N = 4000	47775	45633
trisolv	N = 4000	2142	2079
	N = 6000	4653	4559
	N = 8000	8125	8000

than a corresponding source, even if the two conflicting tiles are not involved in a cycle. As a result, the algorithm always forms an acyclic inter-tile dependence graph without altering the amount of tiles, and tiles can be executed under the lexicographic schedule. In contrast to that approach, tile merging reduces the overall number of tiles, and it does not allow for the lexicographic execution of tiles if a resulting inter-tile dependence relation is not lexicographically-forward.

Paper [7] investigates tiling opportunities enabled by dynamic scheduling of tiles. For the loop nest examined in the paper, the problem of cycles in graphs is solved by means of merging only the first and the last tile of a reduction chain. However, the presented loop nests are tiled manually and the authors do not suggest any way allowing for automatic generation of tiled code.

Paper [1] defines a *relaxed permutability criterion* based on the concept of *live ranges*. The approach enables loop nest tiling by reducing the amount of false dependences between statement instances. This possibility is not considered in our algorithm. We plan to implement it in future.

Paper [15] derives a technique which splits the iteration space of the Nussinov loop nest into "problematic" and "non-problematic" iterations. As a result, tilable iterations are subject to classical tiling, while problematic ones are executed without optimizing the innermost loop.

7   Conclusion

In this paper, we presented an approach for generation of parallel tiled code of loop nests whose original inter-tile dependence graph contains cycles. We examined its applicability on kernels included in the PolyBench suite. In contrast to other studies concerning tile merging as a solution for removing cycles in graphs, we extend the idea by formalizing a complete and automatic procedure. The results of experiments confirm that code produced by means of merging original tiles within each strongly connected component allows for extraction of coarse-grained parallelism and benefits from locality enhancement. However, our technique often enables tiling at the expense of reducing the dimensionality of original tiles, thus wasting optimization opportunities. Furthermore, the experiments have revealed that complex dependence patterns of an inter-tile dependence graph pose challenges on computing its transitive closure. Our future research will therefore focus on developing tiling techniques with lower computational complexity.

References

1. Baghdadi, R., Cohen, A., Verdoolaege, S., Trifunović, K.: Improved loop tiling based on the removal of spurious false dependences. ACM Trans. Archit. Code Optim. **9**, 52:1–52:26 (2013). ACM, New York
2. Bielecki, W., Palkowski, M.: Perfectly nested loop tiling transformations based on the transitive closure of the program dependence graph. In: Wiliński, A., Fray, I., Pejaś, J. (eds.) Soft Computing in Computer and Information Science. AISC, vol. 342, pp. 309–320. Springer, Heidelberg (2015). doi:10.1007/978-3-319-15147-2_26
3. Bielecki, W., Palkowski, M.: Using free scheduling for programming graphic cards. In: Keller, R., Kramer, D., Weiss, J.-P. (eds.) Facing the Multicore - Challenge II. LNCS, vol. 7174, pp. 72–83. Springer, Heidelberg (2012). doi:10.1007/978-3-642-30397-5_7
4. Feautrier, P.: Some efficient solutions to the affine scheduling problem. Part I. One-dimensional time. Int. J. Parallel Prog. **21**, 313–348 (1992). Kluwer Academic Publishers, USA
5. Feautrier, P.: Some efficient solutions to the affine scheduling problem. Part II. Multidimensional time. Int. J. Parallel Prog. **21**, 389–420 (1992). Kluwer Academic Publishers, USA
6. Irigoin, F., Triolet, R.: Supernode partitioning. In: 15th ACM SIGPLAN-SIGACT Symposium on Principles of Programming Languages, pp. 319–329. ACM, San Diego (1988)
7. Mullapudi, R.T., Bondhugula, U.: Tiling for dynamic scheduling. In: 4th International Workshop on Polyhedral Compilation Techniques, Vienna, Austria (2014)
8. OpenMP Architecture Review Board: OpenMP Application ProgramInterface Version 3.0 (2008). http://www.openmp.org/mp-documents/spec30.pdf
9. Palkowski, M., Klimek, T., Bielecki, W.: TRACO: an automatic loop nest parallelizer for numerical applications. In: Federated Conference on Computer Science and Information Systems 2015, pp. 681–686 (2015)
10. PolyBench/C: The Polyhedral Benchmark suite (2015). http://web.cse.ohio-state.edu/~pouchet/software/polybench

11. Pugh, W., Rosser, E.: Iteration space slicing and its application to communication optimization. In: 11th International Conference on Supercomputing, pp. 221–228. ACM (1997)

12. Verdoolaege, S.: *isl*: An integer set library for the polyhedral model. In: Fukuda, K., Hoeven, J., Joswig, M., Takayama, N. (eds.) ICMS 2010. LNCS, vol. 6327, pp. 299–302. Springer, Heidelberg (2010). doi:10.1007/978-3-642-15582-6_49

13. Verdoolaege, S.: Presburger Formulas and Polyhedral Compilation, v0.02. Polly Labs and KU Leuven (2016)

14. Verdoolaege, S., Grosser, T.: Polyhedral extraction tool. In: 2nd International Workshop on Polyhedral Compilation Techniques, Paris, France (2012)

15. Wonnacott, D., Jin, T., Lake, A.: Automatic tiling of mostly-tileable loop nests. In: 5th International Workshop on Polyhedral Compilation Techniques, Amsterdam, The Netherlands (2015)

Performance Evaluation of Impact of State Machine Transformation and Run-Time Library on a C# Application

Anna Derezińska[(⊠)] and Marian Szczykulski

Institute of Computer Science, Warsaw University of Technology,
Nowowiejska 15/19, 00-665 Warsaw, Poland
A.Derezinska@ii.pw.edu.pl

Abstract. State machines are important behavioral models used in Model-Driven Development (MDD) of software applications. UML models are transformed into code and combined with a run-time library realizing state machine notions. Mapping of state machine concepts, including concurrent behavior issues, can be realized in various ways. We discuss several problems of call and time event processing and their impact of an application performance. In experiments, different solutions were evaluated and quantitatively compared. These variants were applied in refactoring of FXU - a framework for C# code generation and application development based on UML classes and state machine models.

Keywords: UML state machines · Statecharts · Model-Driven Software Development · Code generation · Model to code transformation · C#

1 Introduction

In Model-Driven Software Development (MDSD) various kinds of models and their transformations are used [12]. Among dynamic models, behavioral state machines are commonly used in the model to code transformations [8]. However, the scope of supported state machine notions is limited very often. In this paper, we focus on an approach to model-driven development in which a code in a general purpose language is created. Source models cover UML classes and all notions of behavioral state machines [21]. A target application is built from the auto-generated code in C#, a run-time library realizing the state machine concepts, and optionally an additional code refining details of operation bodies.

In realization of state machines of model classes we have to deal with various events, cooperation of parallel machines and their orthogonal regions. Discussion of problems of state machine transformations and their execution is usually aimed at semantic issues and "how to do" [1,6–9,13]. However, different variants of model transformations and realization of model concepts in a run-time environment has an impact on the target application performance. Time performance is of the most interest for applications originated from state machines with extensive usage of events, including time events. These kinds of models are

© Springer International Publishing AG 2017
S. Kobayashi et al. (eds.), *Hard and Soft Computing for Artificial Intelligence, Multimedia and Security*, Advances in Intelligent Systems and Computing 534, DOI 10.1007/978-3-319-48429-7_30

commonly used, e.g., in the reactive system domain. Moreover, using parallel constructs we can benefit from multi-core or other concurrent architectures.

Based on experiences in development, usage, further refactoring and evaluation of the Framework for eXecutable UML (FXU) [14,18] we pose a question "how to do more effectively". Several variant solutions to handling of time and call events in state machines were proposed and implemented in the FXU environment. Their consequences were experimentally evaluated with the assistance of developed benchmark models. Experiments confirmed measurable benefits of the system refactoring, especially for specific kinds of models.

The main contributions of the paper are analysis of different proposals to time and call event strategies that can be realized in practice and quantitative evaluation of the solution impact on target applications. There is lack of measurable evidences addressing this problem in the context of model-based realization of full state machines in C# or analogous technologies.

In the next section, we recall related work. Alternative solutions to state machine transformation and realization issues are discussed in Sect. 3. Experiments and their results are presented in Sect. 4. The paper finishes with conclusions.

2 Background and Related Work

Many MDD tools support transformation of UML state machines into code. State machines are transformed in different ways, including simple attributes depicting states [13], solutions based on the state machine design pattern, and other approaches [8]. Different ideas in development of state machine concepts, e.g. transforming of composite states [1], dealing with history [5], code generation with OCL expressions in state machines [11] are reported. Samek described state machine coding to C++ [16], but the framework does not directly supports concurrency. However, most of the tools do not cover the full specification of UML state machines, e.g., in many cases omitting or limiting the scope of *orthogonal states*, *history*, or event types, especially *time events*.

Many model-driven approaches focus on a direct model interpretation, and not creation of a target code, as discussed in this paper. Interpretation approaches give in general longer execution time [23]. A comprehensive research, together with a benchmark evaluation was given by Höfig [10]. The performance of this interpretation approach was, in dependance of source models, 3 to 460 times slower than of a generated code. Direct execution of models is also a goal of an interpreter in xtUML [2], with an opportunity to transform a tested model to general languages, but not to C#. Moreover, we cannot relate our work to solutions based on Foundation Subset for Executable UML (fUML) [22], as time events are not within the scope of it.

Detailed information concerning performance issues in MDD-originated applications and run-time frameworks for state machines is missing in general. Another survey focuses of selected approaches to flattening of hierarchical state machines [7]. It regards only few tools but admits a lack of detailed performance evaluation of the implemented approaches.

In Framework of eXecutable UML (FXU), UML classes and comprehensive state machine models are transformed [14,18]. At the beginning, FXU was first and the only tool that supported MDD of state machine models towards the C# programming language. In a systematic review [8], no other tools than FXU were referenced as supporting transformation of a full state machine into C#. At present, there are some tools that transform models to C# code. Executable finite state machine [17] seems to cover many of state machine features. However, lack of UML model mapping details hampers possibility to compare our results at the moment. Other tools with C# are mostly limited to class models (e.g. IBM Rational Software Architect [20]), or to simplified state machines (e.g. Sparx Enterprise Architect, SmartState, etc.). An extensive range of state machine notions is taken into account in some MDD frameworks, such as the IBM Rational Rhapsody tool [19], and the Umple tool [9], although they do not support the C# language.

A number of interpretation issues of UML, and especially of state machines, is left open in the UML specification and regarded as *semantic variation points* [21]. Such problems are solved in different ways in various practical solutions, though often without a clear explanation, or are advocated to be decided by a user [3]. In different versions of FXU tool, we resolved some semantic variation points in a manner consistent with the UML specification [5,6]. In the Umple tool [9], a user can apply variant solutions for an event handling and queuing. Prout at all [15] presented a semantically configurable code generator (CGG) that using a set of semantic parameters produces a code generator of state machine-like models to code. The overall execution time was not better than those of commercial tools. However the target is a Java interpreter of models, no time events or history are handled and no data about impact of call event processing on performance are given. Hence, we cannot compare our results with this approach.

3   Event Processing and Multithreading

Behavior of UML state machines is controlled by events that can cause effects in a systems. There are several types of events: *call event, time event, change event, signal event, any-receive-event* [21]. A transition between states can be labeled by an event that triggers traversal via this transition; or traversal is implicitly triggered by a *completion event*. Primary algorithms on event processing in FXU were presented in [4].

Requirements on parallel execution of many state machines, handling of many concurrent regions in orthogonal states, and processing of events can be specified with threads. Multithreading is supported by language environments and operating systems. Threads are executed separately but share common code and date of a process. Realization of a multithreading run-time environment should correspond to a certain interpretation consistent with the UML specification describing state machine behavior [21]. The specification does not decide about the specific implementation. The run-time environment of FXU is using threads that are supported by the C# programming language.

In the UML specification no precise policy of handling arriving events is established. Usually events are stored in an event buffer and processed by an event processing routine. In FXU, each state machine had its own event queue. It was processed by an event scheduling routine of this state machine, which was run in a separate thread. The routine took an event from a queue and elected transitions that are triggered by this event. For each transition, an exit from its source state was completed in a separate thread. After exits from all appropriate substates, the transition was realized. Finally, entry to target states was performed. Transitions were also served in separate threads. Processing of events of a state machine was accomplished during the whole time of the state machine existence.

A variation point in the UML specification concerns also delivering of events to state machines. Assuming a buffer processing of events, an event delivery relies on putting an event into queues of suitable state machines. In refactoring, we considered three methods of an event delivery:

- putting of an event directly into one event queue,
- broadcasting - forwarding an event to all event buffers,
- multicasting - delivering an event to several selected event buffers.

Using those methods we can fulfill all requirements of the UML specification [21] concerning notification of appropriate elements about an event occurrence.

Selected problems of event processing and multithreading in the realization of state machines are presented in the following subsections. The alternative solutions are identified by numbers Alt#.

3.1   Event Priorities - Alt1

According to the UML specification, realization of event priorities is a semantic variant. In the old FXU an event buffer was designed as a modified FIFO queue. All events had the same priority except a completion event that has the highest priority. In the refactored implementation, an event buffer can be designed as a priority queue. Priorities of all kinds of events can be selected by a user. Event priorities are changed using appropriate options and setting a configuration of the run-time environment. This alternative (Alt1) increases a model development flexibility. It can influence performance and security of a designed system.

3.2   Time Events and Multithreading

Time events are initiated just after entering a state before launching an *entry* activity specified for this state. Initialization of handling of a time event corresponds to start of a timer. It counts down a time interval after which the event is placed into an appropriate event buffer. In order to start all time events for a given state, all transitions outgoing the state are visited, as well as states that encompass the state under concern. If a transition is triggered by a time event, a new thread is created. The thread puts the event into the appropriate queue after elapsing of the given time. Two issues of time event processing were alternated. A hypothesis states that *alternative solutions Alt2 and Alt3 can improve time performance of the target application, especially in case of complex state machines with many transitions.*

Creation of Time Events - Alt2. Time events can be created during execution of a state machine. After entering a state, all transitions are visited and checked weather they are triggered by a time event. All time events are added to the appropriate list, and in the next step, processing of all events from this list is started in separate threads. However, time events associated with a given state are known before an application execution. Therefore, the notion of events can be created during the model to code transformation. Visiting of all outgoing transitions can be avoided by adding a list of all time events that can encounter when the automaton is in the state under concern. In this case during the run-time, the time event list is not created but only processed.

Processing of Time Delays - Alt3. Processing of a time event requires calculating of a given time delay. This time passing can be handled by a separate thread devoted to the event. Another alternative is usage of one thread treating all time events in a state. This can be realized in the following steps: (i) At the stage of code generation and initialization of a state machine a list of all time events associated with a state is created, according to the variant solution discussed above (in Alt2). The list should be sorted with the time delay increasing. (ii) While entering the state, all time events are started in one separate thread.

3.3   Call Event Delivery - Alt4

Call events that model initialization of execution of class operations are the mostly used events. An operation of a class of a current state machine can be called, but also a public operation of another class can be triggered. Therefore, it establishes an in- and inter-class communication mechanism within the whole application. One of solutions is *broadcasting*, in which call events are created and placed into event buffers of all state machines of the application. Events are recognized by unique identifiers and the event scheduling routine can check if this event causes a change in the active state configuration of the state machine.

Passing of such events to all state machines is unnecessary in most cases. Therefore, events can be filtered in the event processing routine and distributed to event buffers of appropriate state machines. In result, an event *multicasting* can be applied, or in some cases an event is send to exactly one target buffer. A set of state machines that would be potentially interested in this event can be derived from a source model. In this way, events can be filtered and their set created during the code generation phase. A hypothesis says that *solution (Alt4) improves time performance when there are many call events or when an application includes many classes and their state machines.*

3.4   Creation of New Threads - Alt5

There can be many threads used during execution of state machines. A new thread can be created in the following situations:

(I) A thread of an event processing routine of each state machine.
(II) Threads associated with realization of transitions from a state to another state. They are handling exiting from all current states (a state and sub-states), passing transitions and performing their actions.
(III) Threads managing time passing required by time events (the proposed alternative solution Alt3 was presented in Sect. 3.2).
(IV) A thread performing *do* activities of a state.
(V) Threads handling entering to internal regions of an orthogonal state.

Switching between threads lead to an additional overhead both on time and memory used in the run-time environment. Decreasing a number of threads should be profitable while concerning the application performance.

Another alternative solution (Alt5) concerned transitions between states (case II in the above list). If there is only one transition triggered by a certain event taken from the event queue, it is not necessary to create a new thread to handle this transition. Exit from the present state, execution of an action associated with the transition and entering a new state can be fulfilled in the current thread. A hypothesis concerning Alt5 states that *time performance improves when a number of simple transitions is higher than a number of complex ones*.

4   Performance Evaluation

Alternatives presented in Sect. 3 were implemented in the refactored FXU [18]. The refactoring was performed both in the code generator and the run-time environment. It was aimed at improvement of performance of created applications. Verification of the refactoring took into account two main directions: (i) correctness of the transformation and run-time support, and (ii) performance evaluation. The first area, verified with a set of test models applied to regression testing is beyond the scope of this paper. The second aim was realized with experiments using two benchmark models made available at [18]. The benchmarks were designed to examine the impact of selected variants on realization time of specific parts of state machines, as well as, on the execution time of the whole application. In this paper results of verification based on the benchmarks are discussed.

Several time measures of event processing were examined. They are denoted by letters A-E and illustrated in (Fig. 1).

A. `Transition time` - time interval since finishing exit from a previous state (X) till beginning entry to a next state (Y)
B. `Realization time of an` *entry-exit* `time event` - time interval since entering a state (X) till begin of exiting this state

C. **Realization time of an *entry-entry* time event** - time interval since entering a state (X) till entering a next state (Y), where the transition from X to Y was triggered by a time event

D. **Reaction time to a call event** - time interval since beginning realization of a call event in a state (Y) till starting of exit from the state (Y)

E. **Event handling time** - time interval since beginning realization of a call event in a state (Y) till entering a next state, where the transition outgoing from the state (Y) was triggered by a call event

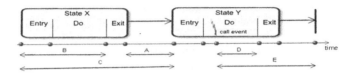

Fig. 1. Time intervals in event processing

While comparing different values measured in modified and non-modified applications a relative difference was observed. The *difference* is calculated as: $(OldValue - NewValue)/OldValue[\%]$, where old and new values are specified in particular experiments.

The benchmark models were prepared with IBM RSA [20] and transformed to C# with the FXU generator. Various modified and non-modified applications were compared. They were created and run within the FXU run-time library in MS VisualStudio. In the experiments, the following exemplary priorities of events were selected: (1) completion event, (2) time event, (3) signal, (4) call event and change event. They could be changed according to the modification facility (Alt1).

4.1 Evaluation of Time Event Processing

Benchmark. To verify strategies of time event processing, a benchmark model was developed, in which many time events are used. The model presents an electronic treadmill. An outline of its state machine is shown in (Fig. 2). Detailed description of time events and guard conditions that label model transitions are not presented in order to keep the figure legible, but is available at [18]. There are two basic states, when the treadmill is turn off or turn on. The latter is a complex state. After the treadmill was activated, a mode is selected. The main execution substate (*ModeSelected*) has two orthogonal regions comprising states of different speeds and different positions. The concurrent behavior is synchronized by a *Join* pseudostate before entering a *Finishing* substate.

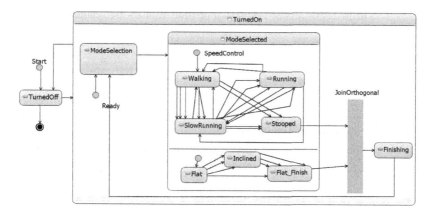

Fig. 2. Benchmark for evaluation of time events - Treadmill state machine

Experiments. We conducted experiments to evaluate an impact of creation strategies of time events (Alt2) and processing of time events in a state (Alt3) on execution time. Measured time could also be influenced by a change of event processing of transitions (Alt5). The modification Alt5 was switched off in these applications in order to separately evaluate its impact. Evaluation of Alt5 was moved on to another benchmark and further experiments (Sect. 4.2).

Two applications aimed at different test scenarios were designed. In the first application, four instances of the benchmark model were created. There were executed in parallel receiving about 10 time events per a state machine. Using this scenario, time overhead of a time event was examined.

Results of the first application are given in Table 1. Column "Old" includes measurements done on a non-modified application, while column "New" with modifications Alt2 and Alt3. Time measures given in rows 2, 4, 5, 6 correspond to definitions B, D, A, and C (Fig. 1). In rows 3 and 7 a time overhead is shown. It was calculated as a relative *difference* between two time intervals: "new"- an actual time of a time event realization, and "old"- a nominal (ideal) time specified by the time event. This kind of difference was calculated for *entry-exit* time event (row 3) and *entry-exit* one (row 7).

In the second application one instance of the Treadmill benchmark model was created. About 200 time events were processed during its run. This test scenario focused on examination of a total execution time of a state machine. The application was configured and executed in three configurations: (I) without modifications, (II) only with Alt2, and (III) with Alt2 and Alt3 modifications.

Results of the experiments are summarized in Table 2. Mean execution time refers to arithmetic mean over five execution runs of the second application in the considered versions. Relative difference of times calculated for all three pairs of the application versions are given in the bottom rows.

Table 1. Results of evaluation of time events - 1st scenario

Measures	Mean values		Difference
	Old	New	[%]
1 Execution time of state machine [ms]	100819.75	100427.45	0.39
2 Realization time of entry-exit time event [ms] (B)	16031.4	16012.08	0.12
3 Overhead of entry-exit time event [%]	0.23	0.09	63.05
4 Reaction time to a call event [ms] (D)	24.45	12.45	49.08
5 Execution time of a transition (no action) [ms] (A)	6.45	0.25	96.12
6 Realization time of entry-entry time event [ms] (C)	20071.73	20038.60	0.17
7 Overhead of entry-entry time event [%]	0.73	0.27	62.40

Table 2. Results of evaluation of time events - 2nd scenario

	Old version	New versions	
	(I)	(II) Alt2 & no Alt3	(III) Alt2 & Alt3
Mean execution time [ms]	167607	166693	156705
Difference (I-II) [%]		0.55	
Difference (II-III) [%]			5.99
Difference (I-III) [%]			6.50

Discussion. According to results of the first scenario (Table 1), modifications Alt2 and Alt3 caused decreasing of an additional overhead on realization of time events (about 62–63%). However, this time overhead was not big, and in the context of the state machine the whole execution time changed very little. Reaction time to call event was lowered almost on 50 % and execution time of a transition about 96 %. The higher impact of these facts on the overall execution time would be in case of the higher number of such events.

In the second scenario more events were generated. Observing results (Table 2), we conclude that modification Alt3 has a visible impact on the execution time improvement (about 6 %). It lowered the number of threads that process time events. Modification Alt2, i.e. moving of a time event searching to the transformation phase, has a smaller impact on the execution time.

4.2 Evaluation of Event Processing and Multithreading

Benchmark. Another benchmark model was devoted to evaluation of passing and handling call events (Fig. 3). The model reflects a simple Robot behavior and has 34 transitions and 33 various events [18]. All transitions are triggered by appropriate call events of different operations. In the diagram, call events

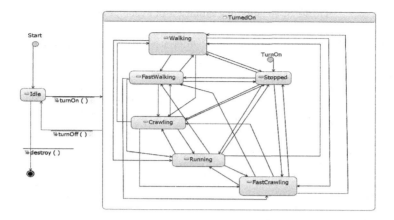

Fig. 3. Benchmark for evaluation of event processing - Robot state machine

that label transitions between sub-states in the complex state *TurnedOn* are not shown to keep the figure legible. The model was used in evaluation of the impact of the thread realizations in a transition between states (Alt5).

Experiments. A robot instance with its state machine was build and many events were generated. Different test scenarios were developed. In the first scenario, a given number of call events was generated and all events were placed into an event buffer of the state machine. In the second scenario, events were grouped into packages, 6 events in each. At first one package of events was created and delivered to the state machine buffer. After a specified time delay, the next package of events reached the state machine, and so on.

Experiment results are shown in Table 3. Test cases are identified by two parameters provided in the first two columns: a number of created events and a time delay after delivering a packet of events. All times in the table are given in [ms]. For each transition, time for handling an event was measured (E in Fig. 1). A mean handling time was calculated as an arithmetic mean over all events encountering in the application run. Execution times of the whole application are also given. A column depicted as "Old" states for results of an application without modification. Column "New" gives results for the application generated and run within the modified environment (Alt4 & 5). Column "Difference" presents a relative change of mean handling time and total time, accordingly.

Discussion. This model was designed to focus on call event processing, hence no time events were included. Moreover, only one instance of a class was created, therefore one event buffer was used. Under these conditions, the experiment results are influenced only by changes of event handling, and not by other factors. It can be seen (Table 3) that the modification had a considerable impact on the deceasing of mean handling time of a single call event. In dependence of a test

Table 3. Results of evaluation of event processing

Test case		Mean handling time		Difference	Total time		Difference
#Event	Delay	Old	New	[%]	Old	New	[%]
600	0	854.20	457.42	46.45	2093.4	1265.4	39.55
6000	0	1291.00	930.82	27.90	22060.6	12080.8	45.24
600	50	6.17	0.13	97.89	5748.4	5655.6	1.61
6000	50	5.59	0.31	94.42	52694.6	52821.0	−0.24
600	20	5.40	1.05	80.62	2616.2	2630.4	−0.54
6000	20	8.72	0.68	92.23	22015.0	21941.8	0.33

case, 600 or 6000 transitions were performed during the application execution. In the old run-time version, twice this number (1200 or 12000) additional threads were created, accordingly. The overall execution time was lowered about 40 % for test cases with no time delay. In the remaining test cases, a group of events was delivered regularly and time delay was big enough to handle all events. Therefore, the whole execution time remained practically the same.

Summing up, the modification Alt4&5 caused a substantial improvement in the application performance. It is especially important when many call events are densely processed in an application. It has no impact of the execution time when events are scarcely generated, although in this case processor resources are less occupied.

5   Conclusions

We presented a beneficial refactoring of an MDE environment for model to code transformation and a run-time support for UML classes and state machines. Presented alternative solutions, consistent with the UML specification, made processing of call and time events in a target application more efficient. An event scheduling was modified in order to perform a transition in more effective way, and avoid creating of additional threads. Time events were also processed in a simpler way with less demands on processor resources. The quantitative advantages depend strongly on the type of source models, giving a smaller overhead (above 60 %) and a shorter total execution time (about few %) in case of many time events, and a faster execution (about 40 %) in case of many call events delivered in the similar time.

References

1. Baderddin, O., Lethbridge, T.C., Forwared, A., Elaasar, M., Aljamaan, H., Garzon, M.A.: Enhanced code generation from UML composite state machines. In: Proceedings of 2nd International Conference on Model-Driven Engineering and Software Development (MODELSWARD). pp. 235–245 (2014)

2. Burden, H., Heldal, R., Siljamaki, T.: Executable and translable UML - how difficult can it be?. In: 18th Asia-Pacific Software Engineering Conference, pp. 5–8 (2011)
3. Chauvel, F., Jézéquel, J.-M.: Code generation from UML models with semantic variation points. In: Briand, L., Williams, C. (eds.) MODELS 2005. LNCS, vol. 3713, pp. 54–68. Springer, Heidelberg (2005). doi:10.1007/11557432_5
4. Derezińska, A., Pilitowski, R.: Event processing in code generation and execution framework of UML state machines. In: Madeyski, L., Ochodek, M., Weiss, D., Zendulka, J. (eds.) Software Engineering in Progress, pp. 80–92. Nakom, Poznań (2007)
5. Derezińska, A., Pilitowski, R.: Interpretation of history pseudostates in orthogonal states of UML state machines. In: Feldman, Y.A., Kraft, D., Kuflik, T. (eds.) NGITS 2009. LNCS, vol. 5831, pp. 26–37. Springer, Heidelberg (2009). doi:10.1007/978-3-642-04941-5_5
6. Derezińska, A., Szczykulski, M.: Interpretation problems in code generation from UML state machines - a case study. In: Kwater, T. (ed.) Computing in Science and Technology 2011: Monographs in Applied Informatics, pp. 36–50. Department of Applied Informatics Faculty of Applied Informatics and Mathematics Warsaw University of Life Sciences (2012)
7. Devroey, X., Perrouin, G., Cordy, M., Schobbens, P.Y., Heymans, P., Legay, A.: State machine flattening, a mapping study and tools assessment. In: 8th IEEE International Conference on Software Testing, Verification and Validation, pp. 1–8 (2015)
8. Dominguez, E., Perez, B., Rubio, A.L., Zapata, M.A.: A systematic review of code generation proposals from state machine specifications. Inf. Softw. Technol. **54**(10), 1045–1066 (2012)
9. Garzon, M., Aljamaan, H., Lethbridge, T.: Umple: a framework for model driven development of object-oriented systems. In: IEEE 22nd International Conference on Software Analysis, Evolution and Reengineering SANER 2015, pp. 494–498, 15 June 2015
10. Höfig, E.: Interpretation of behaviour models at runtime: performancebenchmark and case studies. Ph.D. thesis, Berlin Institute of Technology (2011). http://opus.kobv.de/tuberlin/volltexte/2011/3065/
11. Iqbal, M.Z., Arcun, A., Briand, L.: Environment modeling and simulation for automated testing of soft real-time embedded software. Softw. Syst. Model. **14**(1), 1–42 (2013)
12. Liddle, S.: Model-driven software development. In: Embley, D., Thalheim, B. (eds.) Handbook of Conceptual Modeling, pp. 17–54. Springer, Heidelberg (2011)
13. Niaz, I.A., Tanaka, J.: An object-oriented approach to generate Java code from UML statecharts. Int. J. Comput. Inf. Sci. **6**(2), 83–98 (2005)
14. Pilitowski, R., Derezińska, A.: Code generation and execution framework for UML 2.0 classes and state machines. In: Sobh, T. (ed.) Innovations and Advanced Techniques in Computer and Information Sciences and Engineering, pp. 421–427. Springer, Heidelberg (2007)
15. Prout, A., Atlee, J.M., Day, N.A., Shaker, P.: Code generation for a family of executable modelling notations. Softw. Syst. Model. **11**, 251–272 (2012)
16. Samek, M.: Practical Statecharts in C/C++: Quantum Programming for Embedded Systems. CMP Books, San Francisco (2002)
17. Executable finite state machines. https://state.software/
18. FXU Framework for eXecutable UML. http://galera.ii.pw.edu.pl/~adr/FXU/

19. IBM Rational Rhapsody Developer. http://www-03.ibm.com/software/products/en/ratirhap
20. IBM Rational Software Architect. http://www-03.ibm.com/software/products/en/ratisoftarch
21. Unified Modelling Language (UML) (2015). http://www.omg.org/spec/UML
22. Semantics of a Foundation Subset for Executable UML models (fUML), ptc/2016-01-05 (2016). http://www.omg.org/spec/FUML/
23. Wasowski, A.: Code generation and model driven development for constrained embedded software. Ph.D. thesis, University of Copenhagen (2005)

The Method of Evaluating the Usability of the Website Based on Logs and User Preferences

Luiza Fabisiak[✉]

Faculty of Economics and Management, Institute of IT in Management,
University of Szczecin, Szczecin, Poland
luiza.fabisiak@gmail.com

Abstract. The article includes the development of usability assessment services based on data from the internal structure of the websites. Considered the problem of evaluating the usability of the websites is made concrete on the basis of: selecting appropriate criteria, determine their significance, selection decision support methods and users preferences. Websites user preferences are variable in time and is usually different from those included in the requirement form, applicable to the stage of creating the site. Once created website may lose its usability due to the variable needs of its users. Services aging process, development of software and hardware in computer science, development of civilization and technology, the flow of time, variable fashion, variable conditions related to users' behavior, forcing the company to undertake a new researches and develop new methods to evaluate the usability of websites The research paper is to assess usability of websites based on the data contained in the log, which will take into account changing user preferences based on the history of service use.

Keywords: Usability · Web services · Decision support methods · Rough sets

1 Introduction

Dynamic development of the Internet as well as wider and wider scope of rendered Internet services require even more engagement in adapting the Internet service offer to the requirements of specific user groups. The most often used criteria for Internet service assessment is their web usability and utility. Utility and web usability assessment of a given Internet service is closely related to its popularity among users of the said service. One of the best ways of gaining in popularity, as far as Internet services are concerned, is discovering user preferences. Researching and discovering preferences of website users are applied, i.e., in order to: evaluate the behaviour of users exploring these services, identify priorities for their choice, provide data for monitoring 'traffic' on the web page, improve retention rate or to comply with quality requirements, for example the ISO standard of service.

This paper introduces a synthetic usability method, which is based on historical data as well as on an evaluation mapping current and future user preference; this is due to the fact that none of the above-mentioned need not necessarily constitute a sufficient

© Springer International Publishing AG 2017
S. Kobayashi et al. (eds.), *Hard and Soft Computing for Artificial Intelligence, Multimedia and Security*, Advances in Intelligent Systems and Computing 534, DOI 10.1007/978-3-319-48429-7_31

evaluation (historical evaluation does not need to be in accordance with current user preference, while forecasted preferences generally exhibit little predictability).

Article includes a study evaluation of the usability of website method. The problem of assessment was explained on the basis of the selection of appropriate criteria, determining of its relevance and selection of appropriate multi-criteria method. Defining user preferences based on logs enabled using rough sets to select the relevant criteria in the analyzed structure of the decision problem.

2   Literature Review

In the literature, an attempt was made formalize the evaluation methods for various types of sites [1–3]. Also created a number of organizations to study the wider usability of websites. In practice, there are two types of ratings usability: quality (subjective opinions) and quantitative (objective, measurable facts) [4]. Among the methods for evaluation and usability testing for basic were considered tests with users. They use them to examine, how users perform fundamental for service properly defined sets of tasks. Another method is online questionnaires, which are the most common tools for measurement. Most of the research methods are based on an online questionnaire as a collection of primary data (not yet gathered) and measuring the effects of marketing [5]. Research in the laboratory (user participation) is a technique of assessing the usefulness of websites, applications and devices, which involves the cooperation of representatives of the target group (users) [6]. Besides basic tools for research utilities, methods based on the so-called modern technologies are also used. They include: eyetracking and clictracking. These methods belong to the traffic analysis. Eyetracking is tracking technology clusters eyesight on the website or application. Clictracking is a method used to track user website activity. It is a method that allows verifying the functionality of competing prototypes of the interface. It should be noted, that a major impact on the credibility of the assessment of the usability of websites in the above-mentioned methods is to select a group of experts and their assessment. E. Baharat and S. Nitzan in their works give the conditions that must be met in order to obtain an appropriate assessment [7], in the opinion of experts. D. G. Saari presented methodology to assess the character group [8–10]. Bury H. and D. Wagner provides an extensive discussion D.G. Saari and they drew attention to the difficulty of analyzing the results for a large number of objects [11, 12].

When using the above-described methods to assess the usability of the services there is visible shortage of a multi-faceted and systemic tool. These methods enable usability testing from the point of view of the real and the time-varying user preferences.

3   General Scheme of the Method

Method presented in the article is based on the assessment of the benefit based on historical data (evaluation based on usability log) and data mapping current and future preferences of users (surveys). These results are rating partially. They are marked symbolically as Z_d - assessment of the usability from logs and P_S - evaluation of usability as a result of the survey (user preferences study). Finally, to evaluate the usability of websites it has been created so-called synthetic evaluation of the usability of websites U. The method of synthetic evaluation the usability of websites consists of three stages.

1. In the first stage: Calculate the usability of this service on the basis of the data contained in log (rating partial Z_d – Subsect. 3.1).
2. In the second stage: Evaluate the usability of the service on the basis of the data resulting from the study of the users preferences of this service (rating partial P_S) – (Subsect. 3.2).
3. Final, third stage: Make the final evaluation of service usability U in the form so-called synthetic evaluation of the usability, based on the assessment of partial grade Z_d and P_S. – (Subsect. 3.3)

Methods of partial grade Z_d and P_S as well as synthetic evaluation of the usability U of the Internet services has been presented in subsequent chapters of this study.

3.1 Internet Service Usability on the Basis of Data Included in the Logs

Internet service logs provide data concerning users and their behaviour during visiting web pages. These are automatically collected data while browsing websites.

To make it more general, in order to assess the usability of any service at the time marked t on the basis of the log, it has been assumed that the service in question is service S_0 with a log arbitrarily marked as D_0. This D_0 log provides data on characteristic features concerning the use of service S_0; these features correspond with attributes contained in D_0 which describe the manner in which S_0 service is used. The set of the said attributes in D_0 log for S_0 service is marked with the A_0 symbol, according to formula (1):

$$A_0 = \{a_1^0, a_2^0 \ldots a_m^0\} \tag{1}$$

where:

a_j^0 - attribute describes the j -th property (feature) how to use the website S_0 in logs websites D_0,

$j \in \{1, 2, \ldots, m_o\}$

m_i - number of attributes in the log D_0.

The usability of S_0 service, symbolically marked as Z_{d_0}, is variable in time and its value is the value of an unknown g function, such that (2):

$$Z_{d_0}(t) = g(A_0, \zeta, t) \tag{2}$$

where:

ζ - is unknown and unidentified logs websites D_0 set of factors affecting the usability of the website S_0,

t - at the t moment,

A_0 - a set of attributes defined in the logs D_0 website S_0, defined by formula (1).

It needs to be emphasized that with the g function and the ζ set remaining unknown as well as lack of any identification methods of the above-mentioned, the usability $Z_{d_0}(t)$ of the S_0 service at the t moment may be determined as rank of S_0 service usability value against other services similar to S_0 which has been determined on the basis of the known set of attributes available in the logs. It is then a type of categorical assessment expressing the rank of the value. For this reason, it has been assumed in further deliberations that at the t moment there are other, additionally available services similar to S_0 with logs. A similar service is understood as having the same range of basic functions assigned to the service. Similar services are related to each other not only by means of web usability, but also by content; they usually have similar recipients, though often with individual preferences. Let those services similar to S_0 be services (3):

$$S_1, S_1, \ldots, S_n \tag{3}$$

where:

symbol n indicates the number of websites similar to the site S_0, such that $n \geq 1$.

Assuming that for every S_i service, where $1 \in \{1, 2, \ldots, n\}$ at any t moment the logs, symbolically marked as D_i are readily available and that each of the D_i logs of the S_i service similar to service S_0 (for $i = 1, 2, \ldots n$) contains an m_i set of A_i attributes denoting the manner in which the said services are used, i.e. (4).

$$\forall_{1 <= i <= n} \exists_{m_i \in I} A_i = \left\{ a_1^i, a_2^i \ldots a_{m_i}^i \right\} \tag{4}$$

where:

I - the set of non-negative integers,
n - number of considered additional services,
$i \in \{1, 2, \ldots, n\}$,
m_i - number of attributes in the log D_i
a_j^i - attribute describes the j -th usage of website S_i in logs website D_i,
$j \in \{1, 2, \ldots, m_i\}$

it is possible to consider a set of common Δ attributes defined according to function (5):

$$\Delta = \bigcap_{i=0}^{n} A_i \tag{5}$$

It contains conjoint attributes which characterise services $S_0, S_1, S_2, \ldots, S_n$ from the point of view of their usability. However, Δ set is by no means an empty set. It results from the fact that website logs contain, among other things, the most frequently used attributes such as: number of visits, duration of a given visit on the web page, number of references etc.

Following the assumption that Δ set contains p elements, i.e., let (6):

$$\Delta = \{k_1, k_2, \ldots k_p\} \tag{6}$$

where:

$i \in \{1, 2, \ldots, p\}$ is the number of the item in the collection Δ,

P - cardinality of the set Δ,

k_i - the i-th element of the set Δ.

The elements of Δ are the measurements of $S_0, S_1, S_2, \ldots, S_n$. service usability. A frequent set of criteria (elements of Δ) are usually: number of server references: e.g. k_1,, number of users visiting the service: e.g. k_2, web pages from which users enter the service: e.g. k_3, web pages from which users most frequently tend to leave the service: e.g. k_4, reference number: e.g. k_5, etc.

For services $S_0, S_1, S_2, \ldots, S_n$ and respective $D_0, D_1, D_2, \ldots, D_n$, it is then possible to create a source data matrix $\Omega[\xi \times p]$, where the elements are the values of consecutive criteria in subsequent log entries, where ξ represents the number of total records in all logs beginning with D_0 log and finishing with log D_n, and where p is the count of Δ set.

In further deliberations it has been assumed that the matrix Ω is a matrix with dimensions amounting to $N \times p$, where N a number of lines of the said Ω matrix and p represents the number of its columns (the number of attributes of Δ set). Some of attributes (measurements) of the logs (elements of Δ set) may be relatively dependent from each other and may be significant or insignificant from the point of view of utility assessment. They may also require carrying out calculations (e.g. number of references, the speed at which web pages of a given service are loaded and data are sought or the average number of sessions per user). They may also contain so called information noise, which is a result of various errors that can appear in data collected from the logs. The existence of such noise in data inscribed in the Ω set impedes, from the point of view of credibility of assessment, the process of isolating true, solid and relevant information. Due to the above-mentioned, when having Ω matrix at one's disposal, an adequate method of clearing, organising and validating data contained in Ω needs to be implemented; this reduces the amount of information noise. Generally, methods of clearing, organising and validating as well as normalising are referred to as initial data processing [13]. Irrespective of initial data processing, it is essential to isolate from the criteria included in the Δ set those criteria, which are significant for the usability evaluation as this particular step aims at securing the credibility of the said assessment.

3.1.1   Initial Data Processing Security

Initial processing of data collected from $D_0, D_1, D_2, \ldots, D_n$ (n - number of services similar to S_0) logs involves:

1. Eliminating those lines and columns from Ω matrix, which contain incomplete data;
2. Calculating, on the basis of data contained in Ω matrix, the value of criteria, which belong to Δ set;
3. Normalising data (converting values contained in Ω matrix to values that belong to the same reference field)

In order to obtain a table with unknown values of attributes in relation to fully defined Ω information table three generalisations need to be applied. The first type of initial data processing results from incomplete data with attribute values from Δ set in logs $D_0, D_1, D_2, \ldots, D_n$. The second case is the k_i criterion reduction (i-th element) from

Δ set in Ω information table; it takes place whenever the relation of likelihood for similar cases proves to be asymmetrical. This means that there is a reduction of those k_i (i-th element) from Δ *set* for which there is a dependence between other k_j (j-th element) criteria of this set, where $(j \neq i) \wedge (1 \leq i \leq n) \wedge (1 \leq j \leq n)$. The third instance of generalisation is the normalisation of data. At this point a change of values included in the Ω matrix o values that belong to the same referential field and measurement units takes place. Normalisation is necessary when the values of Ω matrix come from different ranges.

The final outcome of the initial data processing (processing data from Ω matrix (instance 1, 2 and 3) with attributes from Δ set) is a Γ matrix with criteria from $\prod$ set. The set of criteria $\prod$ is then consistent with formula (7).

$$\prod = \{q_1, q_2, \ldots, q_u\} \subseteq \Delta \wedge u \in I \wedge u \leq p \tag{7}$$

where:

  I - the set of positive integers numbers,
  Δ - set of criteria before pre-treatment of the matrix Ω,
  u - number of criteria for evaluation the usability of websites after pre-treatment.

3.1.2   Evaluation of Significance of Assessment Criteria

In order to carry out an assessment of Internet services in respect of their usability and on the basis of attributes contained in a data matrix Γ together with criteria for assessment $\prod$, one needs to choose from $\prod$ set those criteria, which are significant when taking into consideration the aim of the assessment. Additionally, it is necessary to eliminate from the set of distinguishable criteria $\prod$ and data from the table Γ also those elements whose deletion will not result in loss of significance as far as outcomes of assessment process are concerned. The method of eliminating data is referred to as reduct designation method. According to its definition, a reduct P is such a subset of *available data* set, where the objects are prescribed on a set of Φ criteria and for which knowledge dilation algorithm (in this case of evaluating the usability of the website) reaches a generally defined concordance function [14]. Reduction of data may contribute to: an increase of effectiveness of knowledge dilation concerning utility of a given service; reduction of timespan for recognizing the knowledge about the usability; elimination of noise and errors from the set of data; lower requirements concerning computing assets of the method; simplification of complexity of the knowledge representation structure of the problem in question; determination of a compromise between the level of data availability, their coherence and the value of adopted assessment function.

The aim of applying the above-mentioned instances of reducing $\prod$ set of Γ matrix is extracting a set with the lowest count as this makes it possible to differentiate between final assessment variants and, at the same time, to reduce the number of assessment dimensions and leave those attributes from $\prod$ set that are relevant for carrying out (to perform) usability evaluation. The process of data reduction applies methods that involve selecting certain features (attributes - criteria) by choosing such a subset of features (attributes - criteria) Φ that $\Phi \subseteq \prod \subseteq \Delta$. Feature selection may be obtained by eliminating from $\prod$ set those columns of Γ matrix, which present too little variability; this can be done by means of so called statistical measurement of attribute variability or a method, which is based on the similarities of the features. One of reduction methods is also evaluation of data significance.

In order to determine a relative value of significance of attributes (criteria) from $\prod$ set the method of rough sets [15] has been used; it requires completing information matrix Γ with a vector of decision attributes d (usability) defined in function (8).

$$d = \begin{bmatrix} d_o \\ d_1 \\ \ldots \\ d_n \end{bmatrix} \tag{8}$$

where:

d_i - means a usability of websites S_i,
i - number of the next service, wherein $i \in I \wedge 0 \le i \le n$,
n - determines the number of analyzed sites,
I - set of positive integers numbers.

Matrix obtained as a result of the above-mentioned is marked as Ψ. It is compatible with Table 1.

It is table Ψ that allows for carrying out an analysis concerning the significance of criteria from $\prod$ set; the process is possible by means of rough sets method.

Ultimately, as a result of reduces application (assessment of significance of $\prod$ set attributes) a set of independent and significant attributes is derived; the above-mentioned set of attributes is referred to in this paper as log-based service usability assessment criteria. In accordance with function (9) this set is marked with Φ symbol

$$\Phi = \{Q_1, Q_2, \ldots Q_r\} \tag{9}$$

where r represents the count of Φ set. It needs to be pointed out that numbers r and u and p (used in formulas 6 and 7) fulfil the condition (10), such that

$$r \le u \le p \tag{10}$$

Table 1. The matrix Ψ relevant data source of evaluation the usability of the websites decision attribute d (after pre-treatment and reduction of irrelevant criteria)

	q_1	q_2	...	q_u	d
S_0	The value criterion q_1 in D_0	The value criterion q_2 in D_0	...	The value criterion q_u in D_0	d_o
...	...	...	...	...	...
S_n	The value criterion q_1 in D_n	The value criterion q_2 in D_n	...	The value criterion q_u in D_n	d_n

and, that:

$$\Phi \subseteq \prod \subseteq \Delta \tag{11}$$

Determined in the above-mentioned process Φ set of significant criteria described in logs $D_0, D_1, ..., D_n$ of similar services $S_0, S_1, S_2, ..., S_n$ may be used to carry out their of evaluation the usability as well as usability of S_0.

With criteria from Φ known, basing on data from the logs it is possible to create table Θ with significant source data of evaluation the usability of the website. The said table is depicted in Table 2.

A formalised form of Θ matrix is in compliance with function (12),

$$\Theta = \begin{bmatrix} O_{0,1} & O_{0,2} & \cdots & O_{0,r} \\ O_{1,1} & O_{1,2} & \cdots & O_{1,r} \\ \cdots & \cdots & \cdots & \cdots \\ O_{n,1} & O_{n,2} & \cdots & O_{n,r} \end{bmatrix} \tag{12}$$

where:

$n + 1$ - the number of websites tested usability,
r - cardinality of set criteria Φ,
$O_{i,j}$ - table element Θ,
$i \in \{0, ...n\}, j \in \{1, ..., r\}$ are the number of rows and columns of a matrix Θ,

The above-mentioned Θ matrix contains significant criteria to perform an of evaluation the usability of the website $S_0, S_1, ...S_n$ and values of these significant criteria which have been determined on the basis of logs $D_0, D_1, ...D_n$. This matrix constitutes an information table for a multiple-criteria of $S_0, S_1, ...S_n$ service assessment.

3.1.3   Multi-Criteria Internet Service Assessment

In order to assess similar services $S_0, S_1, ..., S_n$ basing on Θ matrix with numerous significant criteria from Φ set, the right method has to be chosen. Let the method sought for carrying out a multi-criteria $S_0, S_1, ...S_n$ service assessment be marked as M. This M method should make it possible to define weights of various criteria from Φ set as there are different components with different influence on user preference. The said

Table 2. The matrix Θ relevant data sources of evaluation the usability of the websites based on the data from the logs websites.

	Q_1	Q_2	...	Q_r
S_0	The value criterion Q_1 in D_0	The value criterion Q_2 in D_0	...	The value criterion Q_r in D_0
S_1	The value criterion Q_1 in D_1	The value criterion Q_2 in D_1	...	The value criterion Q_r in D_1
...	...	...	...	...
S_n	The value criterion Q_1 in D_n	The value criterion Q_2 in D_n	...	The value criterion Q_r in D_n

method should also enable the researcher to compare each of S_0, $S_{1,...,}$ S_n with one another. In order to select the M method from among other potentially suitable methods, Electre I method has been chose [16]. This means that as M method for service S_0 usability evaluation as well as for S_1, S_2,..., S_n service usability evaluation based on data from D_0, D_1,..., D_n service logs, the multi-criteria AHP method constitutes the best choice [17]. Further calculations with Θ table are carried out by means of the AHP method. The method in question involves: 1. Creating a matrix of pairwise comparisons for each of S_0, S_1,..., S_n assessment objects individually within each criterion from Q_1, Q_2,..., Q_r criteria that belong to Φ set and Θ matrix, whereby this comparison leads to creation of r + 1 ($k = r$ in the AHP method) comparison matrices; let the results of the comparisons be marked as matrices $A^{(1)}$, $A^{(2)}$,..., $A^{(r)}$ while the matrix with Q_1, Q_2,..., Q_r criteria comparison as $A^{(0)}$; 2. Determining a ranking for each of $A^{(0)}$, $A^{(1)}$, $A^{(2)}$,..., $A^{(r)}$ matrices individually; 3. Determining a multiple-criteria ranking for services S_0, S_1,..., S_n; according to this ranking each S_i service is given a usability grade Z_{d_i} where $i \in \{0, 1, . . .n\}$.

Procedures for completing steps 1, 2 and 3 have been described in literature [18]. The final outcome of application of the AHP method is S_0, S_1,..., S_n evaluation of service usability (S_0 evaluation of service usability in particular) in the form of Z_d vector which follows function (13):

$$Z_d = \begin{bmatrix} Z_{d_o} \\ Z_{d_i} \\ \dots \\ Z_{d_n} \end{bmatrix} \tag{13}$$

where:

Z_{di} - S_i, service usability,
$i \in \{0, 1, . . .n\}$,
n - number of considered services.

The resulting evaluation Z_d is a partial grade, which was used in the construction of the final model of the synthetic evaluation of the usability of websites method.

3.2   Internet Service Usability Assessment Based on User Preference

The assessment of future and current user preferences (P_s grade) constitutes a crucial step of Internet service usability assessment. In order to solve the problem a multi-dimensional web usability analysis method, i.e., the so-called a point grade method has been applied. It needs to be pointed out that the criteria applied here are treated as equal and that preference range has been assigned coefficient values. A point grade method consists in experts allotting different grades to the criteria. Grades range from 0 to 1, where a 0 grade represents lack of a given feature, grade 0.25 - a low (satisfactory) level of a feature; 0.5 - an average (sufficient) level of a feature; 0.75 - high (fair) level of a feature; 1 - exceptionally high level of a given feature. Studies on experts' assessment may use various scales in order to assess the preferences, there is usually used the min-max normalization, which follows function (14) is applied for the purpose of preference assessment. It adjusts the values obtained in the research in such a way that the set values fall within a given range [19].

$$p''_{Si} = \frac{w - p_{min}}{p_{max} - p_{min}} (p_{newmax} - p_{newmin}) + p_{newmin} \tag{14}$$

where:

p''_{Si} - the new value you want to get after normalization,
P_{newmin} - new minimum value,
P_{newmax} - new maximum value,
P_{min} - previous minimum value,
P_{max} - previous maximum value,
w - the value of assessing the current scale.

After the normalization of data, the scale brings forward values ranging from 0 to 1. A better gradation of the points in expert assessment is achieved by applying grades in natural numbers. It is possible then to apply a preference scale such as one in Table 3.

Once the expert adopts a natural number-based scale, basing on the mapping function presented in Table 3, it is possible to set P_S partial grade for each of S_i services, where $\{i = 0,1,...n\}$.

Table 3. The scale of preferences of the users

Scale expert	<0,1	0,11–0,2	0,21–0,3	0,31–0,4	0,41–0,5	0,51–0,6	0,61–0,7	0,71–0,8	0,81–0,9	0,91–1
Scale preference	1	2	3	4	5	6	7	8	9	10

The final outcome of applying the point grade method are user preferences for S_0, $S_1,...,S_n$ service usability evaluation (S_0 service usability in particular) in the form of P_S vector which follows function (15):

$$P_s = \begin{bmatrix} P_{S_0} \\ P_{S_1} \\ ... \\ P_{S_n} \end{bmatrix} \tag{15}$$

where:

P_{Si} - user preference in the evaluation the usability of websites S_i,
$i \in \{0, 1, ...n\}$
n - number of considered sites.

3.3 Synthetic Method of Evaluate the Usability of the Websites

The ultimate evaluation of Internet service usability is a combination of partial grade Z_d with partial grade P_S (Subsects. 3.1 and 3.2) of examined website. Evaluation derived in this way is called a synthetic evaluation. It takes into account both historical data concerning actual use of services and the assessment of current and future user preference. Final evaluation in the synthetic method is brought forward thanks to the additivity of the usability function, synthesizing contributions of both partial grades with regard to the degree-weights of these contributions and a scale of their reference to the range of derived values. In order to obtain a unified reference scale P_S and Z_d, one needs to carry out a normalisation of these evaluation, for e.g. normalize Z_d against P_S grades. This requires multiplying the value of derived grade Z_d by an n number of examined services.

When applying the synthetic method of usability of the websites evaluation, the log-based data are actual data, while data concerning preferences are treated as unreliable, incomplete and sensitive regarding the choice of a representative user group (experts). For this reason the ultimate synthetic evaluation of usability makes use of a special coefficient β it accounts for weights of both grades. The value of the above mentioned coefficient might be determined by consulting the expert. The same procedure applies when determining the weight's grade β, i.e. again the expert's opinion proves helpful [20]. An expert in the field may argue that the likelihood of an error in received partial grades Z_d against P_S is always known. In case when estimating the likelihood of an error is not possible or is not objective (a subjective or mixed likelihood), the expert uses his knowledge of the subject [21] and decides on β selection (Table 4).

To conclude, in order to determine synthetic evaluation of website usability based on data gathered from logs and users preference evaluation with application of partial grades Z_d and P_S, obtained the usability U_i website S_i (formula 16).

$$U_i = \beta \cdot Z_{d_i} \cdot n + (1 - \beta) \cdot p_{s_i} \tag{16}$$

where:

$i \in \{0, 1, \ldots n\}$
U_i - usability of any website S_i,
p_{s_i} - usability of the service S_i as designated test method preferences of the users
Z_{d_i} - Usability of the service S_i as determined by AHP the websites S_i,
n - the number of rated sites,
β - assessment weight $(0 \leq \beta \leq 1)$.

Ultimately, the evaluation of synthetic websites $S_o, S_1, \ldots, S_n$ are obtained in the form of synthetic usability of the vector U_i according to the following formula (17).

$$U_i = \begin{bmatrix} U_0 \\ U_1 \\ \ldots \\ U_n \end{bmatrix} \tag{17}$$

Table 4. Examples of values β - weight-rating service S_i in the formula (16)

Criterion	Evaluation criterion	Value β	A model of synthetic evaluation of the usability of websites
Hurwitz	Optimism received ratings websites	$\beta = 0,6$ Expert for the evaluation of partial Z_d selects the best of selected decisions, the actual data provide a greater likelihood ratio	$U_i = 0,6 \cdot Z_{d_i} \cdot n + (1 - 0,6) \cdot p_{s_i}$
Laplace	Probability ratings received preferences of users and websites are equal	$\beta = 0,5$ Expert for the evaluation of partial Z_d selects the same lines of action in choosing weight	$U_i = 0,5 \cdot Z_{d_i} \cdot n + (1 - 0,5) \cdot p_{s_i}$
Wald	Not yet rated user preferences	$\beta = 1$ Expert selects one factor optimism as not given the actual subjective evaluation of partial Z_d	$U_i = 1 \cdot Z_{d_i} \cdot n + (1 - 1) \cdot p_{s_i}$
	No similar websites	$\beta = 0$ Expert selects one factor optimism as partial evaluation P_S	$U_i = 1 \cdot Z_{d_i} \cdot n + (1 - 0) \cdot p_{s_i}$

where:

U_i -the evaluation the usability of websites S_i,
$i \in \{0, 1, \ldots n\}$
n - the number of sites considered.

4   Conclusions

The purpose of this paper is to develop means for usability evaluation of any website. It needs to be stressed at this point that literature abounds in various methods and tools that aid website usability evaluation. Due to a fierce competition in the field, it is important to seek such solutions that are truly effective in delivering a credible assessment of examined website. The choice of the right method depends on a specific problem or environment determining the solutions. Internet service usability evaluation presented in this paper takes into account two aspects of service evaluation, mainly user preference, historical and actual data obtained from the logs. It is worth pointing out that this assessment may be applied to any type of website group and for any number of services. The number of web services depends on the problem of evaluation of examined services. Additionally, the evaluation enables gathering log-based data in real-time, so it is possible to create a kind database for the purpose of usability evaluation. The method developed and presented in this paper is a new method in the field the evaluation of usability of the website. The method is classified among methods of software engineering and requirements engineering.

In future work will be carried out advanced tests to evaluate the proposed method. Additionally, directions for further research can be a combination of methods, the evaluation of the usability with the knowledge base in to one tool. This tool will be able to modify the internal structure of websites.

References

1. Krug, S.: Don't Make Me Think, (3rd Edition), New Riders, ISBN-13: 978-0321965516, (2014)
2. Nielsen, J.: Designing Web Usability: The Practice of Simplicity. New Riders Publishing, Indianapolis, ISBN 1-56205-810-X (1999)
3. Nielsen, J., Budiu, R.: Mobile Usability, New Riders Press, ISBN 0-321-88448-5 (2012)
4. Cohen, J.: The Unusually Useful Web Book, 1st edn., New Riders, ISBN-13: 075-2064712060 (2003)
5. Consulting. http://www.amrconsulting.pl. Accessed 4 Mar 2016
6. Symetria. http://www.symetria.pl. Accessed 9 Apr 2016
7. Baharad, E., Nitzan, S.: On the selection of the same winner by all scoring rules. Soc. Choice Welfare **26**, 597–601 (2006)
8. Saari, D.G.: Disposing Dictators, Demystifying Voting Paradoxes, Social Choice Analysis. Cambridge University Press, Cambridge (2008)
9. Saari, D.G.: Complexity and the geometry of voting. Math. Comput. Model. **48**, 1335–1356 (2008)
10. Saari, D.G.: Which is better: the Condorcet or Borda winner. Soc. Choice Welfare **26**, 107–129 (2006)
11. Bury, H., Wagner, D.: Determining group judgment when ties can occur. In: Proceedings of 13th IEEE IFAC International Conference on Methods and Models in Automation and Robotics MMAR, Szczecin, Polska, pp. 779–784 (2007)
12. Bury, H., Wagner, D.: Group judgment with ties. In: Aschemann, H. (ed.) Distance-Based Methods. New Approaches in Automation and Robotics, I-Tech, pp. 153–172 (2008)
13. Pawlak, Z., Skowron, A.: A rough set approach for decision rules generation, ICS Research Report 23/93. University of Technology, Warszawa (1993)
14. Dominik, A.: Analiza danych z zastosowaniem teorii zbiorów przybliżonych. Warszawa (2004, in Polish)
15. Pawlak, Z., Grzymala-Busse, J., Slowinski, R., Ziarko, W.: Rough Sets. Commun. ACM **38** (11), 88–95 (1995)
16. Huang, W.C., Chien-Hua, C., Keelung T.: Using the ELECTRE II method to apply and analyze the differentiation theory. In: Proceedings of the Eastern Asia Society for Transportation Studies (2005)
17. Chmielarz, W.: Metody oceny werbalnych księgarni internetowych Komputerowo: Zintegrowane Zarządzanie obszar: Gospodarka oparta na wiedzy (2010, in Polish)
18. Fabisiak, L., Ziemba, P.: Metody Wielokryterialnego wspomagania decyzji w ocenie użyteczności serwisów internetowych. Wrocław: Wydawnictwo Uniwersytetu Ekonomicznego, nr **20**, 303–312 (2011, in Polish)

19. Jain, A., Nandakumar, K., Ross, A.: Score normalization in multimodal biometric systems. Pattern Recogn. **38**(12), 2270–2285 (2005)
20. Hurwicz, L., Reiter, S.: Designing Economic Mechanisms. Cambridge University Press. Frontmatter (PDF) via Cambridge University Press. ISBN 0-5218-3641-7, (2006)
21. Stąpor, K.: Automatyczna klasyfikacja obiektów. Akademicka Oficyna Wydawnicza EXIT, Warszawa (2005). (in Polish)

Ontology-Based Approaches to Big Data Analytics

Agnieszka Konys[✉]

Faculty of Computer Science and Information Technology,
West Pomeranian University of Technology in Szczecin,
Żołnierska 49, 71-210 Szczecin, Poland
akonys@wi.zut.edu.pl

Abstract. The access to relevant information is one of the determining factors which directly influences on decision-making processes. Huge amounts of data have been accumulated by entities from large variety of sources in many different formats. Due to large amounts of information and continuous processes of generation of new parts, it is necessary to ensure the most effective way of information or data extraction and analysis. The Web of Data provides great opportunities for ontology-based services. The combination of ontology-based approaches and Big Data may help in solving some problems related to extraction of meaningful information from various sources. This paper presents the selected ontology-based approaches to Big Data analytics as well as a proposal of a procedure for ontology-based knowledge discovery.

Keywords: Ontology · Data access · Data analysis · Big Data · Data extraction

1 Introduction

Information can be defined as the creation of meaning on base of analysis, calculation or data exploration. It can encompass data aggregation, data transformation, mapping, correlation or similar operations. Then knowledge constitutes the understanding gained from analyzing information. The processes of information extraction and knowledge extraction can take many forms. The problem is how to make use of knowledge and other technical skills to extract data from different sources.

Due to large amounts of information and continuous processes of generation new parts, it is necessary to ensure the most effective way of information or data extraction and analysis. Huge amounts of data have been accumulated by entities from large variety of sources in many different formats. Data access is one of the determining factors for the potential of value creation processes. It directly influences on decision-making processes, analysis and effective exploitation of the data.

The process of data extraction needs to consider different types of information from diverse sources: structured, semi-structured, and unstructured. The diversity of sources require using variety methods, techniques and tools to manage them in the most effective way. Another inconvenience may concern different formats of data description and scattered data locations. Furthermore, the main problem concerns the increasing time to data access and the quality and actual relevancy of obtained query results.

© Springer International Publishing AG 2017
S. Kobayashi et al. (eds.), *Hard and Soft Computing for Artificial Intelligence, Multimedia and Security*, Advances in Intelligent Systems and Computing 534, DOI 10.1007/978-3-319-48429-7_32

The environment requires quick decisions, on base of current and relevant information, without wasting time for analyzing unnecessary data.

The paper introduces the main problems related to information extraction and analysis from large datasets. It presents the examples of selected ontology-based approaches to Big Data analytics (especially including: two taxonomies for Big Data, abilities offered by conversion, mappings, OBDA solutions, and automatic ontology construction). Moreover, it presents a proposal of a procedure of ontology-based knowledge discovery. The general aim of the procedure is to exploit the combination of ontology and ontology-based solutions and Big Data.

The paper is organized as follows. In Sects. 2 and 3 the main challenges of Big Data and ontology application to knowledge extraction from large datasets are presented. In Sect. 4 the selected ontology-based approaches to Big Data analysis are introduced and described in details. Then, the exemplary taxonomies for Big Data are shown in Sect. 5. Finally, in Sect. 6, a proposal of a procedure for the ontology-based knowledge discovery is presented.

2 Big Data

Big Data refer to large amounts of information, where the traditional solutions are insufficient to ensure the data quality, management and sampling. Big Data are not only generated by traditional information exchange and software use (mobile devices, computers, etc.), but also from the myriads of sensors of various types embedded in various environments [8].

The term of Big Data was first used by Mashey [7], and over the years, it has been variously defined in the literature. In 2001, Laney [18] indicated tree features that characterize Big Data: volume, velocity and variety. Volume refers to the huge amounts of data, then veracity indicates on the messiness and trustworthiness of the data. Velocity concerns the speed of generated new data and the speed at they move around. Variety refers to the different types of data possible to use (structured, semi-structured, and unstructured). In last years, other attributes were added. Nowadays, the term of Big Data very often is defined using 5V's: variety, velocity, volume, veracity and value, by adding two traits: value and veracity to the previous set of V's. Value refers to the quality of our data, then veracity indicates on the problems with uncertainty of data, messiness and trustworthiness, and data quality and accuracy. It is possible to extend a suite of key traits by additional elements, e.g., exhaustivity, resolution, indexicality - fine-grained (in resolution) and uniquely indexical (in identification), relationality, extensionality, and scalability and variability [7, 8, 14, 18].

Regardless of the used traits of Big Data description, this type of data is qualitatively different to traditional, small data. The main differences between small and Big Data was shown by Kitchin [14]. The general problem concerns the diverse forms of characteristics than exist in traditional, small datasets. Kitchin proposed the comparison taxonomy, which is divided into 7 main criteria: variety, velocity, volume, extensionality and scalability, exhaustivity, resolution and indexicality, and relationality. Key differences concern the size (volume), velocity and variety. Big Data are very large, when small data are limited to large. Just comparing the velocity in small and

Big Data, in small data it is slow and bundled, whereas in Big Data it is fast and continuous. Subsequent difference concerns the wide variety in Big Data in contrast to limited to wide in small data. Next, extensionality and scalability in Big Data is very high than small data (low to middling). The exhaustivity refers to entire populations in Big Data in opposite to small data, where samples are enough. The last two criteria: resolution and indexicality, and relationality adopt strong values for Big Data (and additionally tight for resolution and indexicality). In the case of small data, weak to strong values for relationality and course and weak to tight and strong for resolution and indexicality are taken [14]. This short analysis underlines that the traditional methods, techniques, etc. are insufficient to ensure the data quality, relevancy, and management.

Currently, various technologies are available and that are being applied to handle the Big Data (e.g. algorithms and frameworks: Hadoop, DOT, GLADE, Starfish, ODT-MDC, MRAM, CBDMASP, SODSS, BDAF, HACE, CUDA, Storm, Pregel, MLPACK, Mahout, MLAS, PIMRU, Radoop [18]). Many enterprises and private and public organisations struggle to handle distributed information in the most efficient way. The necessity of solving this problem requires using one or more solutions dedicated to Big Data (e.g. machine-learning algorithms; mining algorithms: DBDC, PKM, CloudVista, MSFCUDA, BDCAC, Corest, SOM-MBP, CoS, SVMGA, Quantum SVM, DPSP, DHTRIE, SPC, FPC, DPC, MFPSAM [18]) or applying new ones.

The Web of Data provides great opportunities for ontology-based services, but also poses challenges for tools for editing, using, and reasoning with ontologies, as well as techniques that address bottlenecks for the engineering of large-scale ontologies [12]. The application of Semantic Web technologies may help in solving some problems related to Big Data. Semantic Web technologies are especially dedicated to extract meaningful information from structured, unstructured and semi-structured data from large datasets [6].

3   Ontology Application to Knowledge Extraction from Big Data

In last decade the terms related to Semantic Web become significant elements in the efficient way of information retrieval, processing and supporting availability of machine readable data. A close relation between ontologies and the Semantic Web is noticeable, hence the role of ontologies and their application in knowledge extraction is significant.

From a formal point of view, an ontology can be defined as an explicit representation of a shared conceptualization. It enables defining concepts, relations and instances [17]. One of the advantages of the ontology is that it can cope with the real world complexity and adapt to changes. Furthermore, it can be easily expanded, and it supports knowledge sharing and reuse. The common standard is Ontology Web Language (OWL). Description Logic (DL) is a theoretical base for OWL language for expressing and modelling ontologies.

Generally, an ontology offers a wide spectrum of its application in Big Data context. It can be assumed that the combination of ontologies and Big Data can solve some of the problems identified for large data sets [4]. The main aim of ontology-based solutions is to support the process of dealing with heterogeneous data and the access to relevant data, with regard to increasing number of it [15]. The second area of added value concerns the usage of ontologies for accessing data. The aim is to limit or reduce querying relational databases.

The process of combining the reasoning with Big Data sets with ontological knowledge has made a significant meaning. The main problem concerns the way of ontology storage and effective reasoning, without losing out of sight the need of scalability [4]. The fundamental issue is to ensure scalable reasoners for Big Data. Furthermore, the combinations of the ontology and Big Data may bring some benefits and new opportunities. One of them is the usage of the ontology to combine deep domain knowledge and raw data. The ontology can provide semantics to add raw data. Then, generalized concepts in an ontology can connect data in various concept levels across domains. Another possibility is to use the ontology to bridge datasets across domains [12, 13]. The ontology may provide a common vocabulary to describe or classify datasets. Moreover, the ontology can be used as knowledge to analysis Big Data [16].

Nowadays a number of ontology-based solutions is still rising up. It is possible to indicate many methods, approaches and tools supporting knowledge extraction (e.g., supporting automatic ontology construction from different types of input sources: structured, semi-structured, unstructured, Question Answering Systems (QAS), Ontology-Based Information Extraction (OBIE) systems, Ontology-Based Data Access (OBDA), and different mining techniques) [5]. Ontology-based approaches provide a practical framework to address the semantic challenges presented by Big Data sets. Moreover, they allow to add metadata ontology for annotating data, and they bring benefits for search and retrieval information. The key element is that they offer reasoning processes and understanding of the obtained results. A significant role is assigned to mapping content to defined ontology and then convert content to triples according to defined ontology.

4   Selected Ontology-Based Approaches to Big Data Analysis

4.1   Mapping Ontology to Database

Ontology-based solutions to knowledge extraction support the process of dealing with heterogeneous data, and the access to relevant data, with regard to increasing number of it [16]. The use of ontologies for accessing data is one of the most exciting new applications of description logics in databases and other information systems. The aim is to limit or reduce querying relational databases. It seems that the majority of RDBMS and RDBMS-based tools in predominant use today are usually not designed to search and analyze massive sets containing both structured and unstructured data [13]. Big Data sets require new technique to process, store and make sense of them. In a given ontology, the processes of mapping may refer to mapping classes (concepts) defined in ontology to database schema or mapping classes/instances defined in ontology to data in database.

4.2 Ontology-Based Data Access

One of the most interesting usages of shared conceptualizations is Ontology-Based Data Access (OBDA). OBDA provides a convenient way to deal with large amounts of data spread over heterogeneous data sources [15]. OBDA allows users to formulate queries in a single user-friendly ontology language. These queries are then unfolded and executed on the data sources.

The simplified procedure of OBDA system encompasses three layers: database layer, ontology layer and end-user layer. In many cases an expert participation is required to formulate more complex queries and mappings. OBDA system uses an ontology as a conceptual schema of the subject domain, and as a basis of the user interface for SQL database systems. It offers the direct access to data source [1]. The ontology defines a high-level global schema and provides a vocabulary for user queries. Then, the queries are executed by mappings describing relationship between concepts in the ontologies and their representation. The OBDA system transforms user queries into the vocabulary of the data and then delegates the actual query evaluation to the data sources. Moreover, OBDA system offers the independence of how the data are stored [1, 3, 4].

The main feature of OBDA is to provide the global unified view in terms of conceptualization of a given domain. It is created in an independent way from the representation, preparing for the data stored at the sources. OBDA offers a wide spectrum of functionalities, especially logical/physical independence of the information system. However, OBDA is not deprived of disadvantages. Many efforts have been put in defining the appropriate language for the semantic layer, defining the structure and language of the mappings used to link the data and the semantic layer, studying the complexity of offering a set of useful services such as query answering, database schema extraction, and inconsistency management [3]. Moreover, ontology and mappings are expensive and take some time [2]. The efficiency for translation process and execution queries might be improved.

4.3 Adding Metadata on Data Using Vocabulary Defined in Ontology

Ontologies can also reduce variety in Big Data by aiding the annotation of data and its metadata. Data sets differentiate in completeness of metadata, granularity and vocabulary used. As a consequence, ontologies can mitigate some of this variety by normalizing terms and providing for absent metadata [12]. The processes of data analytics are not possible without explicit understanding of the structures, characteristic or contours of the inquired data sets. Metadata supports operational aspects of data analytics, when the lexicon and ontology helps in the semantic and interpretive aspects of data analytics. Metadata are generally considered to be information about data and are usually formulated and managed to comply with predetermined standards. It is possible to indicate various types of metadata: operational, structural (syntatic), bibliographical, data lineage, and similar. Furthermore, metadata also contain filtering rules that specify the requirements that data must meet to be ontologically useful rather than be excluded as noise [13].

The information quality is determined by the process how well it is understood by a user. Semantic metadata, also called as machine-readable metadata, are treated as a key factor to understanding structured data. To integrate data, model driven techniques use the metadata component. Metadata enabling the current crop of graphical data mapping tools. These tools and techniques provide integration of structured data less costly and error-prone [6]. The standard for metadata is called Ontology Metadata Vocabulary (OMV) [12].

4.4   Conversions to Other Formats

The heterogeneity of documents published on the Web may provides some obstacles. The existence of diverse input sources both structured (e.g. HTML, XML/XML Schema, RDF/OWL, relational data), semi-structured and unstructured (e.g. text, documents, images) poses new challenges for handling data. Another problem concerns the linking between different types of documents available on the Web [4]. Furthermore, the high importance is assigned to the data storage in triples independently of physical scheme. The data are stored in heterogeneous systems, thus the access (by posing queries over the data) is undoubtedly a challenging task. There is no simple way to avoid heterogeneous data sources. To solve this problem, conversion to other formats is applied. OWL format can be easily converted into RDF, XML or other syntax. The RDF and OWL are used to integrate all data formats and standardise existing ontologies (e.g. in bioinformatics domain [13]). Moreover, it is possible to convert database (e.g. RDB) to ontology-based (RDF) database (e.g., linked data such as DBPedia, some bioinformatics databases and similar) [12].

5   Selected Taxonomies for Big Data

5.1   The Taxonomy Proposed by Kitchin and McArdle

The taxonomy for Big Data classification is proposed by Kitchin and McArdle [7]. It encompasses 7 domains and 26 specific types of data. In the taxonomy, the following domains are included: mobile communication, websites, social media/crowdsourcing, sensors, cameras/lasers, transaction process generated data, and administrative. Then, authors mapped 26 sources of data, which were defined in literature as Big Data, against the traits identified by one of the co-author, Kitchin [14]. Applying Kitchin's taxonomy of 7 Big Data traits to 26 datasets drawn from 7 domains, each of which is considered in the literature to constitute Big Data. These 26 types of data are by no means exhaustive of all types of Big Data, for example there are a multitude of Big Data generated within scientific experiments, science computing, and industrial manufacturing. Finally, the authors concluded that two of Vs characteristics, volume and variety, were not key defining characteristics of Big Data [7].

5.2 The 6-Fold Taxonomy Proposed by Murthy

Murthy considers Big Data using a six-fold taxonomy [8]. He presents six-dimensional taxonomy (6-D taxonomy) that includes the following areas: analytics, visualisation, data, security and privacy, storage infrastructure and compute infrastructure. He claimed that these six dimensions arise from the key aspects that are needed to establish Big Data infrastructure classification. Each of dimensions was described in details in [8]. Data can be characterized by two-fold attributes: latency requirements and structure. The second dimension encompasses compute infrastructure, especially including batch or streaming. Storage infrastructure refers to the different database models/types and the ways of used format of description. Then, the analytics processes, especially including machine-learning algorithms can be divided into: supervised, semi-supervised, unsupervised or re-enforcement. The next part concerns visualisation (maps, abstract, interactive, real-time). The last category refers to privacy and security (infrastructure security, data privacy, data management, integrity and reactive security) [8].

5.3 An Ontology-Based Framework for Data Analysis

Kuiler proposed an ontology-based framework for data analysis [13]. The framework relates to the health domain. It is composed of the following components: lexicon and ontology, analytics and reports, metadata (sub-components: identity/security, data ingestion, federation and integration, data anonymization, data distribution), analytic data storage. It provides an IT-neutral conceptualization of a framework that accommodates both Little and Big Data and places them in their proper operational perspective as different kinds of data sources. The lexicon and ontology components provide the semantic foundations for data analytics [13]. The aim is to support the analytical component of the architectural framework. The health domain requires the relevant data analysis from different sources: structured, semi-structured and unstructured. In this domain, the standardized vocabularies exist, but the description analysis depends on the unstructured text data. Hence, it is necessary to process the information to structured form using selected techniques of text mining and natural language processing software. The ontologies and ontology-based solutions have a great impact on a proper data management especially in the health domain, biological domain or genetics.

5.4 Automatic Ontology Construction

The construction of the ontology still remains a hard human task. The process is sometimes assisted by software tools that facilitate the information extraction from a textual corpus. Manually constructing ontology with the help of tools is still practiced to acquire knowledge of many domains. The process of automatic ontology construction may be promising for large datasets.

Current state of the art in automated ontology acquisition typically consists of using existing machine-learning, text-analytic, and natural language processing techniques on annotated or un-annotated data to provide candidate ontology classes, relations, and properties to a human being, who often adjudicates the candidates [12]. Automated generation provides a fundamentally different approach to ontology creation than manual construction by a designer. The general aim of automatic ontology construction is to create ontology from different input sources both structured (e.g., HTML, XML/XML Schema, RDF/OWL, relational data) and unstructured (e.g., text, documents, images) and a possibility to its development in different ways: merging and alignment with other ontology or to use it for a question answering system and create a natural language interface [10]. The resulting knowledge needs to be in a machine-readable and machine-interpretable format and must represent knowledge in a manner that unambiguously defines its meaning and facilitates inferencing [11].

Generally, most of selected approaches offer knowledge extraction from structured and unstructured sources of knowledge, usage a dedicated tool to support mining based approach selection, the process of parsing the data to automate ontology construction, pattern matching, using lexicons, and adaptation to pose question in natural language, validation. It is possible to indicate the exemplary approaches to automatic ontology construction: TERMINAE, SALT, OntoCase, TextOntoEx. The complex analysis is provided in [9].

6 Ontology-Based Knowledge Discovery - A Proposal of a Procedure

The Web of Data provides great opportunities for ontology-based services, but also poses challenges for tools for editing, using, and reasoning with ontologies, as well as techniques that address bottlenecks for the engineering of large-scale ontologies [12]. The analysis of available solutions for Big Data allows to construct a proposal of a procedure for ontology-based knowledge discovery (Fig. 1). The general aim of the procedure is to exploit the combination of ontology and ontology-based solutions and Big Data. The medium level of details is considered. It is worth to emphasize that the final version of this procedure is still elaborated and improved.

The proposed procedure consists of the following steps: (1) replenishment of the knowledge base, (2) data extraction and formatting, (3) data mining processes, (4) result interpretation and analysis. The knowledge base may contain different types of input sources, especially databases, taxonomies or ontologies for Big Data - they provide domain knowledge systematization and provide the lexicon (vocabulary and definitions) specified for a given domain. The relational databases can be easily converted into RDF format to enrich knowledge base. Various techniques are offered by OBDA, enabling user to add mappings between ontology and database, and as a consequence, posing a query. The process of automatic ontology construction is also permitted, but in this case a proper tool should be used to support the whole process of knowledge acquisition. It is worth to mention about semantic enrichment and metadata. Some examples of selected approaches included in the procedure were presented in the previous sections of this paper.

Next, the process of data extraction and formatting takes place. Then, data mining techniques support the processes of data (and information) extraction. In this case, the

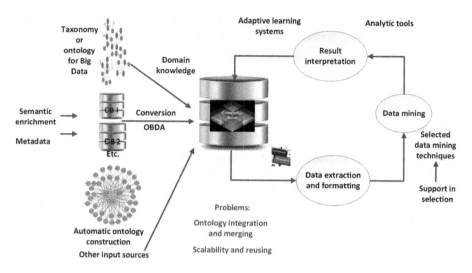

Fig. 1. Ontology-based knowledge discovery

support in selection of a proper mining method or technique is preferable. Thereafter, the obtained results are interpreted. This step should be supported by analytic tools to obtained the relevant set of results.

The combination and exploitation Web of Data and Big Data brings a lot of new opportunities to handle and manage information and data for a large scale. This process will increase in the near future. The environment calls the shots. Requirements for the future may contain the following aspects, especially including the elaboration of tools for engineering large-scale ontologies, ontology and vocabulary mapping and alignment tools, ontological-analytical techniques and hybrid tools [12]. Furthermore, a high importance is assigned to enhance ontology integration, reusing and scalability. These challenges and other prospective requirements will have direct influence of the shape of Web of Data.

7   Conclusions

This paper presented the main ontology-based approaches to Big Data analytics. Ontologies and ontology-based solutions have a wide range of applications, including semantic integration, decision support, search and annotation. They support the process of dealing with heterogeneous data, and the access to relevant data, with regard to increasing number of it. Moreover, they offer a wide spectrum of its application in Big Data context.

The paper presented the examples of selected ontology-based approaches to Big Data analytics (especially including: two taxonomies for Big Data, abilities offered by conversion, mappings, OBDA solutions, and automatic ontology construction). Then, the proposal of the procedure of ontology-based knowledge discovery was introduced. It included some of the described approaches, but the process of the elaboration of presented procedure is still proceeded. It seems that the combination of ontologies and Big Data will yield results in the future.

References

1. Rodríguez-Muro, M., Kontchakov, R., Zakharyaschev, M.: Ontology-based data access: ontop of databases. In: Alani, H., et al. (eds.) ISWC 2013, Part I. LNCS, vol. 8218, pp. 558–573. Springer, Heidelberg (2013). doi:10.1007/978-3-642-41335-3_35
2. Savo, D.F., Lembo, D., Lenzerini, M., Poggi, A., Rodriguez-Muro, M., Romagnoli, V., Ruzzi, M., Stella, G.: MASTRO at work: experiences on ontology-based data access. In: Proceedings of the 23rd International Workshop on Description Logics (DL2010), CEUR WS 573, Waterloo, Canada (2010)
3. Lembo, D., Mora, J., Rosati, R., Savo, D.F., Thorstensen, E.: Mapping analysis in ontology-based data access: algorithms and complexity. In: Arenas, M., et al. (eds.) ISWC 2015, Part I. LNCS, vol. 9366, pp. 217–234. Springer, Heidelberg (2015). doi:10.1007/978-3-319-25007-6_13
4. Heymans, S., et al.: Ontology reasoning with large data repositories. In: Hepp, M., et al. (eds.) Ontology Management. Computing for Human Experience, vol. 7, pp. 89–128. Springer, US (2008)
5. Konys, A.: Knowledge-based approach to question answering system selection. In: Núñez, M., Nguyen, N.T., Camacho, D., Trawiński, B. (eds.) ICCCI 2015, Part I. LNCS (LNAI), vol. 9329, pp. 361–370. Springer, Heidelberg (2015). doi:10.1007/978-3-319-24069-5_34
6. Ajani, S.: An ontology and semantic metadata based semantic search technique for census domain in a big data context. Int. J. Eng. Res. Technol. 3(2), 1–5 (2014)
7. Kitchin, R., McArdle, G.: What makes big data, big data? Exploring the ontological characteristics of 26 datasets. Big Data Soc. 3, 1–10 (2016)
8. Murthy, P., Bharadwaj, A., Subrahmanyam, P.A., et al.: Big Data Taxonomy. Big Data Working Group, Cloud Security Alliance (2014)
9. Konys, A.: A tool supporting mining based approach selection to automatic ontology construction. IADIS J. Comput. Sci. Inf. Syst., 3–10 (2015)
10. Hellmann, S., Auer, S.: Towards web-scale collaborative knowledge extraction. In: Gurevych, I., Kim, J. (eds.) The People's Web Meets NLP, Theory and Applications of Natural Language Processing, pp. 287–313. Springer, Heidelberg (2013)
11. Unbehauen, J., Hellmann, S., Auer, S., Stadler, C.: Knowledge extraction from structured sources. In: Ceri, S., Brambilla, M. (eds.) Search Computing. LNCS, vol. 7538, pp. 34–52. Springer, Heidelberg (2012). doi:10.1007/978-3-642-34213-4_3
12. Gruninger, M., Obst, L.: Semantic web and big data meets applied ontology. Appl. Ontol. 9, 155–170 (2014)
13. Kuiler, E.W.: From big data to knowledge: an ontological approach to big data analytics. Rev. Policy Res. 31(4), 311–318 (2014)
14. Kitchin, R.: The Data Revolution: Big Data, Open Data, Data Infrastructures and Their Consequences. Sage, London (2014)
15. Calvanese, D., et al.: The mastro system for ontology-based data access. Semant. Web J. 2(1), 43–53 (2011)

16. Kozaki K.: Ontology engineering for big data. In: Ontology and Semantic Web for Big Data (ONSD2013) Workshop in the 2013 International Computer Science and Engineering Conference (ICSEC2013), Bangkok, Thailand (2013)
17. Gruber, T.: Toward principles for the design of ontologies used for knowledge sharing. Int. J. Hum Comput Stud. **43**(5–6), 907–928 (1995)
18. Tsai, C.W., et al.: Big data analytics: a survey. J. Big Data **2**, 21 (2015)

Author Index

© Springer International Publishing AG 2017
S. Kobayashi et al. (eds.), *Hard and Soft Computing for Artificial Intelligence, Multimedia and Security*, Advances in Intelligent Systems and Computing 534, DOI 10.1007/978-3-319-48429-7

Printed in the United States
By Bookmasters